The Advanced iOS 6 Developer's Cookbook

Fourth Edition

Erica Sadun

✦Addison-Wesley

Upper Saddle River, NJ • Boston • Indianapolis • San Francisco
New York • Toronto • Montreal • London • Munich • Paris • Madrid
Cape Town • Sydney • Tokyo • Singapore • Mexico City

Many of the designations used by manufacturers and sellers to distinguish their products are claimed as trademarks. Where those designations appear in this book, and the publisher was aware of a trademark claim, the designations have been printed with initial capital letters or in all capitals.

The author and publisher have taken care in the preparation of this book, but make no expressed or implied warranty of any kind and assume no responsibility for errors or omissions. No liability is assumed for incidental or consequential damages in connection with or arising out of the use of the information or programs contained herein.

The publisher offers excellent discounts on this book when ordered in quantity for bulk purchases or special sales, which may include electronic versions and/or custom covers and content particular to your business, training goals, marketing focus, and branding interests. For more information, please contact:

    U.S. Corporate and Government Sales
    1-800-382-3419
    corpsales@pearsontechgroup.com

For sales outside of the U.S., please contact

    International Sales
    international@pearsoned.com

AirPlay, AirPort, AirPrint, AirTunes, App Store, Apple, the Apple logo, Apple TV, Aqua, Bonjour, the Bonjour logo, Cocoa, Cocoa Touch, Cover Flow, Dashcode, Finder, FireWire, iMac, Instruments, Interface Builder, iOS, iPad, iPhone, iPod, iPod touch, iTunes, the iTunes logo, Leopard, Mac, Mac logo, Macintosh, Multi-Touch, Objective-C, Quartz, QuickTime, QuickTime logo, Safari, Snow Leopard, Spotlight, and Xcode are trademarks of Apple, Inc., registered in the U.S. and other countries. OpenGL, or OpenGL Logo, OpenGL is a registered trademark of Silicon Graphics, Inc. The YouTube logo is a trademark of Google, Inc. Intel, Intel Core, and Xeon are trademarks of Intel Corp. in the United States and other countries.

Visit us on the Web: informit.com/aw

Copyright © 2013 Pearson Education, Inc.

All rights reserved. Printed in the United States of America. This publication is protected by copyright, and permission must be obtained from the publisher prior to any prohibited reproduction, storage in a retrieval system, or transmission in any form or by any means, electronic, mechanical, photocopying, recording, or likewise. For information regarding permissions, write to:

    Pearson Education, Inc.
    Rights and Contracts Department
    501 Boylston Street, Suite 900
    Boston, MA 02116
    Fax (617) 671-3447

ISBN-13: 978-0-321-88422-0
ISBN-10: 0-321-88422-1

Editor-in-Chief
Mark Taub

Senior Acquisitions Editor
Trina MacDonald

Senior Development Editor
Chris Zahn

Managing Editor
Kristy Hart

Project Editor
Jovana San Nicolas-Shirley

Copy Editor
Apostrophe Editing Services

Indexer
Brad Herriman

Proofreader
Sarah Kearns

Technical Editors
Mike Shields
Rich Wardwell

Editorial Assistant
Olivia Basegio

Cover Designer
Chuti Prasertsith

Compositor
Nonie Ratcliff

❖

*I dedicate this book with love to my husband, Alberto,
who has put up with too many gadgets and
too many SDKs over the years while remaining
both kind and patient at the end of the day.*

❖

Contents at a Glance

Preface xiii

1 Device-Specific Development 1

2 Documents and Data Sharing 39

3 Core Text 87

4 Geometry 127

5 Networking 167

6 Images 197

7 Camera 229

8 Audio 261

9 Connecting to the Address Book 297

10 Location 339

11 GameKit 371

12 StoreKit 427

13 Push Notifications 447

Index 475

Table of Contents

Preface xiii

1 Device-Specific Development 1
    Accessing Basic Device Information 1
    Adding Device Capability Restrictions 2
    Recipe: Checking Device Proximity and Battery States 5
    Recipe: Recovering Additional Device Information 9
    Recipe: Using Acceleration to Locate "Up" 11
    Working with Basic Orientation 12
    Retrieving the Current Accelerometer Angle Synchronously 13
    Recipe: Using Acceleration to Move Onscreen Objects 16
    Recipe: Accelerometer-Based Scroll View 19
    Recipe: Core Motion Basics 21
    Recipe: Retrieving and Using Device Attitude 26
    Detecting Shakes Using Motion Events 27
    Recipe: Using External Screens 29
    Tracking Users 35
    One More Thing: Checking for Available Disk Space 35
    Summary 36

2 Documents and Data Sharing 39
    Recipe: Working with Uniform Type Identifiers 39
    Recipe: Accessing the System Pasteboard 45
    Recipe: Monitoring the Documents Folder 48
    Recipe: Presenting the Activity View Controller 54
    Recipe: The Quick Look Preview Controller 63
    Recipe: Adding a QuickLook Action 66
    Recipe: Using The Document Interaction Controller 69
    Recipe: Declaring Document Support 75
    Recipe: Creating URL-Based Services 82
    Summary 84

3 Core Text ... 87
Core Text and iOS ... 87
Attributed Strings ... 89
Recipe: Basic Attributed Strings ... 93
Recipe: Mutable Attributed Strings ... 95
The Mystery of Responder Styles ... 98
Recipe: Attribute Stacks ... 100
Recipe: Using Pseudo-HTML to Create Attributed Text ... 105
Drawing with Core Text ... 109
Creating Image Cut-Outs ... 112
Recipe: Drawing Core Text onto a Scroll View ... 114
Recipe: Exploring Fonts ... 116
Adding Custom Fonts to Your App ... 118
Recipe: Splitting Core Text into Pages ... 119
Recipe: Drawing Attributed Text into a PDF ... 120
Recipe: Big Phone Text ... 122
Summary ... 125

4 Geometry ... 127
Recipe: Retrieving Points from Bezier Paths ... 127
Recipe: Thinning Points ... 129
Recipe: Smoothing Drawings ... 132
Recipe: Velocity-Based Stroking ... 135
Recipe: Bounding Bezier Paths ... 137
Recipe: Fitting Paths ... 142
Working with Curves ... 144
Recipe: Moving Items Along a Bezier Path ... 148
Recipe: Drawing Attributed Text Along a Bezier Path ... 151
Recipe: View Transforms ... 154
Recipe: Testing for View Intersection ... 161
Summary ... 166

5 Networking ... 167
Recipe: Secure Credential Storage ... 167
Recipe: Entering Credentials ... 171
Recipe: Handling Authentication Challenges ... 176
Recipe: Uploading Data ... 177

Recipe: Building a Simple Web Server 181
Recipe: OAuth Utilities 184
Recipe: The OAuth Process 188
Summary 196

6 Images 197
Image Sources 197
Reading Image Data 199
Recipe: Fitting and Filling Images 203
Recipe: Rotating Images 208
Recipe: Working with Bitmap Representations 210
Recipe: Basic Image Processing 215
Recipe: Image Convolution 216
Recipe: Basic Core Image Processing 219
Capturing View-Based Screen Shots 221
Drawing into PDF Files 222
Recipe: Reflection 223
Recipe: Emitters 226
Summary 228

7 Cameras 229
Recipe: Snapping Photos 229
Recipe: Enabling a Flashlight 233
Recipe: Accessing the AVFoundation Camera 235
Recipe: EXIF 242
Image Orientations 247
Recipe: Core Image Filtering 249
Recipe: Core Image Face Detection 251
Recipe: Sampling a Live Feed 257
Summary 260

8 Audio 261
Recipe: Playing Audio with AVAudioPlayer 261
Recipe: Looping Audio 269
Recipe: Handling Audio Interruptions 272
Recipe: Recording Audio 274
Recipe: Recording Audio with Audio Queues 280
Recipe: Picking Audio with the MPMediaPickerController 286

Creating a Media Query 288
Recipe: Using the MPMusicPlayerController 290
Summary 294

9 Connecting to the Address Book 297
The AddressBook Frameworks 297
Recipe: Searching the Address Book 322
Recipe: Accessing Contact Image Data 325
Recipe: Picking People 326
Recipe: Limiting Contact Picker Properties 329
Recipe: Adding and Removing Contacts 331
Modifying and Viewing Individual Contacts 334
Recipe: The "Unknown" Person Controller 335
Summary 338

10 Location 339
Authorizing Core Location 339
Recipe: Core Location in a Nutshell 344
Recipe: Geofencing 348
Recipe: Keeping Track of "North" by Using Heading Values 350
Recipe: Forward and Reverse Geocoding 353
Recipe: Viewing a Location 355
Recipe: User Location Annotations 360
Recipe: Creating Map Annotations 363
Summary 369

11 GameKit 371
Enabling Game Center 371
Recipe: Signing In to Game Center 373
Designing Leaderboards and Achievements 375
Recipe: Accessing Leaderboards 378
Recipe: Displaying the Game Center View Controller 380
Recipe: Submitting Scores 381
Recipe: Checking Achievements 382
Recipe: Reporting Achievements to Game Center 383
Recipe: Multiplayer Matchmaking 385
Recipe: Responding to the Matchmaker 387
Recipe: Creating an Invitation Handler 388

Managing Match State ... 390
Recipe: Handling Player State Changes ... 390
Recipe: Retrieving Player Names ... 392
Game Play ... 393
Serializing Data ... 394
Recipe: Synchronizing Data ... 397
Recipe: Turn-by-Turn Matchmaking ... 399
Recipe: Responding to Turn-Based Invitations ... 401
Recipe: Loading Matches ... 402
Recipe: Responding to Game Play ... 403
Recipe: Ending Gameplay ... 407
Recipe: Removing Matches ... 410
Recipe: Game Center Voice ... 411
GameKit Peer Services ... 415
Summary ... 425

12 StoreKit ... **427**
Getting Started with StoreKit ... 427
Creating Test Accounts ... 430
Creating New In-App Purchase Items ... 431
Building a Storefront GUI ... 435
Purchasing Items ... 438
Validating Receipts ... 443
Summary ... 445

13 Push Notifications ... **447**
Introducing Push Notifications ... 447
Provisioning Push ... 451
Registering Your Application ... 454
Recipe: Push Client Skeleton ... 458
Building Notification Payloads ... 465
Recipe: Sending Notifications ... 466
Feedback Service ... 471
Designing for Push ... 473
Summary ... 473

Index ... **475**

Acknowledgments

This book would not exist without the efforts of Chuck Toporek, who was my editor and whipcracker for many years and multiple publishers. He is now at Apple and deeply missed. There'd be no Cookbook were it not for him. He balances two great skill sets: inspiring authors to do what they think they cannot, and wielding the large "reality trout" of whacking[1] to keep subject matter focused and in the real world. There's nothing like being smacked repeatedly by a large virtual fish to bring a book in on deadline and with compelling content.

Thanks go as well to Trina MacDonald (my terrific new editor), Chris Zahn (the awesomely talented development editor), and Olivia Basegio (the faithful and rocking editorial assistant who kept things rolling behind the scenes). Also, a big thank you to the entire Addison-Wesley/Pearson production team, specifically Kristy Hart, Jovana San Nicolas-Shirley, San Dee Phillips, Nonie Ratcliff, and Chuti Prasertsith. Thanks also to the crew at Safari for getting my book up in Rough Cuts and for quickly fixing things when technical glitches occurred.

Thanks go as well to Neil Salkind, my agent of many years, to the tech reviewers Oliver Drobnik, Rich Wardwell, and Duncan Champney, who helped keep this book in the realm of sanity rather than wishful thinking, and to all my colleagues, both present and former, at TUAW, Ars Technica, and the Digital Media/Inside iPhone blog.

I am deeply indebted to the wide community of iOS developers, including Jon Bauer, Tim Burks, Matt Martel, Tim Isted, Joachim Bean, Aaron Basil, Roberto Gamboni, John Muchow, Scott Mikolaitis, Alex Schaefer, Nick Penree, James Cuff, Jay Freeman, Mark Montecalvo, August Joki, Max Weisel, Optimo, Kevin Brosius, Planetbeing, Pytey, Michael Brennan, Daniel Gard, Michael Jones, Roxfan, MuscleNerd, np101137, UnterPerro, Jonathan Watmough, Youssef Francis, Bryan Henry, William DeMuro, Jeremy Sinclair, Arshad Tayyeb, Jonathan Thompson, Dustin Voss, Daniel Peebles, ChronicProductions, Greg Hartstein, Emanuele Vulcano, Sean Heber, Josh Bleecher Snyder, Eric Chamberlain, Steven Troughton-Smith, Dustin Howett, Dick Applebaum, Kevin Ballard, Hamish Allan, Lutz Bendlin, Oliver Drobnik, Rod Strougo, Kevin McAllister, Jay Abbott, Tim Grant Davies, Maurice Sharp, Chris Samuels, Chris Greening, Jonathan Willing, Landon Fuller, Jeremy Tregunna, Christine Reindl, Wil Macaulay, Stefan Hafeneger, Scott Yelich, Mike Kale, chrallelinder, John Varghese, Robert Jen, Andrea Fanfani, J. Roman, jtbandes, Artissimo, Aaron Alexander, Christopher Campbell Jensen, Nico Ameghino, Jon Moody, Julián Romero, Scott Lawrence, Evan K. Stone, Kenny Chan Ching-King, Matthias Ringwald, Jeff Tentschert, Marco Fanciulli, Neil Taylor, Sjoerd van Geffen, Absentia, Nownot, Emerson Malca, Matt Brown, Chris Foresman, Aron Trimble, Paul Griffin, Paul Robichaux, Nicolas Haunold, Anatol Ulrich (hypnocode GmbH), Kristian Glass, Remy "psy" Demarest, Yanik Magnan, ashikase, Shane Zatezalo, Tito Ciuro, Mahipal Raythattha, Jonah Williams of Carbon Five, Joshua Weinberg, biappi, Eric Mock, and everyone at the iPhone developer channels at irc.saurik.com and irc.freenode.net, among many others too numerous to name individually. Their techniques, suggestions, and feedback helped make this book possible. If I have overlooked anyone who helped contribute, please accept my apologies for the oversight.

Special thanks go out to my family and friends, who supported me through month after month of new beta releases and who patiently put up with my unexplained absences and frequent howls of despair. I appreciate you all hanging in there with me. And thanks to my children for their steadfastness, even as they learned that a hunched back and the sound of clicking keys is a pale substitute for a proper mother. My kids provided invaluable assistance over the past few months by testing applications, offering suggestions, and just being awesome people. I try to remind myself on a daily basis how lucky I am that these kids are part of my life.

About the Author

Erica Sadun is the bestselling author, coauthor, and contributor to several dozen books on programming, digital video and photography, and web design, including the widely popular *The iOS 5 Developer's Cookbook*. She currently blogs at TUAW.com and has blogged in the past at O'Reilly's Mac Devcenter, Lifehacker, and Ars Technica. In addition to being the author of dozens of iOS-native applications, Erica holds a Ph.D. in Computer Science from Georgia Tech's Graphics, Visualization, and Usability Center. A geek, a programmer, and an author, she's never met a gadget she didn't love. When not writing, she and her geek husband parent three geeks-in-training, who regard their parents with restrained bemusement, when they're not busy rewiring the house or plotting global dominance.

Preface

Welcome to another iOS Cookbook!

With iOS 6, Apple's mobile device family has reached new levels of excitement and possibility. This Cookbook is here to help you start developing. This revision introduces new features announced at the latest WWDC, showing you how to incorporate them into your applications.

For this edition, my publishing team has sensibly split the Cookbook material into manageable print volumes. This book, *The Advanced iOS 6 Developer's Cookbook,* centers on common frameworks such as StoreKit, GameKit, and Core Location and handy techniques such as image manipulation typesetting. It helps you build applications that leverage special-purpose libraries and move beyond the basics. This volume is for those who have a strong grasp on iOS development and are looking for practical how-to's for specialized areas.

Its companion volume, *The Core iOS 6 Developer's Cookbook,* provides solutions for the heart of day-to-day development. It covers all the classes you need for creating iOS applications using standard APIs and interface elements. It contains the recipes you need for working with graphics, touches, and views to create mobile applications.

Finally, there's *Learning iOS 6: A Hands-On Guide to the Fundamentals of iOS Programming,* which covers much of the tutorial material that used to compose the first several chapters of the Cookbook. There you can find all the fundamental how-to's you need to learn iOS 6 development from the ground up. From Objective-C to Xcode, debugging to deployment, *Learning iOS 6* teaches you how to start with Apple's development tool suite.

As in the past, you can find sample code at github. You'll find the repository for this Cookbook at https://github.com/erica/iOS-6-Cookbook, all of it refreshed for iOS 6 after WWDC 2012.

If you have suggestions, bug fixes, corrections, or anything else you'd like to contribute to a future edition, please contact me at erica@ericasadun.com. Let me thank you all in advance. I appreciate all feedback that helps make this a better, stronger book.

—Erica Sadun, September 2012

What You Need

It goes without saying that, if you plan to build iOS applications, you need at least one iOS device to test your application, preferably a new model iPhone or tablet. The following list covers the basics of what you need to begin:

- **Apple's iOS SDK**—You can download the latest version of the iOS SDK from Apple's iOS Dev Center (http://developer.apple.com/ios). If you plan to sell apps through the App Store, become a paid iOS developer. This costs $99/year for individuals and $299/year for enterprise (that is, corporate) developers. Registered developers receive certificates that enable them to "sign" and download their applications to their iPhone/iPod touch for testing and debugging and to gain early access to prerelease versions of iOS. Free-program developers can test their software on the Mac-based simulator but cannot deploy to devices or submit to the App Store.

> **University Student Program**
>
> Apple also offers a University Program for students and educators. If you are a computer science student taking classes at the university level, check with your professor to see whether your school is part of the University Program. For more information about the iPhone Developer University Program, see http://developer.apple.com/support/iphone/university.

- **A modern Mac running Mac OS X Lion (v 10.7) or, preferably, Mac OS X Mountain Lion (v 10.8)**—You need plenty of disk space for development, and your Mac should have as much RAM as you can afford to put into it.

- **An iOS device**—Although the iOS SDK includes a simulator for you to test your applications in, you really do need to own iOS hardware to develop for the platform. You can tether your unit to the computer and install the software you've built. For real-life App Store deployment, it helps to have several units on hand, representing the various hardware and firmware generations, so you can test on the same platforms your target audience uses.

- **An Internet connection**—This connection enables you to test your programs with a live Wi-Fi connection and with an EDGE or 3G service.

- **Familiarity with Objective-C**—To program for the iPhone, you need to know Objective-C 2.0. The language is based on ANSI C with object-oriented extensions, which means you also need to know a bit of C. If you have programmed with Java or C++ and are familiar with C, you can make the move to Objective-C.

Your Roadmap to Mac/iOS Development

One book can't be everything to everyone. Try as I might, if we were to pack everything you need to know into this book, you wouldn't be able to pick it up. (As it stands, this book offers an excellent tool for upper-body development. Please don't sue if you strain yourself lifting it.) There is, indeed, a lot you need to know to develop for the Mac and iOS platforms. If you are just starting out and don't have any programming experience, your first course of action should be to take a college-level course in the C programming language. Although the alphabet might start with the letter A, the root of most programming languages, and certainly your path as a developer, is C.

When you know C and how to work with a compiler (something you'll learn in that basic C course), the rest should be easy. From there, you can hop right on to Objective-C and learn how to program with that alongside the Cocoa frameworks. The flowchart shown in Figure P-1 shows you key titles offered by Pearson Education that can help provide the training you need to become a skilled iOS developer.

Preface   xv

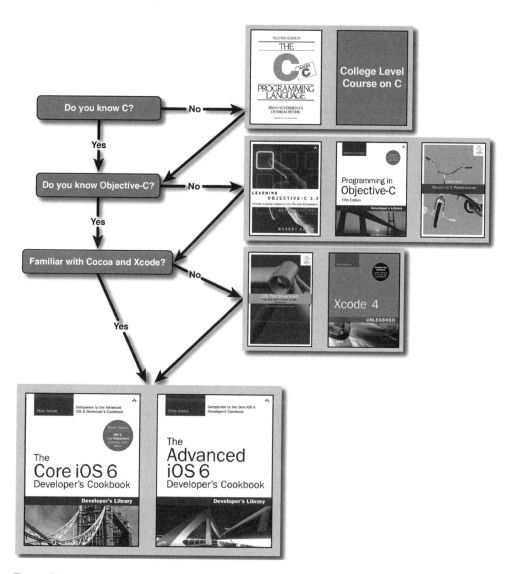

Figure P-1   A roadmap to becoming an iOS developer

When you know C, you have a few options for learning how to program with Objective-C. If you want an in-depth view of the language, you can either read Apple's documentation or pick up one of these books on Objective-C:

- *Objective-C Programming: The Big Nerd Ranch Guide,* by Aaron Hillegass (Big Nerd Ranch, 2012)

- *Learning Objective-C: A Hands-on Guide to Objective-C for Mac and iOS Developers*, by Robert Clair (Addison-Wesley, 2011)
- *Programming in Objective-C 2.0, Fourth Edition*, by Stephen Kochan (Addison-Wesley, 2012)

With the language behind you, next up is tackling Cocoa and the developer tools, otherwise known as Xcode. For that, you have a few different options. Again, you can refer to Apple's documentation on Cocoa and Xcode,[2] or if you prefer books, you can learn from the best. Aaron Hillegass, founder of the Big Nerd Ranch in Atlanta,[3] is the coauthor of *iOS Programming: The Big Nerd Ranch Guide, Second Edition*, and author of *Cocoa Programming for Mac OS X*, soon to be in its fourth edition. Aaron's book is highly regarded in Mac developer circles and is the most-recommended book you'll see on the cocoa-dev mailing list. To learn more about Xcode, look no further than Fritz Anderson's *Xcode 4 Unleashed* from Sams Publishing.

> **Note**
>
> There are plenty of other books from other publishers on the market, including the bestselling *Beginning iPhone 4 Development*, by Dave Mark, Jack Nutting, and Jeff LaMarche (Apress, 2011). Another book that's worth picking up if you're a total newbie to programming is *Beginning Mac Programming*, by Tim Isted (Pragmatic Programmers, 2011). Don't just limit yourself to one book or publisher. Just as you can learn a lot by talking with different developers, you can learn lots of tricks and tips from other books on the market.

To truly master Mac development, you need to look at a variety of sources: books, blogs, mailing lists, Apple's documentation, and, best of all, conferences. If you get the chance to attend WWDC, you'll know what I'm talking about. The time you spend at those conferences talking with other developers, and in the case of WWDC, talking with Apple's engineers, is well worth the expense if you are a serious developer.

How This Book Is Organized

This book offers single-task recipes for the most common issues new iOS developers face: laying out interface elements, responding to users, accessing local data sources, and connecting to the Internet. Each chapter groups together related tasks, enabling you to jump directly to the solution you're looking for without having to decide which class or framework best matches that problem.

The iOS 6 Developer's Cookbook offers you "cut-and-paste convenience," which means you can freely reuse the source code from recipes in this book for your own applications and then tweak the code to suit your app's needs.

Here's a rundown of this book's chapters:

- **Chapter 1, "Device-Specific Development"**—Each iOS device represents a meld of unique, shared, momentary, and persistent properties. These properties include the device's current physical orientation, its model name, its battery state, and its access to onboard hardware. This chapter looks at the device from its build configuration to its active onboard sensors. It provides recipes that return a variety of information items about the unit in use.

- **Chapter 2, "Documents and Data Sharing"**—Under iOS, applications can share information and data as well as move control from one application to another using several system-supplied features. This chapter introduces the ways you can integrate documents and data sharing between applications. You see how to add these features into your applications and use them smartly to make your app a cooperative citizen of the iOS ecosystem.

- **Chapter 3, "Core Text"**—This chapter introduces attributed text processing and explores how you can build text features into your apps. You read about adding attributed strings to common UIKit elements, how to create Core Text-powered views, and how to break beyond lines for freeform text typesetting. After reading this chapter, you'll have discovered the power that Core Text brings to iOS.

- **Chapter 4, "Geometry"**—Although UIKit requires less applied math than, say, Core Animation or Open GL, geometry plays an important role when working with Bezier paths and view transforms. Why do you need geometry? It helps you manipulate views in nonstandard ways, including laying out text along custom paths and performing path-follow types of animation. If your eyes glaze over at the mention of Bezier curves, Convex Hulls, and splines, this chapter can help demystify these terms, enabling you to add some powerful customization options to your toolbox.

- **Chapter 5, "Networking"**—Apple has lavished iOS with a solid grounding in all kinds of network computing and its supporting technologies. The networking chapter in the Core Cookbook introduced network status checks, synchronous and asynchronous downloads, JSON, and XML parsing. This chapter continues that theme by introducing more advanced techniques. These include authentication challenges, using the system keychain, working with OAuth, and so forth. Here are handy approaches that should help with your development.

- **Chapter 6, "Images"**—Images are abstract representations, storing data that makes up pictures. This chapter introduces Cocoa Touch images, specifically the UIImage class, and teaches you all the basic know-how you need for working with image data on iOS. In this chapter, you learn how to load, store, and modify image data in your applications. You discover how to process image data to create special effects, how to access images on a byte-by-byte basis, and more.

- **Chapter 7, "Camera"**—Cameras kick images up to the next level. They enable you to integrate live feeds and user-directed snapshots into your applications, and provide raw data sourced from the real world. In this chapter, you read about image capture. You discover how to take pictures using Apple-sourced classes and how to roll your own from scratch. You learn about controlling image metadata and how to integrate live feeds with advanced filtering. This chapter focuses on image capture from a hardware point of view. Whether you're switching on the camera flash LED or detecting faces, this chapter introduces the ins and outs of iOS image capture technology.

- **Chapter 8, "Audio"**—The iOS device is a media master; its built-in iPod features expertly handle both audio and video. The iOS SDK exposes that functionality to developers. A rich suite of classes simplifies media handling via playback, search, and recording. This chapter introduces recipes that use those classes for audio, presenting media to your users and letting your users interact with that media. You see how to build audio players and audio recorders. You discover how to browse the iPod library and how to choose what items to play.

- **Chapter 9, "Connecting to the Address Book"**—This chapter introduces the Address Book and demonstrates how to use its frameworks in your applications. You read about accessing information on a contact-by-contact basis, how to modify and update contact information, and how to use predicates to find just the contact you're interested in. This chapter also covers the GUI classes that provide interactive solutions for picking, viewing, and modifying contacts.

- **Chapter 10, "Location"**—Where you compute is fast becoming just as important as how you compute and what you compute. iOS is constantly on the go, traveling with its users throughout the course of the day. Core Location infuses iOS with on-demand geopositioning. MapKit adds interactive in-application mapping, enabling users to view and manipulate annotated maps. With Core Location and MapKit, you can develop applications that help users meet up with friends, search for local resources, or provide location-based streams of personal information. This chapter introduces these location-aware frameworks and shows you how you can integrate them into your iOS applications.

- **Chapter 11, "GameKit"**—This chapter introduces various ways you can create connected game play through GameKit. GameKit offers features that enable your applications to move beyond a single-player/single-device scenario toward using Game Center and device-to-device networking. Apple's Game Center adds a centralized service that enables your game to offer shared leaderboards and Internet-based matches. GameKit also provides an ad-hoc networking solution for peer-to-peer connectivity.

- **Chapter 12, "StoreKit"**—StoreKit offers in-app purchasing that integrates into your software. With StoreKit, end users can use their iTunes credentials to buy unlockable features, media subscriptions, or consumable assets, such as fish food or sunlight, from within an application. This chapter introduces StoreKit and shows you how to use the StoreKit API to create purchasing options for users.

- **Chapter 13, "Push Notification"**—When off-device services need to communicate directly with users, push notifications provide a solution. Just as local notifications enable apps to contact users at scheduled times, push notifications deliver messages from web-based systems. Push notifications enable devices to display an alert, play a custom sound, or update an application badge. In this way, off-device services connect with an iOS-based client, enabling them to know about new data or updates. This chapter introduces all the push notification basics you need to know.

About the Sample Code

For the sake of pedagogy, this book's sample code uses a single main.m file. This is not how people normally develop iPhone or Cocoa applications, or, honestly, how they should be developing them, but it provides a great way of presenting a single big idea. It's hard to tell a story when readers must look through five or seven or nine individual files at once. Offering a single file concentrates that story, allowing access to that idea in a single chunk.

These examples are not intended as stand-alone applications. They are there to demonstrate a single recipe and a single idea. One main.m file with a central presentation reveals the implementation story in one place. You can study these concentrated ideas and transfer them into normal application structures, using the standard file structure and layout. The presentation in this book does not produce code in a standard day-to-day best-practices approach. Instead, it offers concise solutions that you can incorporate back into your work as needed.

Contrast that to Apple's standard sample code, where you must comb through many files to build up a mental model of the concepts that are being demonstrated. Those examples are built as full applications, often doing tasks that are related to but not essential to what you need to solve. Finding just those relevant portions is a lot of work. The effort may outweigh any gains.

In this book, you find exceptions to this one-file-with-the-story rule: The Cookbook provides standard class and header files when a class implementation is the recipe. Instead of highlighting a technique, some recipes offer these classes and categories (that is, extensions to a preexisting class rather than a new class). For those recipes, look for separate .m *and* .h files in addition to the skeletal main.m that encapsulates the rest of the story.

For the most part, the examples for this book use a single application identifier: com.sadun.helloworld. This book uses one identifier to avoid clogging up your iOS devices with dozens of examples at once. Each example replaces the previous one, ensuring that your home screen remains relatively uncluttered. If you want to install several examples simultaneously, simply edit the identifier, adding a unique suffix, such as com.sadun.helloworld.table-edits. You can also edit the custom display name to make the apps visually distinct. Your Team Provisioning Profile matches every application identifier, including com.sadun.helloworld. This enables you to install compiled code to devices without having to change the identifier; just make sure to update your signing identity in each project's build settings.

Getting the Sample Code

You'll find the source code for this book at github.com/erica/iOS-6-Cookbook on the open-source GitHub hosting site. There, you can find a chapter-by-chapter collection of source code that provides working examples of the material covered in this book. Recipes are numbered as they are in the book. Recipe 6 in Chapter 5, for example, appears in the C05 folder in the 06 subfolder.

Any project numbered 00 or that has a suffix (such as 05b or 02c) refers to material used to create in-text coverage and figures. Normally I delete these extra projects. Early readers of this manuscript requested that I include them in this edition. You can find a half dozen or so of these extra samples scattered around the repository.

If you do not feel comfortable using git directly, GitHub offers a download button. It was at the right side of the main page at the time this book was written, about halfway down the first page. It enables you to retrieve the entire repository as a ZIP archive or tarball.

Contribute!

Sample code is never a fixed target. It continues to evolve as Apple updates its SDK and the Cocoa Touch libraries. Get involved. You can pitch in by suggesting bug fixes and corrections as well as by expanding the code that's on offer. GitHub enables you to fork repositories and grow them with your own tweaks and features, and share those back to the main repository. If you come up with a new idea or approach, let me know. My team and I are happy to include great suggestions both at the repository and in the next edition of this Cookbook.

Getting Git

You can download this Cookbook's source code using the git version control system. An OS X implementation of git is available at http://code.google.com/p/git-osx-installer. OS X git implementations include both command-line and GUI solutions, so hunt around for the version that best suits your development needs.

Getting GitHub

GitHub (http://github.com) is the largest git-hosting site, with more than 150,000 public repositories. It provides both free hosting for public projects and paid options for private projects. With a custom web interface that includes wiki hosting, issue tracking, and an emphasis on social networking of project developers, it's a great place to find new code or collaborate on existing libraries. You can sign up for a free account at its Web site, enabling you to copy and modify the Cookbook repository or create your own open-source iOS projects to share with others.

Contacting the Author

If you have any comments or questions about this book, please drop me an e-mail message at erica@ericasadun.com, or stop by the github repository and contact me there.

Endnotes

1. No trouts, real or imaginary, were hurt in the development and production of this book. The same cannot be said for countless cans of Diet Coke that selflessly surrendered their contents in the service of this manuscript.

2. See the *Cocoa Fundamentals Guide* (http://developer.apple.com/mac/library/documentation/Cocoa/Conceptual/CocoaFundamentals/CocoaFundamentals.pdf) for a head start on Cocoa, and for Xcode, see *A Tour of Xcode* (http://developer.apple.com/mac/library/documentation/DeveloperTools/Conceptual/A_Tour_of_Xcode/A_Tour_of_Xcode.pdf).

3. Big Nerd Ranch: www.bignerdranch.com.

Editor's Note: We Want to Hear from You!

As the reader of this book, you are our most important critic and commentator. We value your opinion and want to know what we're doing right, what we could do better, what areas you'd like to see us publish in, and any other words of wisdom you're willing to pass our way.

You can e-mail or write me directly to let me know what you did or didn't like about this book—as well as what we can do to make our books stronger.

Please note that I cannot help you with technical problems related to the topic of this book, and that due to the high volume of mail I receive, I might not be able to reply to every message.

When you write, please be sure to include this book's title and author as well as your name and phone or e-mail address. I will carefully review your comments and share them with the author and editors who worked on the book.

E-mail: trina.macdonald@pearson.com

Mail: Trina MacDonald
Senior Acquisitions Editor
Addison-Wesley/Pearson Education, Inc.
75 Arlington St., Ste. 300
Boston, MA 02116

Device-Specific Development

Each iOS device represents a meld of unique, shared, momentary, and persistent properties. These properties include the device's current physical orientation, its model name, its battery state, and its access to onboard hardware. This chapter looks at the device from its build configuration to its active onboard sensors. It provides recipes that return a variety of information items about the unit in use. You read about testing for hardware prerequisites at runtime and specifying those prerequisites in the application's Info.plist file. You discover how to solicit sensor feedback via Core Motion and subscribe to notifications to create callbacks when sensor states change. You read about adding screen mirroring and second-screen output, and about soliciting device-specific details for tracking. This chapter covers the hardware, file system, and sensors available on the iPhone device and helps you programmatically take advantage of those features.

Accessing Basic Device Information

The UIDevice class exposes key device-specific properties, including the iPhone, iPad, or iPod touch model being used, the device name, and the OS name and version. It's a one-stop solution for pulling out certain system details. Each method is an instance method, which is called using the UIDevice singleton, via [UIDevice currentDevice].

The system information you can retrieve from UIDevice includes these items:

- **systemName**—This returns the name of the operating system currently in use. For current generations of iOS devices, there is only one OS that runs on the platform: iPhone OS. Apple has not yet updated this name to match the general iOS rebranding.

- **systemVersion**—This value lists the firmware version currently installed on the unit: for example, 4.3, 5.1.1, 6.0, and so on.

- **model**—The iPhone model returns a string that describes its platform—namely iPhone, iPad, and iPod touch. Should iOS be extended to new devices, additional strings will describe those models. `localizedModel` provides a localized version of this property.
- **userInterfaceIdiom**—This property represents the interface style used on the current device, namely either iPhone (for iPhone and iPod touch) or iPad. Other idioms may be introduced as Apple offers additional platform styles.
- **name**—This string presents the iPhone name assigned by the user in iTunes, such as "Joe's iPhone" or "Binky." This name is also used to create the local hostname for the device.

Here are a few examples of these properties in use:

```
UIDevice *device = [UIDevice currentDevice];
NSLog(@"System name: %@", device.systemName);
NSLog(@"Model: %@", device.model);
NSLog(@"Name: %@", device.name);
```

For current iOS releases, you can use the idiom check with a simple Boolean test. Here's an example of how you might implement an iPad check. Notice the convenience macro. It tests for selector conformance and then returns [UIDevice currentDevice]. userInterfaceIdiom if possible, and UIUserInterfaceIdiomPhone otherwise:

```
#define IS_IPAD (UI_USER_INTERFACE_IDIOM() == UIUserInterfaceIdiomPad)
```

Should this test fail, you may currently assume that you're working with an iPhone/iPod touch. If and when Apple releases a new family of devices, you'll need to update your code accordingly for a more nuanced test.

Adding Device Capability Restrictions

An application's Info.plist property list enables you to specify application requirements when you submit applications to iTunes. These restrictions enable you to tell iTunes what device features your application needs.

Each iOS unit provides a unique feature set. Some devices offer cameras and GPS capabilities. Others don't. Some have onboard gyros, autofocus, and other powerful options. You specify what features are needed to run your application on a device.

When you include the `UIRequiredDeviceCapabilities` key in your Info.plist file, iTunes limits application installation to devices that offer the required capabilities. Provide this list as either an array of strings or a dictionary.

An array specifies each required capability; each item in that array must be present on your device. A dictionary enables you to explicitly require or prohibit a feature. The dictionary keys are the capabilities. The dictionary values set whether the feature must be present (Boolean true) or omitted (Boolean false).

The current keys are detailed in Table 1-1. Only include those features that your application absolutely requires or cannot support. If your application can provide workarounds, do not add restrictions in this way. Table 1-1 discusses each feature in a positive sense. When using a prohibition rather than a requirement, reverse the meaning—for example, that an autofocus camera or gyro cannot be onboard, or that Game Center access cannot be supported.

Table 1-1 **Required Device Capabilities**

Key	Use
telephony	Application requires the Phone application or uses tel:// URLs.
wifi	Application requires local 802.11-based network access. If iOS must maintain that Wi-Fi connection as the app runs, add `UIRequiresPersistentWiFi` as a top-level property list key.
sms	Application requires the Messages application or uses sms:// URLs.
still-camera	Application requires an onboard still camera and can use the image picker interface to capture photos from that still camera.
auto-focus-camera	Application requires extra focus capabilities for macro photography or especially sharp images for in-image data detection.
front-facing-camera	Application requires a front-facing camera on the device.
camera-flash	Application requires a camera flash feature.
video-camera	Application requires a video-capable camera.
accelerometer	Application requires accelerometer-specific feedback beyond simple `UIViewController` orientation events.
gyroscope	Application requires an onboard gyroscope on the device.
location-services	Application uses Core Location of any kind.
gps	Application uses Core Location and requires the additional accuracy of GPS positioning.
magnetometer	Application uses Core Location and requires heading-related events—that is, the direction of travel. (The magnetometer is the built-in compass.)
gamekit	Application requires Game Center access (iOS 4.1 and later).
microphone	Application uses either built-in microphones or (approved) accessories that provide a microphone.
opengles-1	Application requires OpenGL ES 1.1.
opengles-2	Application requires OpenGL ES 2.0.
armv6	Application is compiled only for the armv6 instruction set (3.1 or later).
armv7	Application is compiled only for the armv7 instruction set (3.1 or later).

Key	Use
peer-peer	Application uses GameKit peer-to-peer connectivity over Bluetooth (3.1 or later).
bluetooth-le	Application requires Bluetooth low-energy support (5.0 and later).

For example, consider an application that offers an option for taking pictures when run on a camera-ready device. If the application otherwise works on pre-camera iPod touch units, do not include the still-camera restriction. Instead, use check for camera capability from within the application and present the camera option when appropriate. Adding a still-camera restriction eliminates many early iPod touch (first through third generation) and iPad (first generation) owners from your potential customer pool.

User Permission Descriptions

To protect privacy, the end user must explicitly permit your applications to access reminders, photos, location, contacts, and calendar data. To convince the user to opt-in, it helps to explain how your application can use this data and describe your reason for accessing it. Assign string values to the following keys at the top level of your Info.plist file. When iOS prompts your user for resource-specific permission, it displays these strings as part of its standard dialog box:

- NSRemindersUsageDescription
- NSPhotoLibraryUsageDescription
- NSLocationUsageDescription
- NSContactsUsageDescription
- NSCalendarsUsageDescription

Other Common Info.plist Keys

Here are a few other common keys you may want to assign in your property list, along with descriptions of what they do:

- **UIFileSharingEnabled** (Boolean, defaults to off)—Enables users to access the contents of your app's Documents folder from iTunes. This folder appears at the top level of your app sandbox.

- **UIAppFonts** (Array, strings of font names including their extension)—Specifies custom TTF fonts that you supply in your bundle. When added, you access them using standard UIFont calls.

- **UIApplicationExitsOnSuspend** (Boolean, defaults to off)—Enables your app to terminate when the user clicks the Home button rather than move to the background. When enabled, iOS terminates the app and purges it from memory.

- `UIRequiresPersistentWifi` (Boolean, defaults to off)—Instructs iOS to maintain a Wi-Fi connection while the app is active.
- `UIStatusBarHidden` (Boolean, defaults to off)—If enabled, hides the status bar as the app launches.
- `UIStatusBarStyle` (String, defaults to UIStatusBarStyleDefault)—Specifies the style of the status bar at app launch.

Recipe: Checking Device Proximity and Battery States

The `UIDevice` class offers APIs that enable you to keep track of device characteristics including the states of the battery and proximity sensor. Recipe 1-1 demonstrates how you can enable and query monitoring for these two technologies. Both provide updates in the form of notifications, which you can subscribe to so your application is informed of important updates.

Enabling and Disabling the Proximity Sensor

Proximity is an iPhone-specific feature at this time. The iPod touch and iPad do not offer proximity sensors. Unless you have some pressing reason to hold an iPhone against body parts (or vice versa), using the proximity sensor accomplishes little.

When enabled, it has one primary task. It detects whether there's a large object right in front of it. If so, it switches the screen off and sends a general notification. Move the blocking object away and the screen switches back on. This prevents you from pressing buttons or dialing the phone with your ear when you are on a call. Some poorly designed protective cases keep the iPhone's proximity sensors from working properly.

Siri uses this feature. When you hold the phone up to your ear, it records your query, sending it to be interpreted. Siri's voice interface does not depend on a visual GUI to operate.

Recipe 1-1 also demonstrates how to work with proximity sensing on the iPhone. Its code uses the `UIDevice` class to toggle proximity monitoring and subscribes to `UIDeviceProximity StateDidChangeNotification` to catch state changes. The two states are on and off. When the `UIDevice proximityState` property returns YES, the proximity sensor has been activated.

Monitoring the Battery State

You can programmatically keep track of the battery and device state. These APIs enable you to know the level to which the battery is charged and whether the device is plugged into a charging source. The battery level is a floating-point value that ranges between 1.0 (fully charged) and 0.0 (fully discharged). It provides an approximate discharge level that you can use to query before performing operations that put unusual strain on the device.

For example, you might want to caution your user about performing a large series of mathematical computations and suggest that the user plug in to a power source. You retrieve the battery level via this UIDevice call. The value returned is produced in 5% increments:

```
NSLog(@"Battery level: %0.2f%",
    [UIDevice currentDevice].batteryLevel * 100);
```

The charge state has four possible values. The unit can be charging (that is, connected to a power source), full, unplugged, and a catchall "unknown." Recover the state using the UIDevice batteryState property:

```
NSArray *stateArray = @[
    @"Battery state is unknown",
    @"Battery is not plugged into a charging source",
    @"Battery is charging",
    @"Battery state is full"];

NSLog(@"Battery state: %@",
    stateArray[[UIDevice currentDevice].batteryState]);
```

Don't think of these choices as persistent states. Instead, think of them as momentary reflections of what is actually happening to the device. They are not flags. They are not OR'ed together to form a general battery description. Instead, these values reflect the most recent state change.

You can easily monitor state changes by responding to notifications that the battery state has changed. In this way, you can catch momentary events, such as when the battery finally recharges fully, when the user has plugged in to a power source to recharge, and when the user disconnects from that power source.

To start monitoring, set the batteryMonitoringEnabled property to YES. During monitoring, the UIDevice class produces notifications when the battery state or level changes. Recipe 1-1 subscribes to both notifications. Please note that you can also check these values directly, without waiting for notifications. Apple provides no guarantees about the frequency of level change updates, but as you can tell by testing this recipe, they arrive in a fairly regular fashion.

Recipe 1-1 **Monitoring Proximity and Battery**

```
// View the current battery level and state
- (void) peekAtBatteryState
{
    NSArray *stateArray = [NSArray arrayWithObjects:
                            @"Battery state is unknown",
                            @"Battery is not plugged into a charging source",
                            @"Battery is charging",
                            @"Battery state is full", nil];

    NSString *status = [NSString stringWithFormat:
```

```
            @"Battery state: %@, Battery level: %0.2f%%",
            [stateArray objectAtIndex:[UIDevice currentDevice].batteryState],
            [UIDevice currentDevice].batteryLevel * 100];

    NSLog(@"%@", status);
}

// Show whether proximity is being monitored
- (void) updateTitle
{
    self.title = [NSString stringWithFormat:@"Proximity %@",
        [UIDevice currentDevice].proximityMonitoringEnabled ? @"On" : @"Off"];
}

// Toggle proximity monitoring off and on
- (void) toggle: (id) sender
{
    // Determine the current proximity monitoring and toggle it
    BOOL isEnabled = [UIDevice currentDevice].proximityMonitoringEnabled;
    [UIDevice currentDevice].proximityMonitoringEnabled = !isEnabled;
    [self updateTitle];
}

- (void) loadView
{
    [super loadView];

    // Enable toggling and initialize title
    self.navigationItem.rightBarButtonItem =
        BARBUTTON(@"Toggle", @selector(toggle:));
    [self updateTitle];

    // Add proximity state checker
    [[NSNotificationCenter defaultCenter]
        addObserverForName:UIDeviceProximityStateDidChangeNotification
        object:nil queue:[NSOperationQueue mainQueue]
        usingBlock:^(NSNotification *notification) {
            // Sensor has triggered either on or off
            NSLog(@"The proximity sensor %@",
                [UIDevice currentDevice].proximityState ?
                @"will now blank the screen" : @"will now restore the screen");
    }];

    // Enable battery monitoring
    [[UIDevice currentDevice] setBatteryMonitoringEnabled:YES];
```

```objc
// Add observers for battery state and level changes
[[NSNotificationCenter defaultCenter]
    addObserverForName:UIDeviceBatteryStateDidChangeNotification
    object:nil queue:[NSOperationQueue mainQueue]
    usingBlock:^(NSNotification *notification) {
        // State has changed
        NSLog(@"Battery State Change");
        [self peekAtBatteryState];
}];

[[NSNotificationCenter defaultCenter]
    addObserverForName:UIDeviceBatteryLevelDidChangeNotification
    object:nil queue:[NSOperationQueue mainQueue]
    usingBlock:^(NSNotification *notification) {
        // Level has changed
        NSLog(@"Battery Level Change");
        [self peekAtBatteryState];
}];
}
```

> **Get This Recipe's Code**
>
> To find this recipe's full sample project, point your browser to https://github.com/erica/iOS-6-Advanced-Cookbook and go to the folder for Chapter 1.

Detecting Retina Support

In recent years, Apple introduced the Retina display on its flagship devices. Its pixel density is, according to Apple, high enough so the human eye cannot distinguish individual pixels. Apps shipped with higher-resolution art take advantage of this improved display quality.

The `UIScreen` class offers an easy way to check whether the current device offers a built-in Retina display. Check the screen `scale` property, which provides the factor that converts from the logical coordinate space (points, approximately $1/160^{th}$ of an inch) into a device coordinate space (pixels). It is 1.0 for standard displays, so one point corresponds to one pixel. It is 2.0 for Retina displays (4 pixels per point):

```objc
- (BOOL) hasRetinaDisplay
{
    return ([UIScreen mainScreen].scale == 2.0f);
}
```

The `UIScreen` class also offers two useful display-size properties. The `bounds` returns the screen's bounding rectangle, measured in points. This gives you the full size of the screen, regardless of any onscreen elements such as status bars, navigation bars, or tab bars. The `applicationFrame` property, also measured in points, excludes the status bar, providing the frame for your application's initial window size.

Recipe: Recovering Additional Device Information

Both `sysctl()` and `sysctlbyname()` enable you to retrieve system information. These standard UNIX functions query the operating system about hardware and OS details. You can get a sense of the kind of scope on offer by glancing at the /usr/include/sys/sysctl.h include file on the Macintosh. There you can find an exhaustive list of constants that can be used as parameters to these functions.

These constants enable you to check for core information such as the system's CPU frequency, the amount of available memory, and more. Recipe 1-2 demonstrates this functionality. It introduces a `UIDevice` category that gathers system information and returns it via a series of method calls.

You might wonder why this category includes a platform method, when the standard `UIDevice` class returns device models on demand. The answer lies in distinguishing different types of units.

An iPhone 3GS's model is simply "iPhone," as is the model of an iPhone 4S. In contrast, this recipe returns a platform value of "iPhone2,1" for the 3GS and "iPhone 4,1" for the iPhone 4S. This enables you to programmatically differentiate the 3GS unit from a first-generation iPhone ("iPhone1,1") or iPhone 3G ("iPhone1,2").

Each model offers distinct built-in capabilities. Knowing exactly which iPhone you're dealing with helps you determine whether that unit likely supports features such as accessibility, GPS, and magnetometers.

Recipe 1-2   Extending Device Information Gathering

```
@implementation UIDevice (Hardware)
+ (NSString *) getSysInfoByName:(char *)typeSpecifier
{
    // Recover sysctl information by name
    size_t size;
    sysctlbyname(typeSpecifier, NULL, &size, NULL, 0);

    char *answer = malloc(size);
    sysctlbyname(typeSpecifier, answer, &size, NULL, 0);

    NSString *results = [NSString stringWithCString:answer
        encoding: NSUTF8StringEncoding];
    free(answer);

    return results;
}
```

```objc
- (NSString *) platform
{
    return [UIDevice getSysInfoByName:"hw.machine"];
}

- (NSUInteger) getSysInfo: (uint) typeSpecifier
{
    size_t size = sizeof(int);
    int results;
    int mib[2] = {CTL_HW, typeSpecifier};
    sysctl(mib, 2, &results, &size, NULL, 0);
    return (NSUInteger) results;
}

- (NSUInteger) cpuFrequency
{
    return [UIDevice getSysInfo:HW_CPU_FREQ];
}

- (NSUInteger) busFrequency
{
    return [UIDevice getSysInfo:HW_BUS_FREQ];
}

- (NSUInteger) totalMemory
{
    return [UIDevice getSysInfo:HW_PHYSMEM];
}

- (NSUInteger) userMemory
{
    return [UIDevice getSysInfo:HW_USERMEM];
}

- (NSUInteger) maxSocketBufferSize
{
    return [UIDevice getSysInfo:KIPC_MAXSOCKBUF];
}
@end
```

Get This Recipe's Code

To find this recipe's full sample project, point your browser to https://github.com/erica/iOS-6-Advanced-Cookbook and go to the folder for Chapter 1.

Recipe: Using Acceleration to Locate "Up"

The iPhone provides three onboard sensors that measure acceleration along the iPhone's perpendicular axes: left/right (x), up/down (y), and front/back (z). These values indicate the forces affecting the iPhone, from both gravity and user movement. You can get some neat force feedback by swinging the iPhone around your head (centripetal force) or dropping it from a tall building (freefall). Unfortunately, you might not recover that data after your iPhone becomes an expensive bit of scrap metal.

To subscribe an object to iPhone accelerometer updates, set it as the delegate. The object set as the delegate must implement the UIAccelerometerDelegate protocol:

```
[[UIAccelerometer sharedAccelerometer] setDelegate:self]
```

When assigned, your delegate receives accelerometer:didAccelerate: callback messages, which you can track and respond to. The UIAcceleration structure sent to the delegate method consists of floating-point values for the x, y, and z axes. Each value ranges from –1.0 to 1.0:

```
float x = acceleration.x;
float y = acceleration.y;
float z = acceleration.z;
```

Recipe 1-3 uses these values to help determine the "up" direction. It calculates the arctangent between the X and Y acceleration vectors, returning the up-offset angle. As new acceleration messages are received, the recipe rotates a UIImageView instance with its picture of an arrow, which you can see in Figure 1-1, to point up. The real-time response to user actions ensures that the arrow continues pointing upward, no matter how the user reorients the phone.

Recipe 1-3 **Catching Acceleration Events**

```
- (void)accelerometer:(UIAccelerometer *)accelerometer
    didAccelerate:(UIAcceleration *)acceleration
{
    // Determine up from the x and y acceleration components
    float xx = -acceleration.x;
    float yy = acceleration.y;
    float angle = atan2(yy, xx);
    [arrow setTransform:
        CGAffineTransformMakeRotation(angle)];
}


- (void) viewDidLoad
{
    // Initialize the delegate to start catching accelerometer events
    [UIAccelerometer sharedAccelerometer].delegate = self;
}
```

Figure 1-1 A little math recovers the "up" direction by performing an arctan function using the x and y force vectors. In this example, the arrow always points up, no matter how the user reorients the iPhone.

Get This Recipe's Code

To find this recipe's full sample project, point your browser to https://github.com/erica/iOS-6-Advanced-Cookbook and go to the folder for Chapter 1.

Working with Basic Orientation

The `UIDevice` class uses the built-in `orientation` property to retrieve the physical orientation of the device. iOS devices support seven possible values for this property:

- `UIDeviceOrientationUnknown`—The orientation is currently unknown.
- `UIDeviceOrientationPortrait`—The home button is down.
- `UIDeviceOrientationPortraitUpsideDown`—The home button is up.
- `UIDeviceOrientationLandscapeLeft`—The home button is to the right.
- `UIDeviceOrientationLandscapeRight`—The home button is to the left.

- **UIDeviceOrientationFaceUp**—The screen is face up.
- **UIDeviceOrientationFaceDown**—The screen is face down.

The device can pass through any or all of these orientations during a typical application session. Although orientation is created in concert with the onboard accelerometer, these orientations are not tied in any way to a built-in angular value.

iOS offers two built-in macros to help determine if a device orientation enumerated value is portrait or landscape: namely UIDeviceOrientationIsPortrait() and UIDeviceOrientationIsLandscape(). It is convenient to extend the UIDevice class to offer these tests as built-in device properties:

```
@property (nonatomic, readonly) BOOL isLandscape;
@property (nonatomic, readonly) BOOL isPortrait;

- (BOOL) isLandscape
{
    return UIDeviceOrientationIsLandscape(self.orientation);
}

- (BOOL) isPortrait
{
    return UIDeviceOrientationIsPortrait(self.orientation);
}
```

Your code can subscribe directly to device reorientation notifications. To accomplish this, send beginGeneratingDeviceOrientationNotifications to the currentDevice singleton. Then add an observer to catch the ensuing UIDeviceOrientationDidChangeNotification updates. As you would expect, you can finish listening by calling endGeneratingDeviceOrientationNotification.

Retrieving the Current Accelerometer Angle Synchronously

At times you may want to query the accelerometer without setting yourself up as a full delegate. The following methods, which are meant for use within a UIDevice category, enable you to synchronously return the current device angle along the x/y plane—the front face plane of the iOS device. Accomplish this by entering a new run loop, wait for an accelerometer event, retrieve the current angle from that callback, and then leave the run loop to return that angle:

```
- (void)accelerometer:(UIAccelerometer *)accelerometer
    didAccelerate:(UIAcceleration *)acceleration
{
    float xx = acceleration.x;
    float yy = -acceleration.y;
    device_angle = M_PI / 2.0f - atan2(yy, xx);
```

```
    if (device_angle > M_PI)
        device_angle -= 2 * M_PI;

    CFRunLoopStop(CFRunLoopGetCurrent());
}

- (float) orientationAngle
{
    // Supercede current delegate
    id priorDelegate = [UIAccelerometer sharedAccelerometer].delegate;
    [UIAccelerometer sharedAccelerometer].delegate = self;

    // Wait for a reading
    CFRunLoopRun();

    // Restore delegate
    [UIAccelerometer sharedAccelerometer].delegate = priorDelegate;

    return device_angle;
}
```

This is not an approach to use for continuous polling—use the callbacks directly for that. But for an occasional angle query, these methods provide simple and direct access to the current screen angle.

Calculating Orientation from the Accelerometer

The `UIDevice` class does not report a proper orientation when applications are first launched. It updates the orientation only after the device has moved into a new position or `UIViewController` methods kick in.

An application launched in portrait orientation may not read as "portrait" until the user moves the device out of and then back into the proper orientation. This condition exists on the simulator and on the iPhone device and is easily tested. (Radars for this issue have been closed with updates that the features are working as designed.)

For a workaround, consider recovering the angular orientation directly as just shown. Then, after you determine the device angle, convert from the accelerometer-based angle to a device orientation. Here's how that might work in code:

```
// Limited to the four portrait/landscape options
- (UIDeviceOrientation) acceleratorBasedOrientation
{
    CGFloat baseAngle = self.orientationAngle;
    if ((baseAngle > -M_PI_4) && (baseAngle < M_PI_4))
        return UIDeviceOrientationPortrait;
    if ((baseAngle < -M_PI_4) && (baseAngle > -3 * M_PI_4))
```

```
        return UIDeviceOrientationLandscapeLeft;
    if ((baseAngle > M_PI_4) && (baseAngle < 3 * M_PI_4))
        return UIDeviceOrientationLandscapeRight;
    return UIDeviceOrientationPortraitUpsideDown;
}
```

Be aware that this example looks only at the x-y plane, which is where most user interface decisions need to be made. This snippet completely ignores the z-axis, meaning that you'll end up with vaguely random results for the face-up and face-down orientations. Adapt this code to provide that nuance if needed.

The `UIViewController` class's `interfaceOrientation` instance method reports the orientation of a view controller's interface. Although this is not a substitute for accelerometer readings, many interface layout issues rest on the underlying view orientation rather than device characteristics.

Be aware that, especially on the iPad, a child view controller may use a layout orientation that's distinct from a device orientation. For example, an embedded controller may present a portrait layout within a landscape split view controller. Even so, consider whether your orientation-detection code is satisfiable by the underlying interface orientation. It may be more reliable than device orientation, especially as the application launches. Develop accordingly.

Calculate a Relative Angle

Screen reorientation support means that an interface's relationship to a given device angle must be supported in quarters, one for each possible front-facing screen orientation. As the `UIViewController` automatically rotates its onscreen view, the math needs to catch up to account for those reorientations.

The following method, which is written for use in a `UIDevice` category, calculates angles so that the angle remains in synchrony with the device orientation. This creates simple offsets from vertical that match the way the GUI is currently presented:

```
- (float) orientationAngleRelativeToOrientation:
    (UIDeviceOrientation) someOrientation
{
    float dOrientation = 0.0f;
    switch (someOrientation)
    {
        case UIDeviceOrientationPortraitUpsideDown:
            {dOrientation = M_PI; break;}
        case UIDeviceOrientationLandscapeLeft:
            {dOrientation = -(M_PI/2.0f); break;}
        case UIDeviceOrientationLandscapeRight:
            {dOrientation = (M_PI/2.0f); break;}
        default: break;
    }
```

```
    float adjustedAngle =
        fmod(self.orientationAngle - dOrientation, 2.0f * M_PI);
    if (adjustedAngle > (M_PI + 0.01f))
        adjustedAngle = (adjustedAngle - 2.0f * M_PI);
    return adjustedAngle;
}
```

This method uses a floating-point modulo to retrieve the difference between the actual screen angle and the interface orientation angular offset to return that all-important vertical angular offset.

> **Note**
>
> In iOS 6, use your Info.plist to allow and disallow orientation changes instead of `shouldAutorotateToInterfaceOrientation:`.

Recipe: Using Acceleration to Move Onscreen Objects

With a bit of programming, the iPhone's onboard accelerometer can make objects "move" around the screen, responding in real time to the way the user tilts the phone. Recipe 1-4 builds an animated butterfly that users can slide across the screen.

The secret to make this work lies in adding what a "physics timer" to the program. Instead of responding directly to changes in acceleration, the way Recipe 1-3 did, the accelerometer callback measures the current forces. It's up to the timer routine to apply those forces to the butterfly over time by changing its frame. Here are some key points to keep in mind:

- As long as the direction of force remains the same, the butterfly accelerates. Its velocity increases, scaled according to the degree of acceleration force in the X or Y direction.
- The `tick` routine, called by the timer, moves the butterfly by adding the velocity vector to the butterfly's origin.
- The butterfly's range is bounded. So when it hits an edge, it stops moving in that direction. This keeps the butterfly onscreen at all times. The `tick` method checks for boundary conditions. For example, if the butterfly hits a vertical edge, it can still move horizontally.
- The butterfly reorients itself so it always falling "down." This happens by applying a simple rotation transform in the `tick` method. Be careful when using transforms in addition to frame or center offsets. Always reset the math before applying offsets, and then reapply any angular changes. Failing to do so may cause your frames to zoom, shrink, or skew unexpectedly.

> **Note**
>
> Timers in their natural state do not work with blocks. If you'd rather use a block-based design, check around github to find workarounds that do.

Recipe 1-4 **Sliding an Onscreen Object Based on Accelerometer Feedback**

```
- (void)accelerometer:(UIAccelerometer *)accelerometer
    didAccelerate:(UIAcceleration *)acceleration
{
    // Extract the acceleration components
    float xx = -acceleration.x;
    float yy = acceleration.y;

    // Store the most recent angular offset
    mostRecentAngle = atan2(yy, xx);

    // Has the direction changed?
    float accelDirX = SIGN(xvelocity) * -1.0f;
    float newDirX = SIGN(xx);
    float accelDirY = SIGN(yvelocity) * -1.0f;
    float newDirY = SIGN(yy);

    // Accelerate. To increase viscosity lower the additive value
    if (accelDirX == newDirX) xaccel =
        (abs(xaccel) + 0.85f) * SIGN(xaccel);
    if (accelDirY == newDirY) yaccel =
        (abs(yaccel) + 0.85f) * SIGN(yaccel);

    // Apply acceleration changes to the current velocity
    xvelocity = -xaccel * xx;
    yvelocity = -yaccel * yy;
}

- (void) tick
{
    // Reset the transform before changing position
    butterfly.transform = CGAffineTransformIdentity;

    // Move the butterfly according to the current velocity vector
    CGRect rect = CGRectOffset(butterfly.frame, xvelocity, 0.0f);
    if (CGRectContainsRect(self.view.bounds, rect))
        butterfly.frame = rect;
```

Chapter 1 Device-Specific Development

```
    rect = CGRectOffset(butterfly.frame, 0.0f, yvelocity);
    if (CGRectContainsRect(self.view.bounds, rect))
        butterfly.frame = rect;

    // Rotate the butterfly independently of position
    butterfly.transform =
        CGAffineTransformMakeRotation(mostRecentAngle + M_PI_2);
}

- (void) initButterfly
{
    CGSize size;

    // Load the animation cells
    NSMutableArray *butterflies = [NSMutableArray array];
    for (int i = 1; i <= 17; i++)
    {
        NSString *fileName = [NSString stringWithFormat:@"bf_%d.png", i];
        UIImage *image = [UIImage imageNamed:fileName];
        size = image.size;
        [butterflies addObject:image];
    }

    // Begin the animation
    butterfly = [[UIImageView alloc]
        initWithFrame:(CGRect){.size=size}];
    [butterfly setAnimationImages:butterflies];
    butterfly.animationDuration = 0.75f;
    [butterfly startAnimating];

    // Set the butterfly's initial speed and acceleration
    xaccel = 2.0f;
    yaccel = 2.0f;
    xvelocity = 0.0f;
    yvelocity = 0.0f;

    // Add the butterfly
    butterfly.center = RECTCENTER(self.view.bounds);
    [self.view addSubview:butterfly];


    // Activate the accelerometer
    [[UIAccelerometer sharedAccelerometer] setDelegate:self];

    // Start the physics timer
    [NSTimer scheduledTimerWithTimeInterval: 0.03f
```

```
        target: self selector: @selector(tick)
        userInfo: nil repeats: YES];
}
```

> **Get This Recipe's Code**
> To find this recipe's full sample project, point your browser to https://github.com/erica/iOS-6-Advanced-Cookbook and go to the folder for Chapter 1.

Recipe: Accelerometer-Based Scroll View

Several readers asked me to include a tilt scroller recipe in this edition. A tilt scroller uses the device's built-in accelerometer to control movement around a `UIScrollView`'s content. As the user adjusts the device, the material "falls down" accordingly. Instead of a view being positioned onscreen, the content view scrolls to a new offset.

The challenge in creating this interface lies in determining where the device should have its resting axis. Most people would initially suggest that the display should stabilize when lying on its back, with the Z-direction pointed straight up in the air. It turns out that's actually a fairly bad design choice. To use that axis means the screen must actually tilt away from the viewer during navigation. With the device rotated away from view, the user cannot fully see what is happening onscreen, especially when using the device in a seated position and somewhat when looking at the device while standing overhead.

Instead, Recipe 1-5 assumes that the stable position is created by the Z-axis pointing at approximately 45 degrees, the natural position users holding an iPhone or iPad in their hands. This is halfway between a face-up and a face-forward position. The math in Recipe 1-5 is adjusted accordingly. Tilting back and forward from this slanting position leaves the screen with maximal visibility during adjustments.

The other change in this recipe, compared to Recipe 1-4, is the much lower acceleration constant. This enables onscreen movement to happen more slowly, letting users more easily slow down and resume navigation.

Recipe 1-5   **Tilt Scroller**

```
- (void)accelerometer:(UIAccelerometer *)accelerometer
    didAccelerate:(UIAcceleration *)acceleration
{
    // extract the acceleration components
    float xx = -acceleration.x;
    float yy = (acceleration.z + 0.5f) * 2.0f; // between face-up and face-forward

    // Has the direction changed?
    float accelDirX = SIGN(xvelocity) * -1.0f;
```

```objectivec
    float newDirX = SIGN(xx);
    float accelDirY = SIGN(yvelocity) * -1.0f;
    float newDirY = SIGN(yy);

    // Accelerate. To increase viscosity lower the additive value
    if (accelDirX == newDirX) xaccel = (abs(xaccel) + 0.005f) * SIGN(xaccel);
    if (accelDirY == newDirY) yaccel = (abs(yaccel) + 0.005f) * SIGN(yaccel);

    // Apply acceleration changes to the current velocity
    xvelocity = -xaccel * xx;
    yvelocity = -yaccel * yy;
}

- (void) tick
{
    xoff += xvelocity;
    xoff = MIN(xoff, 1.0f);
    xoff = MAX(xoff, 0.0f);

    yoff += yvelocity;
    yoff = MIN(yoff, 1.0f);
    yoff = MAX(yoff, 0.0f);

    // update the content offset based on the current velocities
    CGFloat xsize = sv.contentSize.width - sv.frame.size.width;
    CGFloat ysize = sv.contentSize.height - sv.frame.size.height;
    sv.contentOffset = CGPointMake(xoff * xsize, yoff * ysize);
}

- (void) viewDidAppear:(BOOL)animated
{
    NSString *map = @"http://maps.weather.com/images/\
        maps/current/curwx_720x486.jpg";
    NSOperationQueue *queue = [[NSOperationQueue alloc] init];
    [queue addOperationWithBlock:
     ^{
         // Load the weather data
         NSURL *weatherURL = [NSURL URLWithString:map];
         NSData *imageData = [NSData dataWithContentsOfURL:weatherURL];

         // Update the image on the main thread using the main queue
         [[NSOperationQueue mainQueue] addOperationWithBlock:^{
             UIImage *weatherImage = [UIImage imageWithData:imageData];
             UIImageView *imageView =
                 [[UIImageView alloc] initWithImage:weatherImage];
             CGSize initSize = weatherImage.size;
             CGSize destSize = weatherImage.size;
```

```
            // Ensure that the content size is significantly bigger
            // than the screen can show at once
            while ((destSize.width < (self.view.frame.size.width * 4)) ||
                   (destSize.height < (self.view.frame.size.height * 4)))
            {
                destSize.width += initSize.width;
                destSize.height += initSize.height;
            }

            imageView.userInteractionEnabled = NO;
            imageView.frame = (CGRect){.size = destSize};
            sv.contentSize = destSize;

            [sv addSubview:imageView];

            // Activate the accelerometer
            [[UIAccelerometer sharedAccelerometer] setDelegate:self];

            // Start the physics timer
            [NSTimer scheduledTimerWithTimeInterval: 0.03f
                target: self selector: @selector(tick)
                userInfo: nil repeats: YES];
        }];
    }];
}
```

> **Get This Recipe's Code**
>
> To find this recipe's full sample project, point your browser to https://github.com/erica/iOS-6-Advanced-Cookbook and go to the folder for Chapter 1.

Recipe: Core Motion Basics

The Core Motion framework centralizes motion data processing. Introduced in the iOS 4 SDK, Core Motion supersedes the direct accelerometer access you've just read about. It provides centralized monitoring of three key onboard sensors. These sensors are composed of the gyroscope, which measures device rotation; the magnetometer, which provides a way to measure compass bearings; and the accelerometer, which detects gravitational changes along three axes. A fourth entry point called *device motion* combines all three of these sensors into a single monitoring system.

Core Motion uses raw values from these sensors to create readable measurements, primarily in the form of force vectors. Measurable items include the following properties:

- **Device attitude (`attitude`)**—The device's orientation relative to some frame of reference. The attitude is represented as a triplet of roll, pitch, and yaw angles, each measured in radians.
- **Rotation rate (`rotationRate`)**—The rate at which the device rotates around each of its three axes. The rotation includes x, y, and z angular velocity values measured in radians per second.
- **Gravity (`gravity`)**—The device's current acceleration vector as imparted by the normal gravitational field. Gravity is measured in g's, along the x, y, and z axes. Each unit represents the standard gravitational force imparted by Earth (namely 32 feet per second per second, or 9.8 meters per second per second).
- **User acceleration (`userAcceleration`)**—The acceleration vector being imparted by the user. Like `gravity`, user acceleration is measured in g's along the x, y, and z axes. When added together, the user vector and the gravity vector represent the total acceleration imparted to the device.
- **Magnetic field (`magneticField`)**—The vector representing the overall magnetic field values in the device's vicinity. The field is measured in microteslas along the x, y, and z axes. A calibration accuracy is also provided, to inform your application of the field measurements quality.

Testing for Sensors

As you read earlier in this chapter, you can use the application's Info.plist file to require or exclude onboard sensors. You can also test an in-app for each kind of possible Core Motion support:

```
if (motionManager.gyroAvailable)
    [motionManager startGyroUpdates];

if (motionManager.magnetometerAvailable)
    [motionManager startMagnetometerUpdates];

if (motionManager.accelerometerAvailable)
    [motionManager startAccelerometerUpdates];

if (motionManager.deviceMotionAvailable)
    [motionManager startDeviceMotionUpdates];
```

Starting updates does not produce a delegate callback mechanism like you encountered with the `UIAccelerometer` class. Instead, you are responsible for polling each value, or you can use a block-based update mechanism that executes a block that you provide at each update (for example, `startAccelerometerUpdatesToQueue:withHandler:`).

Handler Blocks

Recipe 1-6 adapts Recipe 1-4 for use with Core Motion. The acceleration callback has been moved into a handler block, and the x and y values are read from the data's acceleration property. Otherwise, the code remains unchanged. Here, you see the Core Motion basics: A new motion manager is created. It tests for accelerometer availability. It then starts updates using a new operation queue, which persists for the duration of the application run.

The `establishMotionManager` and `shutDownMotionManager` methods enable your application to start up and shut down the motion manager on demand. These methods are called from the application delegate when the application becomes active and when it suspends:

```
- (void) applicationWillResignActive:(UIApplication *)application
{
    [tbvc shutDownMotionManager];
}

- (void) applicationDidBecomeActive:(UIApplication *)application
{
    [tbvc establishMotionManager];
}
```

These methods provide a clean way to shut down and resume motion services in response to the current application state.

Recipe 1-6 **Basic Core Motion**

```
@implementation TestBedViewController
- (void) tick
{
    butterfly.transform = CGAffineTransformIdentity;
    
    // Move the butterfly according to the current velocity vector
    CGRect rect = CGRectOffset(butterfly.frame, xvelocity, 0.0f);
    if (CGRectContainsRect(self.view.bounds, rect))
        butterfly.frame = rect;
    
    rect = CGRectOffset(butterfly.frame, 0.0f, yvelocity);
    if (CGRectContainsRect(self.view.bounds, rect))
        butterfly.frame = rect;
    
    butterfly.transform =
        CGAffineTransformMakeRotation(mostRecentAngle + M_PI_2);
}

- (void) shutDownMotionManager
{
    NSLog(@"Shutting down motion manager");
```

```objc
    [motionManager stopAccelerometerUpdates];
    motionManager = nil;
    
    [timer invalidate];
    timer = nil;
}

- (void) establishMotionManager
{
    if (motionManager)
        [self shutDownMotionManager];
    
    NSLog(@"Establishing motion manager");
    
    // Establish the motion manager
    motionManager = [[CMMotionManager alloc] init];
    if (motionManager.accelerometerAvailable)
        [motionManager
          startAccelerometerUpdatesToQueue:
            [[NSOperationQueue alloc] init]
          withHandler:^(CMAccelerometerData *data, NSError *error)
          {
              // Extract the acceleration components
              float xx = -data.acceleration.x;
              float yy = data.acceleration.y;
              mostRecentAngle = atan2(yy, xx);
              
              // Has the direction changed?
              float accelDirX = SIGN(xvelocity) * -1.0f;
              float newDirX = SIGN(xx);
              float accelDirY = SIGN(yvelocity) * -1.0f;
              float newDirY = SIGN(yy);
              
              // Accelerate. To increase viscosity,
              // lower the additive value
              if (accelDirX == newDirX)
                  xaccel = (abs(xaccel) + 0.85f) * SIGN(xaccel);
              if (accelDirY == newDirY)
                  yaccel = (abs(yaccel) + 0.85f) * SIGN(yaccel);
              
              // Apply acceleration changes to the current velocity
              xvelocity = -xaccel * xx;
              yvelocity = -yaccel * yy;
          }];
    
    // Start the physics timer
    timer = [NSTimer scheduledTimerWithTimeInterval: 0.03f
```

```
            target: self selector: @selector(tick)
            userInfo: nil repeats: YES];
}

- (void) initButterfly
{
    CGSize size;

    // Load the animation cells
    NSMutableArray *butterflies = [NSMutableArray array];
    for (int i = 1; i <= 17; i++)
    {
        NSString *fileName =
            [NSString stringWithFormat:@"bf_%d.png", i];
        UIImage *image = [UIImage imageNamed:fileName];
        size = image.size;
        [butterflies addObject:image];
    }

    // Begin the animation
    butterfly = [[UIImageView alloc]
        initWithFrame:(CGRect){.size=size}];
    [butterfly setAnimationImages:butterflies];
    butterfly.animationDuration = 0.75f;
    [butterfly startAnimating];

    // Set the butterfly's initial speed and acceleration
    xaccel = 2.0f;
    yaccel = 2.0f;
    xvelocity = 0.0f;
    yvelocity = 0.0f;

    // Add the butterfly
    butterfly.center = RECTCENTER(self.view.bounds);
    [self.view addSubview:butterfly];
}

- (void) loadView
{
    [super loadView];
    self.view.backgroundColor = [UIColor whiteColor];
    [self initButterfly];
}
@end
```

> **Get This Recipe's Code**
>
> To find this recipe's full sample project, point your browser to https://github.com/erica/iOS-6-Advanced-Cookbook and go to the folder for Chapter 1.

Recipe: Retrieving and Using Device Attitude

Imagine an iPad sitting on a desk. There's an image displayed on the iPad, which you can bend over and look at. Now imagine rotating that iPad as it lays flat on the desk, but as the iPad moves, the image does not. It maintains a perfect alignment with the world around it. Regardless of how you spin the iPad, the image doesn't "move" as the view updates to balance the physical movement. That's how Recipe 1-7 works, taking advantage of a device's onboard gyroscope—a necessary requirement to make this recipe work.

The image adjusts however you hold the device. In addition to that flat manipulation, you can pick up the device and orient it in space. If you flip the device and look at it over your head, you see the reversed "bottom" of the image. You can also tilt it along both axes: the one that runs from the home button to the camera, and the other that runs along the surface of the iPad, from the midpoints between the camera and home button. The other axis, the one you first explore, is coming out of the device from its middle, pointing to the air above the device and passing through that middle point to behind it. As you manipulate the device, the image responds to create a virtual still world within that iPad.

Recipe 1-7 shows how to do this with just a few simple geometric transformations. It establishes a motion manager, subscribes to device motion updates, and then applies image transforms based on the roll, pitch, and yaw returned by the motion manager.

Recipe 1-7   Using Device Motion Updates to Fix an Image in Space

```
- (void) shutDownMotionManager
{
    NSLog(@"Shutting down motion manager");
    [motionManager stopDeviceMotionUpdates];
    motionManager = nil;
}

- (void) establishMotionManager
{
    if (motionManager)
        [self shutDownMotionManager];

    NSLog(@"Establishing motion manager");

    // Establish the motion manager
```

```objc
    motionManager = [[CMMotionManager alloc] init];
    if (motionManager.deviceMotionAvailable)
        [motionManager
          startDeviceMotionUpdatesToQueue:
            [NSOperationQueue currentQueue]
          withHandler: ^(CMDeviceMotion *motion, NSError *error) {
              CATransform3D transform;
              transform = CATransform3DMakeRotation(
                  motion.attitude.pitch, 1, 0, 0);
              transform = CATransform3DRotate(transform,
                  motion.attitude.roll, 0, 1, 0);
              transform = CATransform3DRotate(transform,
                  motion.attitude.yaw, 0, 0, 1);
              imageView.layer.transform = transform;
        }];
}
```

Get This Recipe's Code

To find this recipe's full sample project, point your browser to https://github.com/erica/iOS-6-Advanced-Cookbook and go to the folder for Chapter 1.

Detecting Shakes Using Motion Events

When the iPhone detects a motion event, it passes that event to the current first responder, the primary object in the responder chain. Responders are objects that can handle events. All views and windows are responders and so is the application object.

The responder chain provides a hierarchy of objects, all of which can respond to events. When an object toward the start of the chain receives an event, that event does not get passed further down. The object handles it. If it cannot, that event can move on to the next responder.

Objects often become the first responder by declaring themselves to be so, via become FirstResponder. In this snippet, a UIViewController ensures that it becomes the first responder whenever its view appears onscreen. Upon disappearing, it resigns the first responder position:

```objc
- (BOOL)canBecomeFirstResponder {
    return YES;
}

// Become first responder whenever the view appears
- (void)viewDidAppear:(BOOL)animated {
```

```
    [super viewDidAppear:animated];
    [self becomeFirstResponder];
}

// Resign first responder whenever the view disappears
- (void)viewWillDisappear:(BOOL)animated {
    [super viewWillDisappear:animated];
    [self resignFirstResponder];
}
```

First responders receive all touch and motion events. The motion callbacks mirror UIView touch callback stages. The callback methods are as follows:

- **motionBegan:withEvent:**—This callback indicates the start of a motion event. At the time of writing this book, there was only one kind of motion event recognized: a shake. This may not hold true for the future, so you might want to check the motion type in your code.

- **motionEnded:withEvent:**—The first responder receives this callback at the end of the motion event.

- **motionCancelled:withEvent:**—As with touches, motions can be canceled by incoming phone calls and other system events. Apple recommends that you implement all three motion event callbacks (and, similarly, all four touch event callbacks) in production code.

The following snippet shows a pair of motion callback examples. If you test this on a device, you can notice several things. First, the began and ended events happen almost simultaneously from a user perspective. Playing sounds for both types is overkill. Second, there is a bias toward side-to-side shake detection. The iPhone is better at detecting side-to-side shakes than the front-to-back and up-down versions. Finally, Apple's motion implementation uses a slight lockout approach. You cannot generate a new motion event until a second or so after the previous one was processed. This is the same lockout used by Shake to Shuffle and Shake to Undo events:

```
- (void)motionBegan:(UIEventSubtype)motion
    withEvent:(UIEvent *)event {

    // Play a sound whenever a shake motion starts
    if (motion != UIEventSubtypeMotionShake) return;
    [self playSound:startSound];
}

- (void)motionEnded:(UIEventSubtype)motion withEvent:(UIEvent *)event
{
    // Play a sound whenever a shake motion ends
     if (motion != UIEventSubtypeMotionShake) return;
    [self playSound:endSound];
}
```

Recipe: Using External Screens

There are many ways to use external screens. Take the newest iPads, for example. The second and third generation models offer built-in screen mirroring. Attach a VGA or HDMI cable and your content can be shown on external displays and on the built-in screen. Certain devices enable you to mirror screens wirelessly to Apple TV using AirPlay, Apple's proprietary cable-free over-the-air video solution. These mirroring features are extremely handy, but you're not limited to simply copying content from one screen to another in iOS.

The `UIScreen` class enables you to detect and write to external screens independently. You can treat any connected display as a new window and create content for that display separate from any view you show on the primary device screen. You can do this for any wired screen, and starting with the iPad 2 (and later) and the iPhone 4S (and later), you can do so wirelessly using AirPlay to Apple TV 2 (and later). A third-party app called Reflector enables you to mirror your display to Mac or Windows computers using AirPlay.

Geometry is important. Here's why. iOS devices currently include the 320×480 old-style iPhone displays, the 640×960-pixel Retina display units, and the 1024×768-pixel iPads. Typical composite/component output is produced at 720×480 pixels (480i and 480p), VGA at 1024×768 and 1280×720 (720p), and then there's the higher quality HDMI output available as well.

Add to this the issues of overscan and other target display limitations, and Video Out quickly becomes a geometric challenge. Fortunately, Apple has responded to this challenge with some handy real-world adaptations. Instead of trying to create one-to-one correspondences with the output screen and your built-in device screen, you can build content based on the available properties of your output display. You just create a window, populate it, and display it.

If you intend to develop Video Out applications, don't assume that your users are strictly using AirPlay. Many users still connect to monitors and projectors using old-style cable connections. Make sure you have at least one of each type of cable on-hand (composite, component, VGA, and HDMI) and an AirPlay-ready iPhone and iPad, so you can thoroughly test on each output configuration. Third-party cables (typically imported from the Far East, not branded with Made for iPhone/iPad) won't work, so make sure you purchase Apple-branded items.

Detecting Screens

The `UIScreen` class reports how many screens are connected. You know that an external screen is connected whenever this count goes above 1. The first item in the screens array is always your primary device screen:

```
#define SCREEN_CONNECTED ([UIScreen screens].count > 1)
```

Each screen can report its bounds (that is, its physical dimensions in points) and its screen scale (relating the points to pixels). Two standard notifications enable you to observe when screens have been connected to and disconnected from the device.

```
// Register for connect/disconnect notifications
[[NSNotificationCenter defaultCenter]
    addObserver:self selector:@selector(screenDidConnect:)
    name:UIScreenDidConnectNotification object:nil];
[[NSNotificationCenter defaultCenter]
    addObserver:self selector:@selector(screenDidDisconnect:)
    name:UIScreenDidDisconnectNotification object:nil];
```

Connection means *any* kind of connection, whether by cable or via AirPlay. Whenever you receive an update of this type, make sure you count your screens and adjust your user interface to match the new conditions.

It's your responsibility to set up windows whenever new screens are attached and tear them down upon detach events. Each screen should have its own window to manage content for that output display. Don't hold onto windows upon detaching screens. Let them release and then re-create them when new screens appear.

> **Note**
>
> Mirrored screens are not represented in the `screens` array. Instead the mirror is stored in the main screen's `mirroredScreen` property. This property is nil when mirroring is disabled, unconnected, or simply not supported by the device's abilities.
>
> Creating a new screen and using it for independent external display always overrides mirroring. So even if the user has enabled mirroring, when your application begins writing to and creating an external display, it takes priority.

Retrieving Screen Resolutions

Each screen provides an `availableModes` property. This is an array of resolution objects ordered from least-to-highest resolution. Each mode has a `size` property indicating a target pixel-size resolution. Many screens support multiple modes. For example, a VGA display might have as many as one-half dozen or more different resolutions it offers. The number of supported resolutions varies by hardware. There will always be at least one resolution available, but you should offer choices to users when there are more.

Setting Up Video Out

After retrieving an external screen object from the `[UIScreens screens]` array, query the available modes and select a size to use. As a rule, you can get away with selecting the last mode in the list to always use the highest possible resolution, or the first mode for the lowest resolution.

To start a Video Out stream, create a new `UIWindow` and size it to the selected mode. Add a new view to that window for drawing on. Then assign the window to the external screen and make it key and visible. This orders the window to display and prepares it for use. After you do that, make the original window key again. This allows the user to continue interacting with the primary screen. Don't skip this step. Nothing makes end users more cranky than discovering their expensive device no longer responds to their touches:

```
self.outputWindow = [[UIWindow alloc] initWithFrame:theFrame];
outputWindow.screen = secondaryScreen;
[outputWindow makeKeyAndVisible];
[delegate.view.window makeKeyAndVisible];
```

Adding a Display Link

Display links are a kind of timer that synchronizes drawing to a display's refresh rate. You can adjust this frame refresh time by changing the display link's `frameInterval` property. It defaults to 1. A higher number slows down the refresh rate. Setting it to 2 halves your frame rate. Create the display link when a screen connects to your device. The `UIScreen` class implements a method that returns a display link object for its screen. You specify the target for the display link and a selector to call.

The display link fires on a regular basis, letting you know when to update the Video Out screen. You can adjust the interval up for less of a CPU load, but you get a lower frame rate in return. This is an important trade-off, especially for direct manipulation interfaces that require a high level of CPU response on the device side.

The code you see in Recipe 1-8 uses common modes for the run loop, providing the least latency. You `invalidate` your display link when you are done with it, removing it from the run loop.

Overscanning Compensation

The `UIScreen` class enables you to compensate for pixel loss at the edge of display screens by assigning a value to the `overscanCompensation` property. The techniques you can assign are described in Apple's documentation but basically correspond to whether you want to clip content or pad it with black space.

VIDEOkit

Recipe 1-8 introduces VIDEOkit, a basic external screen client. It demonstrates all the features needed to get up and going with wired and wireless external screens. You establish screen monitoring by calling `startupWithDelegate:`. Pass it the primary view controller whose job it will be to create external content.

The internal `init` method starts listening for screen attach and detach events and builds and tears down windows as needed. An informal delegate method (`updateExternalView:`) is called each time the display link fires. It passes a view that lives on the external window that the delegate can draw onto as needed.

In the sample code that accompanies this recipe, the view controller delegate stores a local color value and uses it to color the external display:

```
- (void) updateExternalView: (UIImageView *) aView
{
    aView.backgroundColor = color;
}

- (void) action: (id) sender
{
    color = [UIColor randomColor];
}
```

Each time the action button is pressed, the view controller generates a new color. When VIDEOkit queries the controller to update the external view, it sets this as the background color. You can see the external screen instantly update to a new random color.

> **Note**
>
> Reflector App ($15/single license, $50/5-computer license, reflectorapp.com) provides an excellent debugging companion for AirPlay, offering a no-wires/no-Apple TV solution that works on Mac and Windows computers. It mimics an Apple TV AirPlay receiver, letting you broadcast from iOS direct to your desktop and record that output.

Recipe 1-8 VIDEOkit

```
@interface VIDEOkit : NSObject
{
    UIImageView *baseView;
}
@property (nonatomic, weak)   UIViewController *delegate;
@property (nonatomic, strong) UIWindow *outputWindow;
@property (nonatomic, strong) CADisplayLink *displayLink;
+ (void) startupWithDelegate: (id) aDelegate;
@end

@implementation VIDEOkit
static VIDEOkit *sharedInstance = nil;

- (void) setupExternalScreen
{
    // Check for missing screen
    if (!SCREEN_CONNECTED) return;

    // Set up external screen
    UIScreen *secondaryScreen = [UIScreen screens][1];
    UIScreenMode *screenMode =
```

```objc
            [[secondaryScreen availableModes] lastObject];
    CGRect rect = (CGRect){.size = screenMode.size};
    NSLog(@"Extscreen size: %@", NSStringFromCGSize(rect.size));

    // Create new outputWindow
    self.outputWindow = [[UIWindow alloc] initWithFrame:CGRectZero];
    _outputWindow.screen = secondaryScreen;
    _outputWindow.screen.currentMode = screenMode;
    [_outputWindow makeKeyAndVisible];
    _outputWindow.frame = rect;

    // Add base video view to outputWindow
    baseView = [[UIImageView alloc] initWithFrame:rect];
    baseView.backgroundColor = [UIColor darkGrayColor];
    [_outputWindow addSubview:baseView];

    // Restore primacy of main window
    [_delegate.view.window makeKeyAndVisible];
}

- (void) updateScreen
{
    // Abort if the screen has been disconnected
    if (!SCREEN_CONNECTED && _outputWindow)
        self.outputWindow = nil;

    // (Re)initialize if there's no output window
    if (SCREEN_CONNECTED && !_outputWindow)
        [self setupExternalScreen];

    // Abort if encountered some weird error
    if (!self.outputWindow) return;

    // Go ahead and update
    SAFE_PERFORM_WITH_ARG(_delegate,
        @selector(updateExternalView:), baseView);
}

- (void) screenDidConnect: (NSNotification *) notification
{
    NSLog(@"Screen connected");
    UIScreen *screen = [[UIScreen screens] lastObject];

    if (_displayLink)
    {
        [_displayLink removeFromRunLoop:[NSRunLoop currentRunLoop]
            forMode:NSRunLoopCommonModes];
```

```objc
        [_displayLink invalidate];
        _displayLink = nil;
    }

    self.displayLink = [screen displayLinkWithTarget:self
        selector:@selector(updateScreen)];
    [_displayLink addToRunLoop:[NSRunLoop currentRunLoop]
        forMode:NSRunLoopCommonModes];
}

- (void) screenDidDisconnect: (NSNotification *) notification
{
    NSLog(@"Screen disconnected.");
    if (_displayLink)
    {
        [_displayLink removeFromRunLoop:[NSRunLoop currentRunLoop]
            forMode:NSRunLoopCommonModes];
        [_displayLink invalidate];
        self.displayLink = nil;
    }
}

- (id) init
{
    if (!(self = [super init])) return self;

    // Handle output window creation
    if (SCREEN_CONNECTED)
        [self screenDidConnect:nil];

    // Register for connect/disconnect notifications
    [[NSNotificationCenter defaultCenter]
        addObserver:self selector:@selector(screenDidConnect:)
        name:UIScreenDidConnectNotification object:nil];
    [[NSNotificationCenter defaultCenter] addObserver:self
        selector:@selector(screenDidDisconnect:)
        name:UIScreenDidDisconnectNotification object:nil];

    return self;
}

- (void) dealloc
{
    [self screenDidDisconnect:nil];
    self.outputWindow = nil;
}
```

```
+ (VIDEOkit *) sharedInstance
{
    if (!sharedInstance)
        sharedInstance = [[self alloc] init];
    return sharedInstance;
}

+ (void) startupWithDelegate: (id) aDelegate
{
    [[self sharedInstance] setDelegate:aDelegate];
}
@end
```

> **Get This Recipe's Code**
> To find this recipe's full sample project, point your browser to https://github.com/erica/iOS-6-Advanced-Cookbook and go to the folder for Chapter 1.

Tracking Users

Tracking is an unfortunate reality of developer life. Apple deprecated the `UIDevice` property that provided a unique identifier tied to device hardware. It replaced it with two identifier properties. Use `identifierForAdvertising` to return a device-specific string unique to the current device. The `identifierForVendor` property supplies a string that's tied to each app vendor. This should return the same unique string regardless of which of your apps is in use. This is *not* a customer id. The same app on a different device can return a different string, as can an app from a different vendor.

These identifiers are built using the new `NSUUID` class. You can use this class outside of the tracking scenario to create UUID strings that are guaranteed to be globally unique. Apple writes, "UUIDs (Universally Unique Identifiers), also known as GUIDs (Globally Unique Identifiers) or IIDs (Interface Identifiers), are 128-bit values. A UUID is made unique over both space and time by combining a value unique to the computer on which it was generated and a value representing the number of 100-nanosecond intervals since October 15, 1582 at 00:00:00."

The `UUID` class method can generate a new RFC 4122v4 UUID on demand. Use `[NSUUID UUID]` to return a new instance. (Bonus: It's all in uppercase!) From there, you can retrieve the `UUIDString` representation or request the bytes directly via `getUUIDBytes:`.

One More Thing: Checking for Available Disk Space

The `NSFileManager` class enables you to determine how much space is free on the iPhone and how much space is provided on the device as a whole. Listing 1-1 demonstrates how to check for these values and show the results using a friendly comma-formatted string. The values returned represent the free space in bytes.

Listing 1-1 **Recovering File System Size and File System Free Size**

```
- (NSString *) commaFormattedStringWithLongLong: (long long) num
{
    // Produce a properly formatted number string
    // Alternatively use NSNumberFormatter
    if (num < 1000)
        return [NSString stringWithFormat:@"%d", num];
    return    [[self commasForNumber:num/1000]
        stringByAppendingFormat:@",%03d", (num % 1000)];
}

- (void) action: (UIBarButtonItem *) bbi
{
    NSFileManager *fm = [NSFileManager defaultManager];
    NSDictionary *fattributes =
        [fm fileSystemAttributesAtPath:NSHomeDirectory()];
    NSLog(@"System space: %@",
        [self commaFormattedStringWithLongLong:[[fattributes
        objectForKey:NSFileSystemSize] longLongValue]]);
    NSLog(@"System free space: %@",
        [self commasForNumber:[[fattributes
        objectForKey:NSFileSystemFreeSize] longLongValue]]);
}
```

Summary

This chapter introduced core ways to interact with an iPhone device. You saw how to recover device info, check the battery state, and subscribe to proximity events. You learned how to differentiate the iPod touch from the iPhone and iPad and determine which model you're working with. You discovered the accelerometer and saw it in use through several examples, from the simple "finding up" to the more complex shake detection algorithm. You jumped into Core Motion and learned how to create update blocks to respond to device events in real time. Finally, you saw how to add external screen support to your applications. Here are a few parting thoughts about the recipes you just encountered:

- The iPhone's accelerometer provides a novel way to complement its touch-based interface. Use acceleration data to expand user interactions beyond the "touch here" basics and to introduce tilt-aware feedback.

Summary

- Low-level calls can be App Store-friendly. They don't depend on Apple APIs that may change based on the current firmware release. UNIX system calls may seem daunting, but many are fully supported by the iOS device family.
- Remember device limitations. You may want to check for free disk space before performing file-intensive work and for battery charge before running the CPU at full steam.
- Dive into Core Motion. The real-time device feedback it provides is the foundation for integrating iOS devices into real-world experiences.
- Now that AirPlay has cut the cord for external display tethering, you can use Video Out for many more exciting projects than you might have previously imagined. AirPlay and external video screens mean you can transform your iOS device into a remote control for games and utilities that display on big screens and are controlled on small ones.
- When submitting to iTunes, use your Info.plist file to determine which device capabilities are required. iTunes uses this list of required capabilities to determine whether an application can be downloaded to a given device and run properly on that device.

2
Documents and Data Sharing

Under iOS, applications can share information and data as well as move control from one application to another using a variety of system features. Each application has access to a common system pasteboard that enables copying and pasting across apps. Users can transfer documents from one app to another app that supports that format. They can request a number of system-supplied "actions" to apply to a document, such as printing, tweeting, or posting to a Facebook wall. Apps can declare custom URL schemes that can be embedded in text and web pages. This chapter introduces the ways you can integrate documents and data sharing between applications. You see how to add these features into your applications and use them smartly to make your app a cooperative citizen of the iOS ecosystem.

Recipe: Working with Uniform Type Identifiers

Uniform Type Identifiers (UTIs) represent a central component of iOS information sharing. You can think of them as the next generation of MIME types. UTIs are strings that identify resource types such as images and text. UTIs specify what kind of information is being used for common data objects. They do this without relying on older indicators, such as file extensions, MIME types, or file-type metadata such as `OSTypes`. UTIs replace these items with a newer and more flexible technology.

UTIs use a reverse-domain-style naming convention. Common Apple-derived identifiers look like this: `public.html` and `public.jpeg`. These refer, respectively, to HTML source text and JPEG images, which are both specialized types of file information.

Inheritance plays an important role with UTIs. UTIs use an OO-like system of inheritance, where child UTIs have an "is-a" relationship to parents. Children inherit all attributes of their parents but add further specificity of the kind of data they represent. That's because each UTI can assume a more general or more specific role, as needed. Take the JPEG UTI, for example. A JPEG image (`public.jpeg`) is an image (`public.image`), which is in turn a kind of data (`public.data`), which is a kind of user-viewable (or listenable) content (`public.content`), which is a kind of item (`public.item`), the generic base type for UTIs. This hierarchy is called conformance, where child UTIs conform to parent UTIs. For example, the more specific jpeg UTI conforms to the more general image or data UTI.

Figure 2-1 shows part of Apple's basic conformance tree. Any item lower down on the tree must conform to all of its parent data attributes. Declaring a parent UTI implies that you support all of its children. So, an application that can open `public.data` must service text, movies, image files, and more.

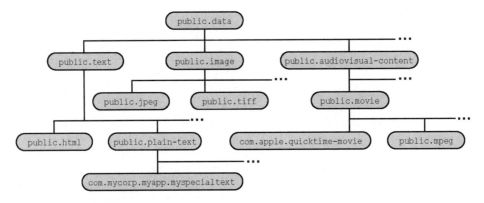

Figure 2-1 Apple's public UTI conformance tree.

UTIs enable multiple inheritance. An item can conform to more than one UTI parent. So, you might imagine a data type that offers both text and image containers, which declares conformance to both.

There is no central registry for UTI items, although each UTI should adhere to conventions. The `public` domain is reserved for iOS-specific types, common to most applications. Apple has generated a complete family hierarchy of public items. Add any third-party company-specific names by using standard reverse domain naming (for example, `com.sadun.myCustomType` and `com.apple.quicktime-movie`).

Determining UTIs from File Extensions

The Mobile Core Services framework offers utilities that enable you to retrieve UTI information based on file extensions. Be sure to include the header files and link your apps to the framework when using these C-based functions. The following function returns a preferred UTI when passed a path extension string. The preferred identifier is a single UTI string:

```
#import <MobileCoreServices/MobileCoreServices.h>

NSString *preferredUTIForExtension(NSString *ext)
{
    // Request the UTI via the file extension
```

```
    NSString *theUTI = (__bridge_transfer NSString *)
        UTTypeCreatePreferredIdentifierForTag(
            kUTTagClassFilenameExtension,
            (__bridge CFStringRef) ext, NULL);
    return theUTI;
}
```

You can pass a MIME type instead of a file extension to `UTTypeCreatePreferredIdentifierForTag()` by using `kUTTagClassMIMEType` as the first argument. This function returns a preferred UTI for a given MIME type:

```
NSString *preferredUTIForMIMEType(NSString *mime)
{
    // Request the UTI via the file extension
    NSString *theUTI = (__bridge_transfer NSString *)
        UTTypeCreatePreferredIdentifierForTag(
            kUTTagClassMIMEType,
            (__bridge CFStringRef) mime, NULL);
    return theUTI;
}
```

Together these functions enable you to move from file extensions and MIME types to the UTI types used for modern file access.

Moving from UTI to Extension or MIME Type

To go the other way, producing a preferred extension or MIME types from a UTI, use `UTTypeCopyPreferredTagWithClass()`. The following functions return `jpeg` and `image/jpeg`, respectively, when passed `public.jpeg`:

```
NSString *extensionForUTI(NSString *aUTI)
{
    CFStringRef theUTI = (__bridge CFStringRef) aUTI;
    CFStringRef results =
        UTTypeCopyPreferredTagWithClass(
            theUTI, kUTTagClassFilenameExtension);
    return (__bridge_transfer NSString *)results;
}

NSString *mimeTypeForUTI(NSString *aUTI)
{
    CFStringRef theUTI = (__bridge CFStringRef) aUTI;
    CFStringRef results =
        UTTypeCopyPreferredTagWithClass(
            theUTI, kUTTagClassMIMEType);
    return (__bridge_transfer NSString *)results;
}
```

You must work at the leaf level with these functions, meaning at the level that declares the type extensions directly. Extensions are declared in property lists, where features like file extensions and default icons are described. So, for example, passing `public.text` or `public.movie` to the extension function returns `nil`, whereas `public.plain-text` and `public.mpeg` return extensions of `txt` and `mpg`, respectively.

The former items live too high up the conformance tree, providing an abstract type rather than a specific implementation. There's no current API function to look down to find items that descend from a given class that are currently defined for the application. You may want to file an enhancement request at bugreport.apple.com. Surely, all the extensions and MIME types are registered somewhere (otherwise, how would the `UTTypeCopyPreferredTagWithClass()` look up work in the first place?), so the ability to map extensions to more general UTIs should be possible.

MIME Helper

Although the extension-to-UTI service is exhaustive, returning UTIs for nearly any extension you throw at it, the UTI-to-MIME results are scattershot. You can usually generate a proper MIME representation for any common item; less common ones are rare.

The following lines show an assortment of extensions, their UTIs (retrieved via `preferredUTIForExtension()`), and the MIME types generated from each UTI (via `mimeTypeForUTI()`). As you can see, there are quite a number of blanks. These functions return nil when they cannot find a match:

```
xlv: dyn.age81u5d0 / (null)
xlw: com.microsoft.excel.xlw / application/vnd.ms-excel
xm: dyn.age81u5k / (null)
xml: public.xml / application/xml
z: public.z-archive / application/x-compress
zip: public.zip-archive / application/zip
zoo: dyn.age81y55t / (null)
zsh: public.zsh-script / (null)
```

To address this problem, the sample code for this recipe includes an extra `MIMEHelper` class. It defines one function, which returns a MIME type for a supplied extension:

```
NSString *mimeForExtension(NSString *extension);
```

Its extensions and MIME types are sourced from the Apache Software Foundation, which has placed its list into the public domain. Out of the 450 extensions in the sample code for this recipe, iOS returned all 450 UTIs but only 88 MIME types. The Apache list ups this number to 230 recognizable MIME types.

Testing Conformance

You test conformance using the `UTTypeConformsTo()` function. This function takes two arguments: a source UTI and a UTI to compare to, returning true if the first UTI conforms to the second. Use this to test whether a more specific item conforms to a more general one. Test equality using `UTTypeEqual()`. Here's an example of how you might use conformance testing, determining if a file path likely points to an image resource:

```
BOOL pathPointsToLikelyUTIMatch(NSString *path, CFStringRef theUTI)
{
    NSString *extension = path.pathExtension;
    NSString *preferredUTI = preferredUTIForExtension(extension);
    return (UTTypeConformsTo(
        (__bridge CFStringRef) preferredUTI, theUTI));
}

BOOL pathPointsToLikelyImage(NSString *path)
{
    return pathPointsToLikelyUTIMatch(path, CFSTR("public.image"));
}

BOOL pathPointsToLikelyAudio(NSString *path)
{
    return pathPointsToLikelyUTIMatch(path, CFSTR("public.audio"));
}
```

Retrieving Conformance Lists

`UTTypeCopyDeclaration()` offers the most general (and most useful) of all UTI functions in the iOS API. It returns a dictionary that includes the following keys:

- **kUTTypeIdentifierKey**—The UTI name, which you passed to the function (for example, `public.mpeg`).

- **kUTTypeConformsToKey**—Any parents that the type conforms to (for example, `public.mpeg` conforms to `public.movie`).

- **kUTTypeDescriptionKey**—A real-world description of the type in question if one exists (for example, "MPEG movie").

- **kUTTypeTagSpecificationKey**—A dictionary of equivalent OSTypes (for example, MPG and MPEG), file extensions (mpg, mpeg, mpe, m75, and m15), and MIME types (video/mpeg, video/mpg, video/x-mpeg, and video/x-mpg) for the given UTI.

In addition to these common items, you encounter more keys that specify imported and exported UTI declarations (`kUTImportedTypeDeclarationsKey` and `kUTExportedType DeclarationsKey`), icon resources to associate with the UTI (`kUTTypeIconFileKey`), a URL that points to a page describing the type (`kUTTypeReferenceURLKey`), and a version key that offers a version string for the UTI (`kUTTypeVersionKey`).

Use the returned dictionary to ascend through the conformance tree to build an array that represents all the items that a given UTI conforms to. For example, the `public.mpeg` type conforms to `public.movie`, `public.audiovisual-content`, `public.data`, `public.item`, and `public.content`. These items are returned as an array from the `conformanceArray` function that follows in Recipe 2-1.

Recipe 2-1 **Testing Conformance**

```
// Build a declaration dictionary for the given type
NSDictionary *utiDictionary(NSString *aUTI)
{
    NSDictionary *dictionary =
        (__bridge_transfer NSDictionary *)
            UTTypeCopyDeclaration((__bridge CFStringRef) aUTI);
    return dictionary;
}

// Return an array where each member is guaranteed unique
// but that preserves the original ordering wherever possible
NSArray *uniqueArray(NSArray *anArray)
{
    NSMutableArray *copiedArray =
        [NSMutableArray arrayWithArray:anArray];

    for (id object in anArray)
    {
        [copiedArray removeObjectIdenticalTo:object];
        [copiedArray addObject:object];
    }

    return copiedArray;
}

// Return an array representing all UTIs that a given UTI conforms to
NSArray *conformanceArray(NSString *aUTI)
{
    NSMutableArray *results =
        [NSMutableArray arrayWithObject:aUTI];
    NSDictionary *dictionary = utiDictionary(aUTI);
    id conforms = [dictionary objectForKey:
        (__bridge NSString *)kUTTypeConformsToKey];

    // No conformance
    if (!conforms) return results;

    // Single conformance
    if ([conforms isKindOfClass:[NSString class]])
    {
        [results addObjectsFromArray:conformanceArray(conforms)];
```

```
        return uniqueArray(results);
    }

    // Iterate through multiple conformance
    if ([conforms isKindOfClass:[NSArray class]])
    {
        for (NSString *eachUTI in (NSArray *) conforms)
            [results addObjectsFromArray:conformanceArray(eachUTI)];
        return uniqueArray(results);
    }

    // Just return the one-item array
    return results;
}
```

Get This Recipe's Code

To find this recipe's full sample project, point your browser to https://github.com/erica/iOS-6-Advanced-Cookbook and go to the folder for Chapter 2.

Recipe: Accessing the System Pasteboard

Pasteboards, also known as clipboards on some systems, provide a central OS feature for sharing data across applications. Users can copy data to the pasteboard in one application, switch tasks, and then paste that data into another application. Cut/copy/paste features are similar to those found in most operating systems. Users can also copy and paste within a single application, when switching between text fields or views, and developers can establish private pasteboards for app-specific data that would not be understood by other apps.

The `UIPasteboard` class offers access to a shared device pasteboard and its contents. This snippet returns the general system pasteboard, which is appropriate for most general copy/paste use:

`UIPasteboard *pb = [UIPasteboard generalPasteboard];`

In addition to the general shared system pasteboard, iOS offers both application-specific pasteboards to better ensure data privacy, which do not extend beyond the application, and custom-named pasteboards that can be used across applications, but only with applications that know and use the pasteboard name key. Create app-specific pasteboards using `pasteboardWith-UniqueName`, which returns an application pasteboard object that persists until the application quits.

Create custom pasteboards using `pasteboardWithName:create:`, which returns a pasteboard with the specified name. Use reverse-DNS naming for the pasteboard (for example, `com.sadun.shared-application-pasteboard`). The create parameter specifies whether the system should create the pasteboard if it does not yet exist. This kind of pasteboard can persist beyond a single application run; set the persistent property to YES after creation. Use `removePasteboardWithName:` to destroy a pasteboard and free up the resources used by it.

Storing Data

Pasteboards can store one or more entries at a time. Each has an associated type, using the UTI to specify what kind of data is stored. For example, you might find `public.text` (and more specifically `public.utf8-plain-text`) to store text data, `public.url` for URL address, and `public.jpeg` for image data. These are among many other common data types used on iOS. The dictionary that stores the type and the data is called an `item`, and you can retrieve an array of all available items via the pasteboard's `items` property.

You can determine the kinds of items currently stored with a simple message. Query a pasteboard for its available types by sending it the `pasteboardTypes` message. This returns an array of types currently stored on the pasteboard:

```
NSArray *types = [pb pasteboardTypes];
```

You can set data on the pasteboard and associate a type by passing an `NSData` object and a UTI that describes a type the data conforms to. Alternatively, for property list objects (that is, string, date, array, dictionary, number, or URL), set an `NSValue` via `setValue:forPasteboardType:`. These property list objects are stored internally somewhat differently than their raw-data cousins, giving rise to the method differentiation:

```
[[UIPasteboard generalPasteboard]
    setData:theData forPasteboardType:theUTI];
```

Storing Common Types

Pasteboards are further specialized for several data types, which represent the most commonly used pasteboard items. These are colors (not a property list "value" object), images (also not a property list "value" object), strings, and URLs. The `UIPasteboard` class provides specialized getters and setters to make it easier to handle these items. You can treat each of these as properties of the pasteboard, so you can set and retrieve them using dot notation. What's more, each property has a plural form, allowing you to access those items as arrays of objects.

Pasteboard properties greatly simplify using the system pasteboard for the most common use cases. The property accessors include the following:

- **string**—Sets or retrieves the first string on the pasteboard.
- **strings**—Sets or retrieves an array of all strings on the pasteboard.
- **image**—Sets or retrieves the first image on the pasteboard.
- **images**—Sets or retrieves an array of all images on the pasteboard.
- **URL**—Sets or retrieves the first URL on the pasteboard.

- **URLs**—Sets or retrieves an array of all URLs on the pasteboard.
- **color**—Sets or retrieves the first color on the pasteboard.
- **colors**—Sets or retrieves an array of all colors on the pasteboard.

Retrieving Data

When using one of the four special classes, simply use the associated property to retrieve data from the pasteboard. Otherwise, you can fetch data using the `dataForPasteboardType:` method. This method returns the data from the first item whose type matches the UTI sent as a parameter. Any other matching items in the pasteboard are ignored.

Should you need to retrieve all matching data, recover an `itemSetWithPasteboardTypes:` and then iterate through the set to retrieve each dictionary. Recover the data type for each item from the single dictionary key and the data from its value.

As mentioned, `UIPasteboard` offers two approaches for pasting to the pasteboard depending on whether the information being pasted is a property list object or raw data. Use `setValueForPasteboardType:` for property list objects, which include strings, dates, numbers, dictionaries, arrays, and URLs. For general data, use `setData:forPasteboardType:`.

When pasteboards are changed, they issue a `UIPasteboardChangedNotification`, which you can listen to via a default `NSNotificationCenter` observer. You can also watch custom pasteboards and listen for their removal via `UIPasteboardRemovedNotification`.

> **Note**
> If you want to successfully paste text data to Notes or Mail, use `public.utf8-plain-text` as your UTI of choice when storing information to the pasteboard. Using the `string` or `strings` properties automatically enforces this UTI.

Passively Updating the Pasteboard

iOS's selection and copy interface are not, frankly, the most streamlined elements of the operating system. There are times when you want to simplify matters for your user while preparing content that's meant to be shared with other applications.

Consider Recipe 2-2. It enables the user to use a text view to enter and edit text, while automating the process of updating the pasteboard. When the watcher is active (toggled by a simple button tap), the text updates the pasteboard on each edit. This is accomplished by implementing a text view delegate method (`textViewDidChange:`) that responds to edits by automatically assigning changes to the pasteboard (`updatePasteboard`).

This recipe demonstrates the relative simplicity involved in accessing and updating the pasteboard.

Recipe 2-2 Creating an Automatic Text-Entry to Pasteboard Solution

```
- (void) updatePasteboard
{
    // Copy the text to the pasteboard when the watcher is enabled
    if (enableWatcher)
        [UIPasteboard generalPasteboard].string = textView.text;
}

- (void) textViewDidChange: (UITextView *) textView
{
    // Delegate method calls for an update
    [self updatePasteboard];
}

- (void) toggle: (UIBarButtonItem *) bbi
{
    // switch between standard and auto-copy modes
    enableWatcher = !enableWatcher;
    bbi.title = enableWatcher ? @"Stop Watching" : @"Watch";
}
```

> **Get This Recipe's Code**
>
> To find this recipe's full sample project, point your browser to https://github.com/erica/iOS-6-Advanced-Cookbook and go to the folder for Chapter 2.

Recipe: Monitoring the Documents Folder

iOS documents aren't trapped in their sandboxes. You can and should share them with your users. Offer users direct control over their documents and access to any material they may have created on-device. A simple Info.plist setting enables iTunes to display the contents of a user's Documents folder and enables those users add and remove material on demand.

At some point in the future, you may use a simple `NSMetadataQuery` monitor to watch your Documents folder and report updates. At the time this book was written, that metadata surveillance was not yet extended beyond iCloud for use with other folders. Code ported from OS X fails to work as expected on iOS. Currently, there are precisely two available search domains for iOS: the ubiquitous data scope and the ubiquitous documents scope (that is, iCloud and iCloud).

Until general functionality arrives in iOS, use kqueue. This older technology provides scalable event notification. With kqueue, you can monitor add and clear events. This roughly equates to looking for files being added and deleted, which are the primary kinds of updates you want to react to. Recipe 2-3 presents a kqueue implementation for watching the Documents folder.

Enabling Document File Sharing

To enable file sharing, add a `UIFileSharingEnabled` key to the application's Info.plist and set its value to YES, as shown in Figure 2-2. When working with nonraw keys and values, this item is called Application supports iTunes file sharing. iTunes lists all applications that declare file-sharing support in each device's Apps tab, as shown in Figure 2-3.

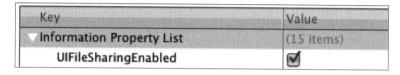

Figure 2-2 Enable `UIFileSharingEnabled` to allow user access to the Documents folder via iTunes.

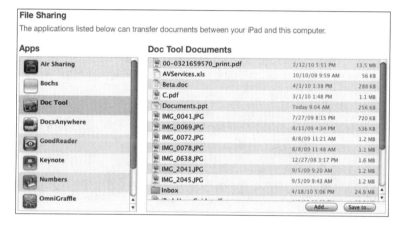

Figure 2-3 Each installed application that declares `UIFileSharingEnabled` is listed in iTunes in the device's Apps tab.

User Control

You cannot specify which kinds of items are allowed to be in the Documents folder. Users can add any materials they like, and remove any items they want to. What they cannot do, however, is navigate through subfolders using the iTunes interface. Notice the Inbox folder in Figure 2-3. This is an artifact left over from application-to-application document sharing, and it should not be there. Users cannot manage that data directly, and you should not leave the subfolder there to confuse them.

Users cannot delete the Inbox in iTunes the way they can delete other files and folders. Nor should your application write files directly to the Inbox. Respect the Inbox's role, which is to capture any incoming data from other applications. When you implement file-sharing support, always check for an Inbox on resuming active status and process that data to clear out the Inbox and remove it whenever your app launches and resumes. Best practices for handling incoming documents are discussed later in this chapter.

Xcode Access

As a developer, you have access not only to the Documents folder, but also to the entire application sandbox. Use the Xcode Organizer (Command-2) > Devices tab > *Device* > Applications > *Application Name* to browse, upload, and download files to and from the sandbox.

Test basic file sharing by enabling the `UIFileSharingEnabled` property to an application and loading data to your Documents folder. After those files are created, use Xcode and iTunes to inspect, download, and delete them.

Scanning for New Documents

Recipe 2-3 works by requesting kqueue notifications in its `beginGeneratingDocumentNotificationsInPath:` method. Here, it retrieves a file descriptor for the path you supply (in this case, the Documents folder) and requests notifications for add and clear events. It adds this functionality to the current runloop, enabling notifications whenever the monitored folder updates.

Upon receiving that callback, it posts a notification (my custom `kDocumentChanged`, in the `kqueueFired` method) and continues watching for new events. This all runs in the primary run loop on the main thread, so the GUI can respond and update itself upon receiving the notification.

The following snippet demonstrates how you might use Recipe 2-3's watcher to update a file list in your GUI. Whenever the contents change, an update notification allows the app to refresh those directory contents listings:

```
- (void) scanDocuments
{
    NSString *path = [NSHomeDirectory()
        stringByAppendingPathComponent:@"Documents"];
    items = [[NSFileManager defaultManager]
        contentsOfDirectoryAtPath:path error:nil];
    [self.tableView reloadData];
}

- (void) loadView
{
    [self.tableView registerClass:[UITableViewCell class]
        forCellReuseIdentifier:@"cell"];
    [self scanDocuments];
```

```
    // React to content changes
    [[NSNotificationCenter defaultCenter]
        addObserverForName:kDocumentChanged
        object:nil queue:[NSOperationQueue mainQueue]
        usingBlock:^(NSNotification *notification){
        [self scanDocuments];
     }];

    // Start the watcher
    NSString *path = [NSHomeDirectory()
        stringByAppendingPathComponent:@"Documents"];
    helper = [DocWatchHelper watcherForPath:path];
}
```

Test this recipe by connecting a device to iTunes. Add and remove items using the iTunes App tab interface. The device's onboard file list updates to reflect those changes in real time.

There are some cautions to be aware of when using this recipe. First, for larger documents, you shouldn't be reading immediately after you're notified of their creation. You may want to poll file sizes to determine when data has stopped being written. Second, iTunes File Sharing transfer can, upon occasion, stall. Code accordingly.

Recipe 2-3 Using a kqueue File Monitor

```
#import <fcntl.h>
#import <sys/event.h>

#define kDocumentChanged \
    @"DocumentsFolderContentsDidChangeNotification"

@interface DocWatchHelper : NSObject
{
    CFFileDescriptorRef kqref;
    CFRunLoopSourceRef  rls;
}
@property (strong) NSString *path;
+ (id) watcherForPath: (NSString *) aPath;
@end

@implementation DocWatchHelper
@synthesize path;

- (void)kqueueFired
{
    int           kq;
    struct kevent event;
```

```
    struct timespec timeout = { 0, 0 };
    int             eventCount;

    kq = CFFileDescriptorGetNativeDescriptor(self->kqref);
    assert(kq >= 0);

    eventCount = kevent(kq, NULL, 0, &event, 1, &timeout);
    assert( (eventCount >= 0) && (eventCount < 2) );

    if (eventCount == 1)
        [[NSNotificationCenter defaultCenter]
            postNotificationName:kDocumentChanged
            object:self];

    CFFileDescriptorEnableCallBacks(self->kqref,
        kCFFileDescriptorReadCallBack);
}

static void KQCallback(CFFileDescriptorRef kqRef,
    CFOptionFlags callBackTypes, void *info)
{
    DocWatchHelper *helper =
        (DocWatchHelper *)(__bridge id)(CFTypeRef) info;
    [helper kqueueFired];
}

- (void) beginGeneratingDocumentNotificationsInPath:
    (NSString *) docPath
{
    int             dirFD;
    int             kq;
    int             retVal;
    struct kevent   eventToAdd;
    CFFileDescriptorContext context =
        { 0, (void *)(__bridge CFTypeRef) self,
            NULL, NULL, NULL };

    dirFD = open([docPath fileSystemRepresentation], O_EVTONLY);
    assert(dirFD >= 0);

    kq = kqueue();
    assert(kq >= 0);

    eventToAdd.ident  = dirFD;
    eventToAdd.filter = EVFILT_VNODE;
    eventToAdd.flags  = EV_ADD | EV_CLEAR;
    eventToAdd.fflags = NOTE_WRITE;
```

```
    eventToAdd.data   = 0;
    eventToAdd.udata  = NULL;

    retVal = kevent(kq, &eventToAdd, 1, NULL, 0, NULL);
    assert(retVal == 0);

    self->kqref = CFFileDescriptorCreate(NULL, kq,
        true, KQCallback, &context);
    rls = CFFileDescriptorCreateRunLoopSource(
        NULL, self->kqref, 0);
    assert(rls != NULL);

    CFRunLoopAddSource(CFRunLoopGetCurrent(), rls,
        kCFRunLoopDefaultMode);
    CFRelease(rls);

    CFFileDescriptorEnableCallBacks(self->kqref,
        kCFFileDescriptorReadCallBack);
}

- (void) dealloc
{
    self.path = nil;
    CFRunLoopRemoveSource(CFRunLoopGetCurrent(), rls,
        kCFRunLoopDefaultMode);
    CFFileDescriptorDisableCallBacks(self->kqref,
        kCFFileDescriptorReadCallBack);
}

+ (id) watcherForPath: (NSString *) aPath
{
    DocWatchHelper *watcher = [[self alloc] init];
    watcher.path = aPath;
    [watcher beginGeneratingDocumentNotificationsInPath:aPath];
    return watcher;
}
@end
```

Get This Recipe's Code

To find this recipe's full sample project, point your browser to https://github.com/erica/iOS-6-Advanced-Cookbook and go to the folder for Chapter 2.

Recipe: Presenting the Activity View Controller

Newly introduced, iOS 6's activity view controller integrates data activities into the interface shown in Figure 2-4. With minimal development cost on your part, this new controller enables your users to copy items to the pasteboard, post to social media, share via e-mail and texting, and more. Built-in activities include Facebook, Twitter, Weibo, SMS, mail, printing, copying to pasteboard, and assigning data to a contact. Apps can define their own custom services as well, which you can read about later in this section:

- `UIActivityTypePostToFacebook`
- `UIActivityTypePostToTwitter`
- `UIActivityTypePostToWeibo`
- `UIActivityTypeMessage`
- `UIActivityTypeMail`
- `UIActivityTypePrint`
- `UIActivityTypeCopyToPasteboard`
- `UIActivityTypeAssignToContact`

Figure 2-4 The `UIActivityViewController` class offers system and custom services.

Significantly missing from this list are two important activities, namely Open in... for sharing documents between applications and QuickLook for previewing files. This functionality is discussed later in this chapter, with recipes that show you how to support these features independently and, in the case of QuickLook, how to integrate into the Activity View Controller.

> **Note**
>
> You do not need to own an AirPrint printer to test print activities. Ecamm's Printopia (http://www.ecamm.com/mac/printopia/, $19.95) creates a virtual printer on your local network, which you can use from both devices *and* the Simulator. Choose to print to any local printers, to a file on your Mac or a file on Dropbox. It's a terrific investment for any developer working with the new Activity View Controller. Netputing makes a similar product called handyPrint (http://www.netputing.com/handyprint). handyPrint accepts PayPal donations.

Presenting the Activity View Controller

How you present the controller varies by device. Show it modally on members of the iPhone-family and in a popover on tablets. The `UIBarButtonSystemItemAction` icon provides the perfect way to populate bar buttons linking to this controller.

Best of all, there's almost no work required on your end. After users select an activity, the controller handles all further interaction such as presenting a mail or Twitter compose sheet, adding a picture to the onboard library, or assigning it to a contact.

Activity Item Sources

Recipe 2-4 creates and presents the view controller from code. This implementation has its main class adopt the `UIActivityItemSource` protocol and adds `self` to the items array passed to the controller. Adopting the source protocol helps the controller understand how to use callbacks when retrieving data items. This represents the first of two ways to create and present the controller.

The protocol's two mandatory methods supply the item to process (the data that will be used for the activity) and a placeholder for that item. The item corresponds to an object that's appropriate for a given activity type. You can vary which item you return based on the kind of activity that's passed to the callback. For example, you might tweet "I created a great song in *App Name*" but you might send the actual sound file through e-mail.

The placeholder for an item is typically the same data returned as the item unless you have objects that you must process or create. In that case, you can create a placeholder object without real data.

Both callbacks (item and placeholder) run on the main thread, so keep your data small. If you need to process your data extensively, consider using a provider instead.

Item Providers

The `UIActivityItemProvider` class enables you to delay passing data. It's a type of operation (`NSOperation`) that offers you the flexibility to manipulate data before sharing. For example, you might need to process a large video file before it can be uploaded to a social sharing site or subsample some audio from a larger sequence.

Subclass the provider class and implement the `item` method. This takes the place of the `main` method you normally use with operations. Generate the processed data, safe in the knowledge that it will run asynchronously without blocking your user's interactive experience.

Item Source Callbacks

The callbacks methods enable you to vary your data based on each one's intended use. Use the activity types (such as Facebook, or Add to Contacts; they're listed earlier in this section) to choose the exact data you want to provide. This is especially important when selecting from resolutions for various uses. When printing, keep your data quality high. When tweeting, a low-res image may do the job instead.

If your data is invariant, that is, you'll be passing the same data to e-mail as you would to Facebook, you can directly supply an array of data items (typically strings, images, and URLs). For example, you could create the controller like this. This uses a single image:

```
UIActivityViewController *activity = [[UIActivityViewController alloc]
    initWithActivityItems:@[imageView.image]
    applicationActivities:nil];
```

This direct approach is far simpler. Your primary class doesn't need to declare the item source protocol; you do not need to implement the extra methods. It's a quick-and-easy way to manage activities for simple items.

You're not limited to passing single items, either. Include additional elements in the activity items array as needed. The following controller might add its two images to an e-mail or save both to the system camera roll, depending on the user's selection. Broadening activities to use multiple items enables users to be more efficient while using your app:

```
UIImage *secondImage = [UIImage imageNamed:@"Default.png"];
UIActivityViewController *activity = [[UIActivityViewController alloc]
    initWithActivityItems:@[imageView.image, secondImage]
    applicationActivities:nil];
```

Recipe 2-4 **The Activity View Controller**

```
- (void) presentViewController:
    (UIViewController *)viewControllerToPresent
{
    if (IS_IPHONE)
    {
```

```objc
        [self presentViewController:viewControllerToPresent
            animated:YES completion:nil];
    }
    else
    {
        popover = [[UIPopoverController alloc]
            initWithContentViewController:viewControllerToPresent];
        popover.delegate = self;
        [popover presentPopoverFromBarButtonItem:
                self.navigationItem.leftBarButtonItem
            permittedArrowDirections:UIPopoverArrowDirectionAny
            animated:YES];
    }
}

// Return the item to process
- (id)activityViewController:
        (UIActivityViewController *)activityViewController
    itemForActivityType:(NSString *)activityType
{
    return imageView.image;
}

// Return a thumbnail version of that item
- (id)activityViewControllerPlaceholderItem:
    (UIActivityViewController *)activityViewController
{
    return imageView.image;
}

// Create and present the view controller
- (void) action
{
    UIActivityViewController *activity =
        [[UIActivityViewController alloc]
            initWithActivityItems:@[self] applicationActivities:nil];
    [self presentViewController:activity];
}
```

Get This Recipe's Code

To find this recipe's full sample project, point your browser to https://github.com/erica/iOS-6-Advanced-Cookbook and go to the folder for Chapter 2.

Adding Services

Each app can provide application-specific services by subclassing the `UIActivity` class and presenting a custom view controller. The view controller enables the user to process the passed data in some way. Listing 2-1 introduces a skeletal activity that presents a simple text view. This view lists the items passed to it by the activity controller. It displays each item's class and description.

This listing includes details for two distinct classes. The first class implements a simple text controller and is intended for use within a navigation hierarchy. It includes a view for presenting text and a handler that updates the calling `UIActivity` instance by sending `activityDidFinish:` when the user taps *Done*.

Adding a way for your activity to complete is important, especially when your controller doesn't have a natural endpoint. When your action uploads data to an FTP server, you know when it completes. If it Tweets, you know when the status posts. In this example, it's up to the user to determine when this activity finishes. Make sure your view controller contains a weak property pointing back to the activity, so you can send the did-finish method after your work concludes.

The activity class contains a number of mandatory and optional items. You should implement *all* the methods shown in this listing. The methods to support a custom activity include the following:

- **activityType**—Returns a unique string describing the type of activity. One of this string's counterparts in the system-supplied activities is `UIActivityTypePostToFacebook`. Use a similar naming scheme. This string identifies a particular activity type and what it does. In this listing, I return `@"CustomActivityTypeListItemsAndTypes"`, which describes the activity.

- **activityTitle**—Supply the text you want to show in the activity controller. The custom text in Figure 2-5 was returned by this method. Use active descriptions when describing your custom action. Follow Apple's lead, for example, "Save to Camera Roll", "Print", "Copy". Your title should finish the phrase, "I want to..." For example, "I want to print," "I want to copy," or, in this example, "I want to list items." Use header case and capitalize each word except for minor ones like "to" or "and."

- **activityImage**—Returns an image for the controller to use. The controller adds a backsplash and converts your image to a one-value bitmap, layering it on top. Use simple art on a transparent background (57x57 iPhone, 72x72 iPad, double for Retina screen scale) to build the contents of your icon image. You'll want to inset your art at least 15% from the sides to allow space to inset from the controller-supplied rounded rectangle that frames it.

- **canPerformWithActivityItems:**—Scan the passed items and decide if your controller can process them. If so, return YES.

- **prepareWithActivityItems:**—Store the passed items for later use (here, they're assigned to a local instance variable) and perform any necessary pre-processing.

- **`activityViewController`**—Return a fully initialized presentable view controller using the activity items passed to you earlier. This controller is automatically presented to the user, where she can customize options before performing the promised action.

Figure 2-5 Add your own custom application activities.

Adding custom activities allows your app to expand its data handling possibilities while integrating features into a consistent system-supplied interface. It's a powerful iOS feature. The strongest activity choices will integrate with system services (such as copying to the pasteboard, or saving to the photo album) or provide a connection to off-device APIs, such as Facebook, Twitter, Dropbox, and FTP.

This example, which simply lists items, represents a weak use case. There's no reason the same feature couldn't be provided as a normal in-app screen. When you think "actions," try to project outside the app. Connect your user's data with sharing and processing features that expand beyond the normal GUI.

Listing 2-1 **Application Activities**

```
// All activities present a view controller. This custom controller
// provides a full-sized text view.
@interface TextViewController : UIViewController
  @property (nonatomic, readonly) UITextView *textView;
  @property (nonatomic, weak) UIActivity *activity;
@end
```

```
@implementation TextViewController

// Make sure you provide a done handler of some kind, such as this
// or an integrated button that finishes and wraps up
- (void) done
{
    [_activity activityDidFinish:YES];
}

// Just a super-basic text view controller
- (id) init
{
    if (!(self = [super init])) return nil;
    _textView = [[UITextView alloc] init];
    _textView.font = [UIFont fontWithName:@"Futura" size:16.0f];
    _textView.editable = NO;

    [self.view addSubview:_textView];
    PREPCONSTRAINTS(_textView);
    STRETCH_VIEW(self.view, _textView);

    // Prepare a Done button
    self.navigationItem.rightBarButtonItem =
        BARBUTTON(@"Done", @selector(done));

    return self;
}
@end

@interface MyActivity : UIActivity
@end
@implementation MyActivity
{
    NSArray *items;
}

// A unique type name
- (NSString *)activityType
{
    return @"CustomActivityTypeListItemsAndTypes";
}

// The title listed on the controller
- (NSString *) activityTitle
{
    return @"Cookbook";
}
```

```objc
// A custom image, displayed as a bitmap over a textured background
// This one says "iOS" in a rounded rect edge
- (UIImage *) activityImage
{
    CGRect rect = CGRectMake(0.0f, 0.0f, 75.0f, 75.0f);
    UIGraphicsBeginImageContext(rect.size);
    rect = CGRectInset(rect, 15.0f, 15.0f);
    UIBezierPath *path = [UIBezierPath
        bezierPathWithRoundedRect:rect cornerRadius:4.0f];
    [path stroke];
    rect = CGRectInset(rect, 0.0f, 10.0f);
    [@"iOS" drawInRect:rect
        withFont:[UIFont fontWithName:@"Futura" size:18.0f]
        lineBreakMode:NSLineBreakByWordWrapping
        alignment:NSTextAlignmentCenter];
    UIImage *image = UIGraphicsGetImageFromCurrentImageContext();
    UIGraphicsEndImageContext();

    return image;
}

// Specify if you can respond to these items
- (BOOL)canPerformWithActivityItems:(NSArray *)activityItems
{
    return YES;
}

// Store the items locally for later use
- (void)prepareWithActivityItems:(NSArray *)activityItems
{
    items = activityItems;
}

// Return a view controller, in this case one that lists
// its items and their classes
- (UIViewController *) activityViewController
{
    TextViewController *tvc = [[TextViewController alloc] init];
    tvc.activity = self;
    UITextView *textView = tvc.textView;

    NSMutableString *string = [NSMutableString string];
    for (id item in items)
        [string appendFormat:
            @"%@: %@\n", [item class], [item description]];
    textView.text = string;
```

```
    // Make sure to provide some kind of done: handler in
    // your main controller.
    UINavigationController *nav = [[UINavigationController alloc]
        initWithRootViewController:tvc];
    return nav;
}
@end
```

Items and Services

The services presented for each item vary by the kind of data you pass. Table 2-1 lists offered activities by source data type. As you see in this chapter, preview controller support expands beyond these foundation types:

- iOS's Quick Look framework integrates activity controllers into its file previews. The Quick Look-provided activity controller can print and e-mail many kinds of documents. Some document types support other activities as well.
- Document Interaction Controllers offer "open in" features that enable you to share files between application. It adds activities into its "options" style presentation, combining activities with "open in" choices.

Table 2-1 **Activity Types for Data Types**

Source	Offered Activity
NSString String, single or multiple	Mail, Message, Twitter, Facebook, Weibo, and Copy
UIImage Image, single	Mail, Twitter, Facebook, Weibo, Assign to Contact, Save to Camera Roll, and Print, Copy
UIImage Image, multiple	Mail, Facebook, Assign to Contact, Save to Camera Roll, Print, and Copy
UIColor Colors	Copy
NSURL URLs	Mail, Message, Twitter, Facebook, Weibo, and Copy
NSDictionary Dictionaries	Any supported activities for objects contained with in the dictionary. Sadly, the same does not hold true for arrays, which are unsupported.

Source	Offered Activity
Unsupported items	For example, `NSData`, `NSArray`, `NSDate`, or `NSNumber`: Nothing, a blank view controller.
Various items	Union of all supported types; for example, for string + image, you get: Mail, Message, Twitter, Facebook, Weibo, Assign to Contact, Save to Camera Roll, and Print Copy.

Supporting HTML E-mail

If you want to send HTML using an E-mail activity, make sure your item's text string starts with `@"<html>"`. Differentiate your HTML-based e-mail text from your plain Twitter contents by implementing the item source protocol and returning appropriate items according to the activity selected by the user.

Excluding Activities

You can specifically exclude activities by supplying a list of activity types to the excludedActivityTypes property:

```
UIActivityViewController *activity = [[UIActivityViewController alloc]
    initWithActivityItems:items applicationActivities:@[appActivity]];
activity.excludedActivityTypes = @[UIActivityTypeMail];
```

Recipe: The Quick Look Preview Controller

The Quick Look preview controller class enables users to preview many document types. This controller supports text, images, PDF, RTF, iWork file, Microsoft Office documents (Office 97 and newer, including doc, ppt, xls, and so on), and comma-separated value (csv) files. You supply a supported file type and the Quick Look controller displays it for the user. An integrated system-supplied activity view controller helps share the previewed document, as shown in Figure 2-6.

Either push or present your preview controllers. The controller adapts to both situations, working with navigation stacks and with modal presentation. Recipe 2-5 demonstrates both approaches.

Figure 2-6 This Quick Look controller was presented modally and shows the screen after the user has tapped the action button. Quick Look handles a wide range of document types, enabling users to see the file contents before deciding on an action to apply to them. Most Quick Look types support Mail and Print. Many support Copy, and image files offer even more options.

Implementing QuickLook

Quick Look support requires a few simple steps:

1. Declare the `QLPreviewControllerDataSource` protocol in your primary controller class.
2. Implement the `numberOfPreviewItemsInPreviewController:` and `previewController:previewItemAtIndex:` data source methods. The first of these methods returns a count of items to preview. The second returns the preview item referred to by the index.
3. Preview items must conform to the `QLPreviewItem` protocol. This protocol consists of two required properties: a preview title and an item URL. Recipe 2-5 creates a conforming `QuickItem` class. This class implements an absolutely minimal approach to support the data source.

After these requirements are met, your code is ready to create a new preview controller, sets its data source, and present or push it.

Recipe 2-5 **Quick Look**

```
@interface QuickItem : NSObject <QLPreviewItem>
@property (nonatomic, strong) NSString *path;
@property (readonly) NSString *previewItemTitle;
@property (readonly) NSURL *previewItemURL;
@end

@implementation QuickItem

// Title for preview item
- (NSString *) previewItemTitle
{
    return [_path lastPathComponent];
}

// URL for preview item
- (NSURL *) previewItemURL
{
    return [NSURL fileURLWithPath:_path];
}
@end

#define FILE_PATH    [NSHomeDirectory() \
    stringByAppendingPathComponent:@"Documents/PDFSample.pdf"]

@interface TestBedViewController : UIViewController
    <QLPreviewControllerDataSource>
@end

@implementation TestBedViewController
- (NSInteger) numberOfPreviewItemsInPreviewController:
    (QLPreviewController *) controller
{
    return 1;
}

- (id <QLPreviewItem>) previewController:
        (QLPreviewController *) controller
    previewItemAtIndex: (NSInteger) index;
{
    QuickItem *item = [[QuickItem alloc] init];
    item.path = FILE_PATH;
    return item;
}
```

```
// Push onto navigation stack
- (void) push
{
    QLPreviewController *controller =
        [[QLPreviewController alloc] init];
    controller.dataSource = self;
    [self.navigationController
        pushViewController:controller animated:YES];
}

// Use modal presentation
- (void) present
{
    QLPreviewController *controller =
        [[QLPreviewController alloc] init];
    controller.dataSource = self;
    [self presentViewController:controller
        animated:YES completion:nil];
}

- (void) loadView
{
    self.view.backgroundColor = [UIColor whiteColor];

    self.navigationItem.rightBarButtonItem =
        BARBUTTON(@"Push", @selector(push));
    self.navigationItem.leftBarButtonItem =
        BARBUTTON(@"Present", @selector(present));
}
@end
```

Get This Recipe's Code

To find this recipe's full sample project, point your browser to https://github.com/erica/iOS-6-Advanced-Cookbook and go to the folder for Chapter 2.

Recipe: Adding a QuickLook Action

QuickLook is significant in its absence from the standard set of actions presented by the system-supplied activity view controller. You can easily add a custom action that provides this feature, offering the same preview created by Recipe 2-5. That's what Recipe 2-6 does, wrapping Recipe 2-5's functionality into a custom `QLActivity` class.

Recipe 2-6 performs a more thorough search for compliant items than you've seen yet in this chapter. It scans the item array passed to it until it finds a local file URL, which it can use for a document preview. If it does not find one, it returns NO from the `canPerformWithActivity-Items:` method and will not be listed on the activity controller.

Recipe 2-6 **Quick Look**

```
@implementation QLActivity
{
    NSArray *items;
    NSArray *qlitems;
    QLPreviewController *controller;
}

// Activity Customization
- (NSString *)activityType
{
    return @"CustomQuickLookActivity";
}

- (NSString *) activityTitle
{
    return @"QuickLook";
}

- (UIImage *) activityImage
{
    return [UIImage imageNamed:@"QL.png"];
}

// Items must include at least one file URL
- (BOOL)canPerformWithActivityItems:(NSArray *)activityItems
{
    for (NSObject *item in activityItems)
        if ([item isKindOfClass:[NSURL class]])
        {
            NSURL *url = (NSURL *)item;
            if (url.isFileURL) return YES;
        }
    return NO;
}

// QuickLook callbacks
- (NSInteger) numberOfPreviewItemsInPreviewController:
    (QLPreviewController *) controller
{
    return qlitems.count;
}
```

```objc
- (id <QLPreviewItem>) previewController: (QLPreviewController *)
    controller previewItemAtIndex: (NSInteger) index;
{
    return qlitems[index];
}

// Item preparation
- (void)prepareWithActivityItems:(NSArray *)activityItems
{
    items = activityItems;

    controller = [[QLPreviewController alloc] init];
    controller.dataSource = self;
    controller.delegate = self;

    NSMutableArray *finalArray = [NSMutableArray array];

    for (NSObject *item in items)
    {
        if ([item isKindOfClass:[NSURL class]])
        {
            NSURL *url = (NSURL *)item;
            if (url.isFileURL)
            {
                QuickItem *item = [[QuickItem alloc] init];
                item.path = url.path;
                [finalArray addObject:item];
            }
        }
    }

    qlitems = finalArray;
}

- (void) previewControllerDidDismiss:
    (QLPreviewController *)controller
{
    [self activityDidFinish:YES];
}

- (UIViewController *) activityViewController
{
    return controller;
}
@end
```

> **Get This Recipe's Code**
>
> To find this recipe's full sample project, point your browser to https://github.com/erica/iOS-6-Advanced-Cookbook and go to the folder for Chapter 2.

Recipe: Using The Document Interaction Controller

The `UIDocumentInteractionController` class enables applications to present a menu of options to users, enabling them to use document files in a variety of ways. With this class, users can take advantage of the following:

- iOS application-to-application document sharing (that is, "open this document in... *some app*")
- Document preview using QuickLook
- Activity controller options such as printing, sharing, and social networking

You've already seen the latter two features in action earlier in this chapter. The document interaction class adds app-to-app sharing on top of these, as shown in Figure 2-7. The controller is presented as a menu, which enables users to specify how they want to interact with the given document.

In iOS 6, the number of "open in" options is no longer limited the way it was in earlier OS releases. That's why you see the page indicator at the bottom of the menu. Users can swipe between screens to see a full complement of "open in" options.

The controller offers two basic menu styles. The "open" style offers only "open in" choices, using the menu space to provide as many destination choices as possible. The "options" style (refer to Figure 2-7) provides a list of all interaction options, including "open in," quick-look, and any supported actions. It's essentially all the good stuff you get from a standard Actions menu, along with "open in" extras. You do have to explicitly add quick-look callbacks, but it takes a little work to do so.

Creating Document Interaction Controller Instances

Each document interaction controller is specific to a single document file. This file is typically stored in the user's Documents folder:

```
dic = [UIDocumentInteractionController
    interactionControllerWithURL:fileURL];
```

Figure 2-7 The `UIDocumentInteractionController` shown in its options style for iPhone (left) and iPad (right). Both presentations include two pages of icons, as indicated by the page controller near the bottom of the display.

You supply a local file URL and present the controller using either the options variation (basically the Action menu) or the open one (just the "open in" items). The two presentation styles are from a bar button or an onscreen rectangle:

- `presentOptionsMenuFromRect:inView:animated:`
- `presentOptionsMenuFromBarButtonItem:animated:`
- `presentOpenInMenuFromRect:inView:animated:`
- `presentOpenInMenuFromBarButtonItem:animated:`

The iPad uses the bar button or rect you pass to present a popover. On the iPhone, the implementation presents a modal controller view. As you would expect, more bookkeeping takes place on the iPad, where users may tap on other bar buttons, may dismiss the popover, and so forth.

You'll want to disable each iPad bar button item after presenting its associated controller, and re-enable it after dismissal. That's important because you don't want your user to re-tap an in-use bar button, and need to handle situations where a different popover needs to take over. Basically, there are a variety of unpleasant scenarios that can happen if you don't carefully monitor which buttons are active and what popover is in play. Recipe 2-7 guards against these scenarios.

Document Interaction Controller Properties

Each document interaction controller offers a number of properties, which can be used in your delegate callbacks:

- The `URL` property enables you to query the controller for the file it is servicing. This is the same URL you pass when creating the controller.

- The `UTI` property is used to determine which apps can open the document. It uses the system-supplied functions discussed earlier in the chapter to find the most preferred UTI match based on the filename and metadata. You can override this in code to set the property manually.

- The `name` property provides the last path component of the URL, offering a quick way to provide a user-interpretable name without having to manually strip the URL yourself.

- Use the `icons` property to retrieve an icon for the file type that's in play. Applications that declare support for certain file types provide image links in their declaration (as you'll see shortly, in the discussion about declaring file support). These images correspond to the values stored for the `kUTTypeIconFileKey` key, as discussed earlier in this chapter.

- The `annotation` property provides a way to pass custom data along with a file to any application that will open the file. There are no standards for using this property; although, the item must be set to some top-level property list object, namely dictionaries, arrays, data, strings, numbers, and dates. Because there are no community standards, use of this property tends to be minimal except where developers share the information across their own suite of published apps.

Providing Document Quick Look Support

Add Quick Look support to the controller by implementing a trio of delegate callbacks. These methods declare which view controller will be used to present the preview, which view will host it, and the frame for the preview size. You may have occasional compelling reasons to use a child view controller with limited screen presence on tablets (such as in a split view, with the preview in just one portion), but for the iPhone family, there's almost never any reason not to allow the preview to take over the entire screen:

```
#pragma mark QuickLook
- (UIViewController *)
    documentInteractionControllerViewControllerForPreview:
        (UIDocumentInteractionController *)controller
{
    return self;
}

- (UIView *) documentInteractionControllerViewForPreview:
    (UIDocumentInteractionController *)controller
{
    return self.view;
}

- (CGRect) documentInteractionControllerRectForPreview:
    (UIDocumentInteractionController *)controller
{
    return self.view.frame;
}
```

Checking for the Open Menu

When using a document interaction controller, the Options menu will almost always provide valid menu choices, especially if you implement the Quick Look callbacks. You may or may not, however, have any open-in options to work with. Those options depend on the file data you provide to the controller and the applications users install on their device.

A no-open-options scenario happens when there are no applications installed on a device that support the file type you are working with. This may be caused by an obscure file type, but more often it occurs because the user has not yet purchased and installed a relevant application.

Always check whether to offer an "Open" menu option. Recipe 2-7 performs a rather ugly test to see if external apps will offer themselves as presenters and editors for a given URL. This is what it does: It creates a new, temporary controller and attempts to present it. If it succeeds, conforming file destinations exist and are installed on the device. If not, there are no such apps, and any "Open In" buttons should be disabled.

On the iPad, you must run this check in `viewDidAppear:` or later—that is, after a window has been established. The method immediately dismisses the controller after presentation. It should not be noticeable by your end user, and none of the calls use animation.

This is obviously a rather dreadful implementation, but it has the advantage of testing as you lay out your interface or when you start working with a new file. I encourage you to file an enhancement request at bugreporter.apple.com.

One further caution: Although this test works on primary views (as in this recipe), it can cause headaches in nonstandard presentations in popovers on the iPad.

> **Note**
>
> You rarely offer users both "option" and "open" items in the same application. Recipe 2-7 uses the system-supplied Action item for the Options menu. You may want to use this in place of "Open in…" text for apps that exclusively use the open style.

Recipe 2-7   **Document Interaction Controllers**

```
@implementation TestBedViewController
{
    NSURL *fileURL;
    UIDocumentInteractionController *dic;
    BOOL canOpen;
}

#pragma mark QuickLook
- (UIViewController *)
    documentInteractionControllerViewControllerForPreview:
        (UIDocumentInteractionController *)controller
{
    return self;
}

- (UIView *) documentInteractionControllerViewForPreview:
    (UIDocumentInteractionController *)controller
{
    return self.view;
}

- (CGRect) documentInteractionControllerRectForPreview:
    (UIDocumentInteractionController *)controller
{
    return self.view.frame;
}

#pragma mark Options / Open in Menu

// Clean up after dismissing options menu
- (void) documentInteractionControllerDidDismissOptionsMenu:
    (UIDocumentInteractionController *) controller
```

```objc
{
    self.navigationItem.leftBarButtonItem.enabled = YES;
    dic = nil;
}

// Clean up after dismissing open menu
- (void) documentInteractionControllerDidDismissOpenInMenu:
    (UIDocumentInteractionController *) controller
{
    self.navigationItem.rightBarButtonItem.enabled = canOpen;
    dic = nil;
}

// Before presenting a controller, check to see if there's an
// existing one that needs dismissing
- (void) dismissIfNeeded
{
    if (dic)
    {
        [dic dismissMenuAnimated:YES];
        self.navigationItem.rightBarButtonItem.enabled = canOpen;
        self.navigationItem.leftBarButtonItem.enabled = YES;
    }
}

// Present the options menu
- (void) action: (UIBarButtonItem *) bbi
{
    [self dismissIfNeeded];
    dic = [UIDocumentInteractionController interactionControllerWithURL:fileURL];
    dic.delegate = self;
    self.navigationItem.leftBarButtonItem.enabled = NO;
    [dic presentOptionsMenuFromBarButtonItem:bbi animated:YES];
}

// Present the open-in menu
- (void) open: (UIBarButtonItem *) bbi
{
    [self dismissIfNeeded];
    dic = [UIDocumentInteractionController interactionControllerWithURL:fileURL];
    dic.delegate = self;
    self.navigationItem.rightBarButtonItem.enabled = NO;
    [dic presentOpenInMenuFromBarButtonItem:bbi animated:YES];
}

#pragma mark Test for Open-ability
-(BOOL)canOpen: (NSURL *) aFileURL
```

```objc
{
    UIDocumentInteractionController *tmp =
        [UIDocumentInteractionController
            interactionControllerWithURL:aFileURL];
    BOOL success = [tmp presentOpenInMenuFromRect:CGRectMake(0,0,1,1)
        inView:self.view animated:NO];
    [tmp dismissMenuAnimated:NO];
    return success;
}

- (void) viewDidAppear:(BOOL)animated
{
    // Only enable right button if the file can be opened
    canOpen = [self canOpen:fileURL];
    self.navigationItem.rightBarButtonItem.enabled = canOpen;
}

#pragma mark View management
- (void) loadView
{
    self.view.backgroundColor = [UIColor whiteColor];
    self.navigationItem.rightBarButtonItem =
        BARBUTTON(@"Open in...", @selector(open:));
    self.navigationItem.leftBarButtonItem =
        SYSBARBUTTON(UIBarButtonSystemItemAction,
            @selector(action:));

    NSString *filePath = [NSHomeDirectory()
        stringByAppendingPathComponent:@"Documents/DICImage.jpg"];
    fileURL = [NSURL fileURLWithPath:filePath];
}
@end
```

Get This Recipe's Code

To find this recipe's full sample project, point your browser to https://github.com/erica/iOS-6-Advanced-Cookbook and go to the folder for Chapter 2.

Recipe: Declaring Document Support

Application documents are not limited to files they create or download from the Internet. As you discovered in the previous recipe, applications may handle certain file types. They may open items passed from other apps. You've already seen document sharing from the sending point of view, using the "open in" controller to export files to other applications. Now it's time to look at it from the receiver's end.

Applications declare their support for certain file types in their Info.plist property list. The Launch Services system reads this data and creates the file-to-app associations used by the document interaction controller.

Although you can edit the property list directly, Xcode 4 offers a simple form as part of the Project > Target > Info screen. Locate the Document Types section. You'll find it below the Custom iOS Target Properties. Open the section and click + to add a new supported document type. Figure 2-8 shows what this looks like for an app that accepts JPEG image documents.

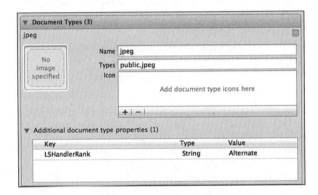

Figure 2-8   Declare supported document types in Xcode's Target > Info pane.

This declaration contains three minimal details: a name, one or more UTIs, and a handler rank, in this case alternate.

- The name is both required and arbitrary. It should be descriptive of the kind of document in play, but it's also somewhat of an afterthought on iOS. This field makes more sense when used on a Macintosh (it's the "kind" string used by Finder), but it is not optional.
- Specify one or more UTIs as your types. This example specifies only public.jpeg. Add commas between items when listing several items. For example, you might have an "image" document type that opened public.jpeg, public.tiff, and public.png. Enumerate specific types when you need to limit file support. Although declaring public.image would cover all three types, it might allow unsupported image styles to be opened as well.
- The launch services handler rank describes how the app views itself among the competition for handling this file type. An "owner" says that this is a native app that creates files of this type. An "alternate," as in Figure 2-8, offers a secondary viewer. You add the `LSHandlerRank` key manually in the additional document type properties.

You may optionally specify icon files. These are used in OS X as document icons and have minimal overlap with the iOS world. In the only case I can think of, you might see these icons in iTunes' Apps tab, when using the File Sharing section to add and remove items. Icons are typically 320x320 (`UTTypeSize320IconFile`) and 64x64 (`UTTypeSize64IconFile`) and are normally limited to files that your app creates and defines a custom type for.

Under the hood, Xcode uses this interactive form to build a `CFBundleDocumentTypes` array in your application's Info.plist. The following snippet shows the information from Figure 2-8 in its Info.plist form:

```
<key>CFBundleDocumentTypes</key>
<array>
    <dict>
        <key>CFBundleTypeIconFiles</key>
        <array/>
        <key>CFBundleTypeName</key>
        <string>jpg</string>
        <key>LSHandlerRank</key>
        <string>Alternate</string>
        <key>LSItemContentTypes</key>
        <array>
            <string>public.jpeg</string>
        </array>
    </dict>
</array>
```

Creating Custom Document Types

When your application builds new kinds of documents, you should declare them in the Exported UTI section of the Target > Info editor, which you see in Figure 2-9. This registers support for this file type with the system and identifies you as the owner of that type.

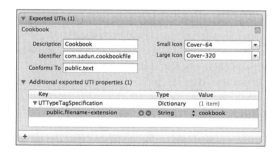

Figure 2-9 Declare custom file types in the Exported UTIs section of the Target > Info editor.

Chapter 2 Documents and Data Sharing

To define the new type, you supply a custom UTI (here, com.sadun.cookbookfile), document art (at 64 and 320 sizes), and specify a filename extension that identifies your file type. As with declaring document support, Xcode builds an exported declaration array into your project's Info.plist file. Here is what that material might look like for the declaration shown in Figure 2-9:

```
<key>UTExportedTypeDeclarations</key>
<array>
    <dict>
        <key>UTTypeConformsTo</key>
        <array>
            <string>public.text</string>
        </array>
        <key>UTTypeDescription</key>
        <string>Cookbook</string>
        <key>UTTypeIdentifier</key>
        <string>com.sadun.cookbookfile</string>
        <key>UTTypeSize320IconFile</key>
        <string>Cover-320</string>
        <key>UTTypeSize64IconFile</key>
        <string>Cover-64</string>
        <key>UTTypeTagSpecification</key>
        <dict>
            <key>public.filename-extension</key>
            <string>cookbook</string>
        </dict>
    </dict>
</array>
```

If added to your project in this way, your app should open any files with the cookbook extension, using the com.sadun.cookbookfile UTI.

Implementing Document Support

When your application provides document support, you should check for an "inbox" folder each time it becomes active. Specifically, see if an Inbox folder has appeared in the Documents folder. If so, you should move elements out of that inbox and to where they belong, typically in the main Documents directory. After the inbox has been cleared, delete it. This provides the best user experience, especially in terms of any file sharing through iTunes where the Inbox and its role may confuse users:

```
- (void)applicationDidBecomeActive:(UIApplication *)application
{
    // perform inbox test here
}
```

When moving items to Documents, check for name conflicts, and use an alternative path name (typically by appending a hyphen followed by a number) to avoid overwriting any existing file. The following method helps find an alternative name for a destination path. It gives up after a thousand attempts, but seriously, none of your users should be hosting that many duplicate document names. If they do, there's something deeply wrong with your overall application design.

Recipe 2-8 walks through the ugly details of scanning for the Inbox and moving files into place. It removes the Inbox after it is emptied. As you can see, any method like this is file manager-intensive. It primarily involves handling all the error combination possibilities that might pop up throughout the task. This should run quickly for small file support. If you must handle large files, such as video or audio, make sure to perform this processing on its own operation queue.

If you plan to support `public.data` files (that is, will open anything), you may want to display those files using `UIWebView` instances. Refer to Technical Q&A QA1630 (http://developer.apple.com/library/ios/#qa/qa1630) for details about which document types iOS can and cannot display in those views. Web views can present most audio and video assets, as well as Excel, Keynote, Numbers, Pages, PDF, PowerPoint, and Word resources in addition to simple HTML.

Recipe 2-8   Handling Incoming Documents

```
#define DOCUMENTS_PATH   [NSHomeDirectory() \
    stringByAppendingPathComponent:@"Documents"]
#define INBOX_PATH       [DOCUMENTS_PATH \
    stringByAppendingPathComponent:@"Inbox"]

@implementation InboxHelper
+ (NSString *) findAlternativeNameForPath: (NSString *) path
{
    NSString *ext = path.pathExtension;
    NSString *base = [path stringByDeletingPathExtension];

    for (int i = 1; i < 999; i++)
    {
        NSString *dest =
            [NSString stringWithFormat:@"%@-%d.%@", base, i, ext];

        // if the file does not yet exist, use this destination path
        if (![[NSFileManager defaultManager]
            fileExistsAtPath:dest])
            return dest;
    }

    NSLog(@"Exhausted possible names for file %@. Bailing.",
        path.lastPathComponent);
    return nil;
}
```

```objc
- (void) checkAndProcessInbox
{
    // Does the Inbox exist? If not, we're done
    BOOL isDir;
    if (![[NSFileManager defaultManager]
        fileExistsAtPath:INBOX_PATH isDirectory:&isDir])
        return;

    NSError *error;
    BOOL success;

    // If the Inbox is not a folder, remove it.
    if (!isDir)
    {
        success = [[NSFileManager defaultManager]
            removeItemAtPath:INBOX_PATH error:&error];
        if (!success)
        {
            NSLog(@"Error deleting Inbox file (not directory): %@",
                error.localizedFailureReason);
            return;
        }
    }

    // Retrieve a list of files in the Inbox
    NSArray *fileArray = [[NSFileManager defaultManager]
        contentsOfDirectoryAtPath:INBOX_PATH error:&error];
    if (!fileArray)
    {
        NSLog(@"Error reading contents of Inbox: %@",
            error.localizedFailureReason);
        return;
    }

    // Remember the number of items
    NSUInteger initialCount = fileArray.count;

    // Iterate through each file, moving it to Documents
    for (NSString *filename in fileArray)
    {
        NSString *source =
            [INBOX_PATH stringByAppendingPathComponent:filename];
        NSString *dest = [DOCUMENTS_PATH
            stringByAppendingPathComponent:filename];

        // Is the file already there?
        BOOL exists =
            [[NSFileManager defaultManager] fileExistsAtPath:dest];
        if (exists) dest = [self findAlternativeNameForPath:dest];
```

```objc
        if (!dest)
        {
            NSLog(@"Error. File name conflict not resolved");
            continue;
        }
        
        // Move file into place
        success = [[NSFileManager defaultManager]
            moveItemAtPath:source toPath:dest error:&error];
        if (!success)
        {
            NSLog(@"Error moving file from Inbox: %@",
                error.localizedFailureReason);
            continue;
        }
    }
    
    // Inbox should now be empty
    fileArray = [[NSFileManager defaultManager]
        contentsOfDirectoryAtPath:INBOX_PATH error:&error];
    if (!fileArray)
    {
        NSLog(@"Error reading contents of Inbox: %@",
            error.localizedFailureReason);
        return;
    }
    
    if (fileArray.count)
    {
        NSLog(@"Error clearing Inbox. %d items remain",
            fileArray.count);
        return;
    }
    
    // Remove the inbox
    success = [[NSFileManager defaultManager]
        removeItemAtPath:INBOX_PATH error:&error];
    if (!success)
    {
        NSLog(@"Error removing inbox: %@",
            error.localizedFailureReason);
        return;
    }
    
    NSLog(@"Moved %d items from the Inbox", initialCount);
}
@end
```

> **Get This Recipe's Code**
> To find this recipe's full sample project, point your browser to https://github.com/erica/iOS-6-Advanced-Cookbook and go to the folder for Chapter 2.

Recipe: Creating URL-Based Services

Apple's built-in applications offer a variety of services that can be accessed via URL calls. You can ask Safari to open web pages, Maps to show a map, or use the `mailto:` style URL to start composing a letter in Mail. A URL scheme refers to the first part of the URL that appears before the colon, such as `http` or `ftp`.

These services work because iOS knows how to match URL schemes to applications. A URL that starts with `http:` opens in Mobile Safari. The `mailto:` URL always links to Mail. What you may not know is that you can define your own URL schemes and implement them in your applications. Not all standard schemes are supported on iOS. The FTP scheme is not available for use.

Custom schemes enable applications to launch whenever Mobile Safari or another application opens a URL of that type. For example, should your application register xyz, any `xyz:` links go directly to your application for handling, where they're passed to the application delegate's URL opening method. You do not have to add any special coding there. If all you want to do is run an application, adding the scheme and opening the URL enables cross-application launching.

Handlers extend launching to allow applications to do something with the URL that's been passed to it. They might open a specific data file, retrieve a particular name, display a certain image, or otherwise process information included in the call.

Declaring the Scheme

To declare your URL scheme, edit the URL Types section of the Target > Info editor (see Figure 2-10) and list the URL schemes you will use. The Info.plist section created by this declaration will look like this:

```
<key>CFBundleURLTypes</key>
<array>
    <dict>
        <key>CFBundleURLName</key>
        <string>com.sadun.urlSchemeDemonstration</string>
        <key>CFBundleURLSchemes</key>
        <array>
            <string>xyz</string>
        </array>
    </dict>
</array>
```

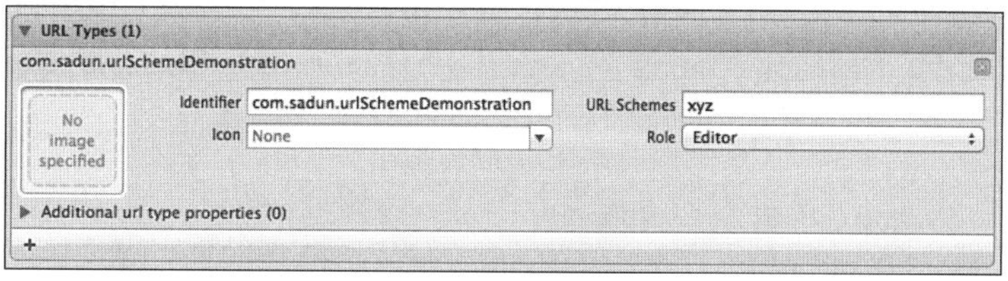

Figure 2-10 Add custom URL schemes in the URL Types section of the Target > Info editor.

The `CFBundleURLTypes` entry consists of an array of dictionaries that describe the URL types the application can open and handle. Each dictionary is quite simple. They contain two keys: a `CFBundleURLName` (defining an arbitrary identifier) and an array of `CFBundleURLSchemes`.

The Schemes array provides a list of prefixes that belong to the abstract name. You can add one scheme or many. This following example declares just one. You may want to prefix your name with an x (for example, `x-sadun-services`). Although the iOS family is not part of any standards organization, the x prefix indicates that this is an unregistered name. A draft specification for x-callback-url is under development at http://x-callback-url.com.

Former iOS developer (and current Apple employee) Emanuele Vulcano has started an informal registry at the CocoaDev Web site (http://cocoadev.com/index.pl?ChooseYourOwniPhoneURLScheme). iOS developers can share their schemes in a central listing, so you can discover services you want to use and promote services that you offer. The registry lists services and their URL schemes, and describes how these services can be used by other developers. Other registries include http://handleopenurl.com, http://wiki.akosma.com/IPhone_URL_Schemes, and http://applookup.com/Home.

Testing URLs

You can test whether a URL service is available. If the `UIApplication`'s `canOpenURL:` method returns `YES`, you are guaranteed that `openURL:` can launch another application to open that URL. You are not guaranteed that the URL is valid, only that its scheme is registered properly to an existing application:

```
if ([[UIApplication sharedApplication] canOpenURL:aURL])
    [[UIApplication sharedApplication] openURL:aURL];
```

Adding the Handler Method

To handle URL requests, you implement the URL-specific application delegate method shown in Recipe 2-9. Unfortunately, this method is guaranteed only to trigger when the application is already running. If it is not, and the app were launched by the URL request, control first goes to the launching methods (will- and did-finish).

You want to ensure that your normal application:didFinishLaunchingWithOptions: method returns YES. This allows control to pass to application:openURL:sourceApplication: annotation:, so the incoming URL can be processed and handled.

Recipe 2-9 Providing URL Scheme Support

```
// Called if the app is open or if the launch returns YES
- (BOOL)application:(UIApplication *)application
    openURL:(NSURL *)url
    sourceApplication:(NSString *)sourceApplication
    annotation:(id)annotation
{
    NSString *logString = [NSString stringWithFormat:
        @"DID OPEN: URL[%@] App[%@] Annotation[%@]\n",
        url, sourceApplication, annotation];
    tbvc.textView.text =
        [logString stringByAppendingString:tbvc.textView.text];
    return YES;
}

// Make sure to return YES
- (BOOL)application:(UIApplication *)application
    didFinishLaunchingWithOptions:(NSDictionary *)launchOptions
{
    window = [[UIWindow alloc]
        initWithFrame:[[UIScreen mainScreen] bounds]];
    tbvc = [[TestBedViewController alloc] init];

    UINavigationController *nav = [[UINavigationController alloc]
        initWithRootViewController:tbvc];
    window.rootViewController = nav;
    [window makeKeyAndVisible];
    return YES;
}
```

> **Get This Recipe's Code**
>
> To find this recipe's full sample project, point your browser to https://github.com/erica/iOS-6-Advanced-Cookbook and go to the folder for Chapter 2.

Summary

Want to share data across applications and leverage system-supplied actions? This chapter showed you how. You read about UTIs and how they are used to specify data roles across applications. You saw how the pasteboard worked, and how you could share files with iTunes. You read about monitoring folders and discovered how to implement custom URLs. You dived deep into the document interaction controller and saw how to add support for everything from printing to copying to previews. Here are a few thoughts to take with you before leaving this chapter:

- You are never limited to the built-in UTIs that Apple provides, but you should follow its lead when you decide to add your own. Be sure to use custom reverse domain naming and add as many details as possible (public URL definition pages, typical icons, and file extensions) in your exported definitions. Precision matters.

- Conformance arrays help you determine what kind of thing you're working with. Knowing that it's an image and not, say, a text file or movie, can help you better process the data associated with any file.

- The general pasteboard offers a terrific way to work with shared data, but if you have application-specific needs for cross-communication, there's no reason you can't use a custom pasteboard to share information. Just be aware that pasteboard data will not persist across reboots.

- The Documents folder belongs to the user and not to you. Remember that and provide respectful management of that directory.

- The documents interaction controller supercedes a lot of the reasons many developers used custom URL schemes. Use the controller to provide the app-to-app interaction your users demand, and don't be afraid of introducing annotation support to help ease transition between apps.

- Don't offer an "Open In" menu option unless there are onboard apps ready to back up that button. The solution you read about in this chapter is crude, but it's better than dealing with angry, frustrated, or confused users through customer support. Consider providing an alert, backed by this method, which explains when there are no other apps available.

3
Core Text

Nuanced text layout has been an important part of the computing experience as long as there have been personal computers. Even the earliest consumer operating systems included word processing with attributed text features. Decades ago, interactive document preparation tools offered font selection, text traits such as bolding, underlining, and italics, indentation, and more.

These same elements form a big part of the iOS experience. Developers use attributed strings and the Core Text framework to offer sophisticated text elements with a minimum of development effort. This chapter introduces attributed text processing and explores how you can build text features into your apps. You read about adding attributed strings to common UIKit elements, how to create Core Text-powered views, and how to break beyond lines for freeform text typesetting. Reading this chapter, you can discover the power that Core Text brings to iOS.

Core Text and iOS

Core Text is a Mac and iOS technology that programmatically supports typesetting tasks including attribute specification and layout. Its APIs enable you to define fonts, colors, alignment, word wrapping, and other features that move strings beyond being a simple collection of characters. Originally a Mac OS X 10.5 framework, Core Text has been arriving slowly at the iOS platform over the last few releases. As you'll discover, even in iOS 6, the process of migration continues.

Attributes

Much of the typesetting story consists of attributes, specifically attributed strings. Attributes are a set of features, like font choice or text color, applied to text within a certain range. Attributed strings, as the name implies, add characteristics to select substrings. An attributed string contains both text information and range-specific attributes applied to that string.

To get a sense of how attributes work and can combine, consider the following string:

This **is a** sample *string* that demonstrates how attributes can combine.

In this non-iOS example, constrained by the realities of book publishing, the preceding string uses a bold attribute from its 6th through 16th characters, and italic from the 11th through 23rd characters.

On iOS, the kinds of attributes applied to a string differ from those used when writing a book. For example, you do not add emphasis or bolding. Those are set by changing the font face—for example, from Courier to Courier-Bold or to Courier-BoldOblique. You specify the range where that font face should apply, just as you can specify ranges to set a given stroke color or a text alignment.

Other attributes do work in a similar manner. You can update the color for a range of characters, add a shadow, or change the font size.

C Versus Objective C

The first C-based Core Text features debuted in iOS 3.2. They used a Core Foundation pattern with little in the way of Objective C-friendliness. Here's an example of what raw Core Text looks like. What follows is a snippet from the iOS 5 Cookbook. It demonstrates how to create a CTParagraphStyleRef that defines a custom alignment and line break mode:

```
uint8_t theAlignment = [self ctAlignment];
CTParagraphStyleSetting alignSetting = {
    kCTParagraphStyleSpecifierAlignment, sizeof(uint8_t),
    &theAlignment};

uint8_t theLineBreak = [self ctBreakMode];
CTParagraphStyleSetting wordBreakSetting = {
    kCTParagraphStyleSpecifierLineBreakMode,
    sizeof(uint8_t),&theLineBreak};

CTParagraphStyleSetting settings[2] = {alignSetting, wordBreakSetting};
CTParagraphStyleRef paraStyle = CTParagraphStyleCreate(settings, 2);
```

Contrast this with the following iOS 6 code that performs the same task. This snippet creates an NSMutableParagraphStyle instance and customizes its properties with the same alignment and line break mode values used in the preceding example. This code is far simpler because it takes advantage of the new Objective-C class:

```
NSMutableParagraphStyle *paraStyle =
    [[NSMutableParagraphStyle alloc] init];
paraStyle.alignment = [self ctAlignment];
paraStyle.lineBreakMode = [self ctBreakMode];
```

iOS 6 introduced Objective-C implementations for nearly all of Core Text, enabling you to skip the library's Core Foundation calls. There are still several notable exceptions, which are explored later in this chapter. That said, iOS is now entering the golden age of customized text. You can add fonts, colors, spacing, and more to nearly everything in UIKit.

UIKit

In iOS 6, UIKit has been updated to become far more text-aware. Many UIKit classes, including text fields, text views, labels, and buttons, enable you to assign attributed (Core Text-style) strings to their `attributedText` property, just as you assign plain `NSStrings` to their `text` properties:

```
textField.attributedText = myAttributedString;
```

In addition, iOS now provides a new (and small) vocabulary of UIKit-specific text attributes such as font, color, and shadow. These are used with navigation bars, segmented controls, and bar items (that is, bar-style elements). You set attributes by calling `setTitleText Attributes:` (navigation bar) and `setTitleTextAttributes:forState:` (segmented control and bar items). Pass an attribute dictionary using the following dictionary keys and values:

- **UITextAttributeFont**—Provides a `UIFont` instance.
- **UITextAttributeTextColor**—Provides a `UIColor` instance.
- **UITextAttributeTextShadowColor**—Provides a `UIColor` instance.
- **UITextAttributeTextShadowOffset**—Provides an `NSValue` instance wrapping a `UIOffset` struct. Offsets include two floats, horizontal and vertical. Use `UIOffsetMake()` to construct the struct from a pair of floating point values.

For example, this snippet sets a segmented control's text color to light gray for its selected state. Whenever the control is selected, the text color changes from white to gray:

```
NSDictionary *attributeDictionary =
    @{UITextAttributeTextColor : [UIColor lightGrayColor]};
[segmentedControl setTitleTextAttributes:attributeDictionary
    forState:UIControlStateSelected];
```

This UIKit vocabulary will grow over time as Apple continues blending typesetting features into its controls and views.

Attributed Strings

You define traits by working with members of the `NSAttributedString` class or its mutable cousin `NSMutableAttributedString`. The mutable version offers far more flexibility, enabling you to layer attributes individually rather adding everything all at once.

To create an attributed string, you allocate it and initialize it with text and an attribute dictionary. The following snippet demonstrates how to build a "Hello World" string that displays with a large, gray font. Figure 3-1 shows the attributed string created here:

```
NSMutableDictionary *attributes = [NSMutableDictionary dictionary];
attributes[NSFontAttributeName] =
    [UIFont fontWithName:@"Futura" size:36.0f];
attributes[NSForegroundColorAttributeName] =
    [UIColor grayColor];

attributedString = [[NSAttributedString alloc]
    initWithString:@"Hello World" attributes: attributes];
textView.attributedText = attributedString;
```

Figure 3-1 Attributed strings store information such as font selection and color.

Typesetting centers on creating attributes dictionaries. You populate the dictionaries using the following keys and pass them to the attributed string. For mutable instances, you specify a character range to apply them to:

- **NSFontAttributeName**—A `UIFont` object, used to set the text font.
- **NSParagraphStyleAttributeName**—An `NSParagraphStyle` object, used to specify a number of paragraph settings including alignment, line break mode, indentation, and more.
- **NSForegroundColorAttributeName** and **NSBackgroundColorAttributeName**— `UIColor` objects that set the color of the text and the color shown behind the text. `NSStrokeColorAttributeName` is synonymous with the foreground color in the current iOS release.
- **NSStrokeWidthAttributeName**—An `NSNumber` storing a floating point value that defines the stroke width as a percentage of the font point size. Negative numbers both stroke and fill the text. Positive numbers create a "hollow" presentation, stroking the edges of each character glyph, as shown in Figure 3-2:

    ```
    attributes[NSStrokeWidthAttributeName] = @(3.0);
    ```

Attributed Strings | 91

Figure 3-2   Positive stroke values outline character glyphs but do not fill the interior.

- **NSStrikethroughStyleAttributeName** and **NSUnderlineStyleAttributeName**—These keys specify whether an item uses a strike-through or underline. In practice, these are essentially Boolean NSNumber instances, storing either 0 or 1. In theory, they may extend in future iOS releases to additional values, so Apple has defined the attribute values to NSUnderlineStyleNone (0) or NSUnderlineStyleSingle (1).

- **NSShadowAttributeName**—An NSShadow object, that sets the shadow's color, offset, and blur radius, as shown in Figure 3-3. This snippet uses a CGSize offset per Apple's documentation, but surely this will soon update to UIOffset support:

```
NSShadow *shadow = [[NSShadow alloc] init];
shadow.shadowBlurRadius = 3.0f;
shadow.shadowOffset = CGSizeMake(4.0f, 4.0f);
attributes[NSShadowAttributeName] = shadow;
```

Figure 3-3   Add shadows on iOS using the new NSShadow class.

- **NSLigatureAttributeName**—Ligatures refer to the way that individual glyphs (character pictures) can be bound together, such as "f" and "I," as shown in Figure 3-4. This key references an NSNumber that selects from "use no ligatures" (0), "use the default ligature" (1), or "use all ligatures" (2). iOS 6 does not support a value of 2.

Be aware that UIKit views may not support ligatures. I had to create Figure 3-4 using direct Core Text drawing. The UITextView I initially assigned the attributed string to did not properly render the "fi" ligature.

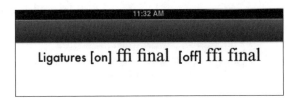

Figure 3-4 Ligatures combine certain letter combinations such as "f" and "i" into a single character glyph. Notice the separate dot on the "i" when ligatures are disabled.

- **NSBaselineOffsetAttributeName**—An `NSNumber`. The floating point number indicates a displacement from the normal text baseline, the point at which letters rest. Some letters such as "j" and "q" fall below the baseline during layout. Negative numbers will likely offset the text above the baseline, positive below it. (As this book was being written, this feature was broken.)
- **NSKernAttributeName**—An `NSNumber`. Not used in iOS 6, it eventually indicates whether kerning is enabled (1) or disabled (0). Kerning allows typesetters to adjust the space between letters, so they naturally overlap, such as when placing a capital A next to a capital V, for example, AV.
- **NSVerticalGlyphFormAttributeName**—An `NSNumber`. Not used in iOS 6, it will eventually support horizontal (0) and vertical (1) text layout.

Paragraph Styles

Paragraph styles are stored in their own objects, members of the `NSParagraphStyle` class. Use the mutable version of the class `NSMutableParagraphStyle` to set style specifics. Assign values to an instance's properties, making sure to match the required data types. The following snippet creates the presentation shown in Figure 3-5, using an extra large spacing between paragraphs and a generous first line indent:

```
NSMutableParagraphStyle *paragraphStyle = [[NSMutableParagraphStyle alloc] init];
paragraphStyle.alignment = NSTextAlignmentLeft;
paragraphStyle.lineBreakMode = NSLineBreakByWordWrapping;
paragraphStyle.firstLineHeadIndent = 36.0f;
paragraphStyle.lineSpacing = 8.0f;
paragraphStyle.paragraphSpacing = 24.0f;
attributes[NSParagraphStyleAttributeName] = paragraphStyle;
```

Most of these values refer to points, such as the spacing between lines or paragraphs, and indentation. If you're careful, you can control these features on a paragraph-by-paragraph basis; assign new attribute dictionaries for each paragraph.

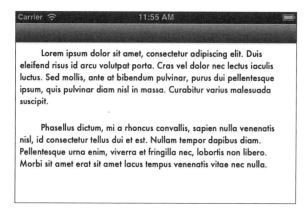

Figure 3-5   Paragraph style attributes include indentation, paragraph-to-paragraph spacing, alignment, and more.

> **Note**
>
> Make sure to create copies of your paragraph style object for each attribute update. If you do not, as I discovered the hard way, you may create an attributed result all of whose paragraph styles point to the same object. Update one and you update them all. Ouch.

Recipe: Basic Attributed Strings

Recipe 3-1 demonstrates building attributed strings by creating a user-controlled paragraph style and font color. Built to support the interface shown in Figure 3-6, the user selects a justification (from left, center, right, and justified) and a color (black, red, green, and blue) from the top navigation bar. The setupText method in Recipe 3-1 applies those choices as the user selects them, so the display updates in real time. It builds an attributes dictionary and uses that dictionary to create an attributed string, which it then loads into the central text view.

Because this recipe creates a simple NSAttributedString instance, the attributes apply across the entire string. As you'll see in the next section, using mutable instances helps you build more complex presentations.

Recipe 3-1   **Using Basic Attributed Strings with a Text View**

```
- (void) setupText
{
    // Establish a new dictionary
    NSMutableDictionary *attributes = [NSMutableDictionary dictionary];

    // Create the paragraph style
    NSMutableParagraphStyle *paragraphStyle =
```

```
            [[NSMutableParagraphStyle alloc] init];
    paragraphStyle.alignment = alignment;
    paragraphStyle.paragraphSpacing = 12.0f;
    attributes[NSParagraphStyleAttributeName] = paragraphStyle;

    // Load up the attributes dictionary
    attributes[NSFontAttributeName] =
        [UIFont fontWithName:@"Futura" size:14.0f];
    attributes[NSForegroundColorAttributeName] = color;

    // Build the attributed string
    attributedString = [[NSAttributedString alloc]
        initWithString:lorem attributes: attributes];

    // Display the string
    textView.attributedText = attributedString;
}
```

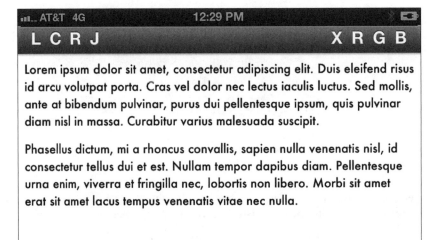

Figure 3-6   This recipe builds a basic interface that enables users to set font coloring and text alignment.

Get This Recipe's Code

To find this recipe's full sample project, point your browser to https://github.com/erica/iOS-6-Advanced-Cookbook and go to the folder for Chapter 3.

Recipe: Mutable Attributed Strings

As the name suggests, the `NSMutableAttributedString` class offers ways to evolve an already-created attributed string. Among the many methods provided by this class, are ones to replace, delete, insert, and append content:

- Use `replaceCharactersInRange:withString:` to substitute characters with new ones. New characters inherit whatever attributes were active at the first character (at `range.location`) of the replaced range. In contrast, the `replaceCharactersInRange:withAttributedString:` brings along whatever attributes and characters were used in the passed string.

- To remove characters, apply `deleteCharactersInRange:`.

- The `insertAttributedString:atIndex:` and `appendAttributedString:` do exactly what their names suggest. The newly added content provides its own characters and attributes.

The class also enables you to adjust attributes by applying traits to a range, often the selected range for a text view instance:

- `setAttributes:range:` entirely replaces the attribute dictionary for any characters that fall within the specified range.

- `addAttributes:range:` layers the attributes in the passed dictionary over any existing attributes in a given range.

- `addAttribute:value:range:` adds a single attribute to the range.

- `removeAttribute:range:` removes a single attribute from the characters within the range.

Setting an attribute dictionary is not without danger; it completely wipes any existing attributes, fully replacing them with whatever attributes have been included in the passed dictionary. Any attributes left unspecified may return to iOS defaults. For example, you might attempt to change the color of a selected range and notice that the font unexpectedly snaps back to 12-point Helvetica.

For modifications, use the `addAttribute:` methods instead. The second and third methods layer new attributes over existing ones, allowing you to replace items in the active attributes with new values. With these methods, you can add attributes that have not yet been specified, or modify existing ones.

The final method enables you to remove single attributes, such as colors, shadows, or font selections, without affecting any other attributes currently in effect.

In real-world use, user-created text selections often cross attribute boundaries. Figure 3-7 offers an example. The words eleifend, risus, and id each display with different font sizes, 36 point, 24 point, and 18 point, respectively. Imagine the user selecting fend ris, and requesting a tint change. To apply this update, the `addAttribute:value:range:` method must individually process each grouping of uniform traits. In this case, the selection breaks down to fend in a large font, and ris in a small font.

Figure 3-7   Applying traits and crossing range boundaries.

Recipe 3-2 re-creates the `addAttribute:value:range:` functionality. This recipe is included because it is extremely useful to iterate through attribute ranges, which is a great example. This code showcases how attribute ranges break down, and how you can query these from your apps. The recipe iterates through a text view's selected range. It queries its attributed string for the attributes found at a certain point, retrieving the full effective range of that attribute dictionary.

In the view's initial state, that range contains all the text stored to the view, a range that's normally much larger than whatever selection the user has made. To handle the situation in which an effective attribute range is bigger than the user selection, Recipe 3-2 performs an intersection, returning a range that's limited to the user selection.

It copies the existing attributes at the current point to a mutable dictionary and sets the new attribute passed to the method. These attributes are then applied back to the attributed string, and the current location moves forward—either to the end of the current effective range, or to the end of the selection.

In Figure 3-7, the fend ris selection contains two attribute ranges based on font size. When run on this example, Recipe 3-2 first tinted the large fend red, and then iterated to tint the remaining ris. In real-world use, selections can prove even more complicated, containing any number of effective ranges to update.

The important points to take away from this recipe are as follows:

- Attributed strings contain both characters and attributes.
- Each range of characters is associated with one set of attributes. You can retrieve a full range by querying for attributes at any character index. These ranges do not overlap.
- Mutable attributed strings enable you to layer new attributes over or remove attributes from parts of your string. Doing so often creates new ranges, so that each range remains limited to one set of attributes.
- You can replace characters at any point in your string; they inherit whatever attributes are active at that range. When you insert or append an attributed string, the added material retains the attributes it was defined with.

Recipe 3-2 **Layering Attributes Via Iterated Ranges**

```
- (void) applyAttribute: (id) attributeValue withName: (NSString *) keyName
{
    // Replicating this approach through custom code
    // [attributedString addAttribute:keyName
    //     value:attributeValue range:range];

    NSRange range = textView.selectedRange;
    if (range.location == NSNotFound) return;

    // Keep track of attribute range effects
    CGFloat fullExtent = range.location + range.length;
    CGFloat currentLocation = range.location;

    // Iterate through each effective range within the selection
    while (currentLocation < fullExtent)
    {
        // Extract the current effective range
        NSRange effectiveRange;
        NSDictionary *currentAttributes =
            [attributedString attributesAtIndex:currentLocation
                effectiveRange:&effectiveRange];

        // Limit that range to the current selection
        NSRange intersection = NSIntersectionRange(range, effectiveRange);

        // Extract and modify the current attributes
        NSMutableDictionary *attributes = [NSMutableDictionary
```

```
            dictionaryWithDictionary:currentAttributes];
        attributes[keyName] = attributeValue;

        // Apply those attributes back. This uses "set attributes"
        // instead of "apply attributes" for demonstration
        [attributedString setAttributes:attributes range:intersection];

        // Update the current location to move past the effective range
        currentLocation = effectiveRange.location + effectiveRange.length;
    }

    [self setAttrStringFrom:textView.attributedText to:attributedString];
}
```

> **Get This Recipe's Code**
> To find this recipe's full sample project, point your browser to https://github.com/erica/iOS-6-Advanced-Cookbook and go to the folder for Chapter 3.

The Mystery of Responder Styles

Introduced in iOS 6, a trio of responder methods enables objects to apply styles to selected text. The methods are defined in `NSObject` but meant for `UIResponder` use. The three methods are `toggleBoldface:`, `toggleItalics:`, and `toggleUnderline:`, and they're a bit curious, as you'll see. They apply attributes to whatever text has been selected in the view.

These methods seem to be limited to text views and text fields. You enable them by setting the `allowsEditingTextAttributes` property to `YES`. The standard iOS 6 contextual text menus update to include new BIU options, as shown in Figure 3-8. The view handles the rest of the implementation details, and there's nothing further you need to do programmatically. You can access the results via the text view or the text field's `attributedText` property.

This new feature is puzzling on a number of levels. To start, `UIFont` instances do not offer direct access to any of these three attributes: bold, italics, or underline.

The underline attribute ties into a fairly complex Core Text system, of which only a single line (`NSUnderlineStyleSingle`) is currently exposed in string attributes (`NSUnderlineStyleAttributeName`). Other native Core Text underlining styles include double lines, tick lines, and a variety of patterns, such as the ones used in Figure 3-8 to accentuate possible misspellings.

The bold and italic traits represent font family variations. Managing these requires a trip to C-based APIs, as shown in Listing 3-1. The listing details how you can retrieve a font descriptor from a family name such as Trebuchet MS and a traits dictionary. You use that descriptor to extract a trait-specific font name such as TrebuchetMS-Italic, or Trebuchet-BoldItalic, and then build a `UIFont` instance with that name.

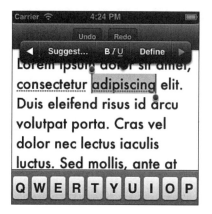 

Figure 3-8 The iOS contextual menu updates to include BIU options after enabling `allowsEditingTextAttributes` in text views and text fields.

Adding to the mystery is that text view delegates do not receive updates to their contents when users apply these toggles. Regular delegate callbacks such as `textViewDidChange:` aren't sent. If you want to monitor changes, and update Undo/Redo capabilities (as shown in Figure 3-8), you can add a timer to poll for items on the undo manager's stack.

The final bit of the mystery is this: Text views do not store mutable attributed string instances. Therefore, whatever mechanism Apple uses to update view contents with attribution updates changes and replaces the text.

This mystery probably adds up to a fundamental redesign that wasn't finished in time for a full iOS 6 debut. What it promises is exciting. For now, you're left with whatever made the cut in iOS 6. For the most part, it's a vast improvement over the Core Foundation text calls like the ones used in Listing 3-1.

Listing 3-1 **Returning a Font from Its Traits**

```
// Handle bold and italics traits
- (UIFont *) fontWithTraits
{
    // Core font elements
    NSString *familyName = self.font.familyName;
    CGFloat newFontSize = self.font.pointSize;
    
    // Return core font
    if (!self.bold && !self.italic)
        return [UIFont fontWithName:familyName size:newFontSize];
    
    // Create traits value
    NSUInteger appliedTraits = 0;
```

```
    if (self.bold) appliedTraits = kCTFontBoldTrait;
    if (self.italic) appliedTraits =
        appliedTraits | kCTFontItalicTrait;
    NSNumber *traitsValue = @(appliedTraits);

    // Build dictionary from family name and traits
    NSDictionary *traitDictionary =
        @{(NSString *)kCTFontSymbolicTrait:traitsValue};
    NSDictionary *dict =
    @{
        (NSString *)kCTFontFamilyNameAttribute:familyName,
        (NSString *)kCTFontTraitsAttribute:traitDictionary,
    };

    // Extract font descriptor
    CFDictionaryRef dictRef = CFBridgingRetain(dict);
        CTFontDescriptorRef desc =
            CTFontDescriptorCreateWithAttributes(dictRef);
    CFRelease(dictRef);

    // If this failed, return core font
    if (!desc)
        return [UIFont fontWithName:familyName size:newFontSize];

    // Otherwise, extract the new font name e.g. whatever-bold
    CTFontRef ctFont = CTFontCreateWithFontDescriptor(
        desc, self.font.pointSize, NULL);
    NSString *newFontName = CFBridgingRelease(
        CTFontCopyName(ctFont, kCTFontPostScriptNameKey));

    // Create font with trait-name
    return [UIFont fontWithName:newFontName size:newFontSize];
}
```

Recipe: Attribute Stacks

During the writing of the iOS 5 version of this cookbook, I developed a string helper class. This class hid the C-based Core Foundation lurking behind attributed string creation, adding an Objective C wrapper. It also added properties that defined attributes for new text appended to the string. For example, you could set a new foreground color and added text would adopt that color.

Fast forward to iOS 6. When iOS 6 debuted, it introduced Objective C classes and constants to support attributed strings. About one-half of the reason for the custom class disappeared. But something else happened along the way to my ditching the recipe: Readers and early testers responded strongly to the stored state features. They liked how they could set attributes, pack on text, update the attributes, and pack on more text.

From there grew the notion of an attribute stack, which could provide a saved state feature similar to `CGContextSaveGState()`. It would enable you to push a set of attributes, such as a new color or font size, and then after adding new text pop back to the previous conditions.

Recipe 3-3 evolves the original wrapper to introduce stack support. Like its inspiration, it exposes properties that can be transformed into an attributes dictionary. However, it introduces functionality to move in the other direction. This implementation retrieves property settings from stored dictionaries, so the object can move back through its saved stack and its properties update to match.

The stack itself is a simple mutable array. Pushing state (via `saveContext`) duplicates the topmost dictionary and stores it on the stack. Restoring state (via `popContext`) pops the stack and exports its attributes out to the object properties.

Recipe 3-3 also provides transient attributes. The `performTransientAttributeBlock:` method automates a save, execute, pop sequence, so any changes do not permanently affect the stored attribute stack. Here's how this method can operate. Inside the embedded transient blocks, properties set the attributes for newly added text:

```
string.font = [UIFont fontWithName:@"Arial" size:24.0f];
[string performTransientAttributeBlock:^(){
    string.foregroundColor = [UIColor redColor];
    string.bold = YES;
    [string appendFormat:@"This is red and bold. "];
    [string performTransientAttributeBlock:^(){
        string.font = [UIFont fontWithName:@"Georgia" size:24.0f];
        string.bold = NO;
        string.underline = YES;
        string.strokeColor = [UIColor greenColor];
        string.strokeWidth = 2.0f;
        [string appendFormat:
            @"This is Green Georgia Outline, Underlined Not Bolded. "];
    }];
    string.bold = NO;
    string.foregroundColor = COOKBOOK_PURPLE_COLOR;
    [string appendFormat:@"This is not green *or* bolded.\n\n"];
}];
[string appendFormat:@"Back to the original attributes."];
```

When the outer block ends, the string returns to a basic 24-point Arial state. No other property updates persist beyond the block's life. As each block executes synchronously, any commands placed after the block are performed after the block has completely finished. Figure 3-9 shows this code in action.

102   Chapter 3   Core Text

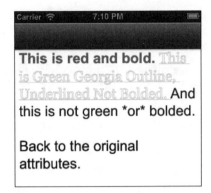

Figure 3-9   Transient attribute blocks in action.

Recipe 3-3   Building Attributed Strings with an Objective-C Wrapper

```
{
    NSMutableArray *stack;
    NSMutableDictionary *top;
}

// Initialize with a new string, a stack, and a current dictionary (top)
- (id) init
{
    if (!(self = [super init])) return self;

    _string = [[NSMutableAttributedString alloc] init];
    stack = [NSMutableArray array];
    top = [NSMutableDictionary dictionary];

    graphStyle = [[NSMutableParagraphStyle alloc] init];
    _font = [UIFont fontWithName:@"Helvetica" size:12.0f];

    return self;
}

// Execute a block with temporary traits
- (void) performTransientAttributeBlock: (AttributesBlock) block
{
    [self saveContext];
    block();
    [self popContext];
}
```

```objc
// Transform object properties into an attributes dictionary
- (NSDictionary *) attributes
{
    NSMutableDictionary *attributes = [NSMutableDictionary dictionary];
    
    // Font and Para Style
    self.font = [self fontWithTraits];
    [attributes setObject:self.font forKey:NSFontAttributeName];
    [attributes setObject:self.paragraphStyle
        forKey:NSParagraphStyleAttributeName];
    
    // Colors
    if (self.foregroundColor)
        [attributes setObject:self.foregroundColor
            forKey:NSForegroundColorAttributeName];
    if (self.backgroundColor)
        [attributes setObject:self.backgroundColor
            forKey:NSBackgroundColorAttributeName];
    if (self.strokeColor)
        [attributes setObject:self.strokeColor
            forKey:NSStrokeColorAttributeName];
    
    // Other Styles
    [attributes setObject:@(self.strokeWidth)
        forKey:NSStrokeWidthAttributeName];
    [attributes setObject:@(self.underline)
        forKey:NSUnderlineStyleAttributeName];
    [attributes setObject:@(self.strikethrough)
        forKey:NSStrikethroughStyleAttributeName];
    if (self.shadow)
        [attributes setObject:self.shadow
            forKey:NSShadowAttributeName];
    [attributes setObject:@(self.useLigatures)
        forKey:NSLigatureAttributeName];
    
    return attributes;
}

// Expose the current attribute set through properties
- (void) setAttributesFromDictionary: (NSDictionary *) dictionary
{
    
    // Update active properties
    graphStyle = dictionary[NSParagraphStyleAttributeName];
    
    // Establish font
    _font = dictionary[NSFontAttributeName];
```

```objc
    CTFontSymbolicTraits traits =
        CTFontGetSymbolicTraits((__bridge CTFontRef)(self.font));
    self.bold = (traits & kCTFontBoldTrait) != 0;
    self.italic = (traits & kCTFontItalicTrait) != 0;

    // Colors
    _foregroundColor = dictionary[NSForegroundColorAttributeName];
    _backgroundColor = dictionary[NSBackgroundColorAttributeName];
    _strokeColor = dictionary[NSStrokeColorAttributeName];

    // Other
    _strokeWidth = ((NSNumber *)dictionary[
        NSStrokeWidthAttributeName]).floatValue;
    _underline = ((NSNumber *)dictionary[
        NSUnderlineStyleAttributeName]).boolValue;
    _strikethrough = ((NSNumber *)dictionary[
        NSStrikethroughStyleAttributeName]).boolValue;
    _useLigatures = ((NSNumber *)dictionary[
        NSLigatureAttributeName]).boolValue;
    _shadow = dictionary[NSShadowAttributeName];
}

#pragma mark - Stack Operations -

// Push the top dictionary onto the stack
- (void) saveContext
{
    [stack addObject:self.attributes];
    top = [NSMutableDictionary dictionaryWithDictionary:top];

    // Create a copy of the style, to point to a distinct object
    graphStyle = [graphStyle mutableCopy];
    top[NSParagraphStyleAttributeName] = graphStyle;
}

// Pop the top dictionary off the stack, and set its attributes
- (BOOL) popContext
{
    if (!stack.count) return NO;

    // Pop it
    top = [stack lastObject];
    [stack removeLastObject];
    [self setAttributesFromDictionary:top];

    return (stack.count > 0);
}
```

Get This Recipe's Code

To find this recipe's full sample project, point your browser to https://github.com/erica/iOS-6-Advanced-Cookbook and go to the folder for Chapter 3.

Recipe: Using Pseudo-HTML to Create Attributed Text

Adding attributes to a string one-by-one, even with the assistance of helper classes, soon becomes tedious. That's why using HTML, a familiar standard built for text markup, can transform and simplify your layout tasks. It's easier to produce basic HTML content than to iteratively implement the Core Text calls that create a similar presentation. Consider the following string:

```
<h1>Using Core Text with Markup</h1>
<p>This is an example <i>of some text that uses <b>HTML styling</b></i>.</p>
```

ISO-compliant HTML layout for this simple string would require quite a bit of Core Text coding to match the requested markup, but why expend all that effort? If you're willing to be flexible, you can automate an HTML-like solution that creates Core Text coding for you.

The `UIWebView` class intrinsically supports typesetting through HTML, and does so in a rigorously standards-supported way. Core Text views, however, can be produced with far lighter weight code, without the memory overhead. For example, you might not want to add a dozen web views to your `UITableView` cells, but Core Text–backed alternatives could work much better, with little memory overhead.

The challenge lies in transforming an HTML (or HTML-like) annotated string into an attributed string. Recipe 3-4 tackles this problem, offering an enhanced HTML subset that parses its contents into an attributed string.

Note

For a superior HTML solution, check out Oliver Drobnik's excellent attributed strings extensions library. You can find his repository at https://github.com/Cocoanetics/DTCoreText.

This class brings more to the equation than standard HTML. Its simple scanning and matching approach means that it's just a matter of minutes to add tags of your own design. I've added support for a couple of convenience tags that don't rely on standard HTML—namely the `color` and `size` tags found toward the end of the `stringFromMarkup:` method. Figure 3-10 juxtaposes the running app with the pseudo markup file that's driving its presentation.

The class works by scanning for tags enclosed in angle brackets. The tags set text attributes using the same stack-based string class discussed in Recipe 3-3. The code appends each run of text using the current set of string attributes.

Chapter 3 Core Text

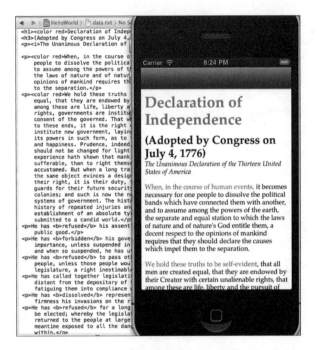

Figure 3-10 Pseudo-markup and its attributed string presentation.

Utilities like these ultimately exist to serve individual development needs, not necessarily to support a standard such as HTML. If you can add new tags with little work for your own use, then a custom solution like this recipe may save you a lot of effort in your application development. Don't feel you must be tied to the standard.

After you build (and rigorously debug, of course) a parser like Recipe 3-4, it can save you a lot of time in future projects. It enables you to reuse Core Text routines without having to bother with a lot of the Core Text overhead.

Recipe 3-4 Pseudo HTML Markup

```
+ (NSAttributedString *) stringFromMarkup: (NSString *) aString
{
    // Core Fonts
    UIFont *baseFont = [UIFont fontWithName:@"Palatino" size:14.0f];
    UIFont *headerFont = [UIFont fontWithName:@"Palatino" size:14.0f];
    
    // Prepare to scan
    NSScanner *scanner = [NSScanner scannerWithString:aString];
    [scanner setCharactersToBeSkipped:[
        NSCharacterSet newlineCharacterSet]];
```

```objc
// Initialize a string helper
FancyString *string = [FancyString string];
string.font = baseFont;

// Iterate through the string
while (scanner.scanLocation < aString.length)
{
    NSString *contentText = nil;

    // Scan until the next tag
    [scanner scanUpToString:@"<" intoString:&contentText];

    // Process entities and append the text
    contentText = [contentText
        stringByReplacingOccurrencesOfString:@"&lt;"
        withString:@"<"];
    if (contentText)
        [string appendFormat:@"%@", contentText];

    // Scan through the tag
    NSString *tagText = nil;
    [scanner scanUpToString:@">" intoString:&tagText];
    if (scanner.scanLocation < aString.length)
        scanner.scanLocation += 1;
    tagText = [tagText stringByAppendingString:@">"];

    // -- PROCESS TAGS --

    // Header Tags
    if (STRMATCH(tagText, @"</h")) // finish any headline
    {
        [string popContext];
        [string appendFormat:@"\n"];
        continue;
    }

    if (STRMATCH(tagText, @"<h"))
    {
        int hlevel = 0;
        if (STRMATCH(tagText, @"<h1>")) hlevel = 1;
        else if (STRMATCH(tagText, @"<h2>")) hlevel = 2;
        else if (STRMATCH(tagText, @"<h3>")) hlevel = 3;

        [string performTransientAttributeBlock:^(){
            // add a wee spacer
            string.font = [UIFont boldSystemFontOfSize:8.0f];
```

```objc
            [string appendFormat:@"\n"];
        }];
        
        [string saveContext];
        string.bold = YES;
        string.font = [UIFont fontWithName:headerFont.fontName
            size:20.0f + MAX(0, (4 - hlevel)) * 4.0f];
    }
    
    // Bold and Italics
    if (STRMATCH(tagText, @"<b>")) string.bold = YES;
    if (STRMATCH(tagText, @"</b>")) string.bold = NO;
    if (STRMATCH(tagText, @"<i>")) string.italic = YES;
    if (STRMATCH(tagText, @"</i>")) string.italic = NO;
    
    // Paragraph and line break tags
    if (STRMATCH(tagText, @"<br>")) [string appendFormat:@"\n"];
    if (STRMATCH(tagText, @"</p>")) [string appendFormat:@"\n\n"];
    
    // Color
    if (STRMATCH(tagText, @"<color"))
    {
        if STRMATCH(tagText, @"blue")
            string.foregroundColor = [UIColor blueColor];
        if STRMATCH(tagText, @"red")
            string.foregroundColor = [UIColor redColor];
        if STRMATCH(tagText, @"green")
            string.foregroundColor = [UIColor greenColor];
    }
    if (STRMATCH(tagText, @"</color>"))
        string.foregroundColor = nil;
    
    // Size
    if (STRMATCH(tagText, @"<size"))
    {
        // Scan the value for the new font size
        NSScanner *newScanner = [NSScanner scannerWithString:tagText];
        NSCharacterSet *cs =
            [[NSCharacterSet decimalDigitCharacterSet] invertedSet];
        [newScanner setCharactersToBeSkipped:cs];
        
        CGFloat fontSize;
        [newScanner scanFloat:&fontSize];
        [string saveContext];
        string.font = [UIFont fontWithName:string.font.fontName
            size:fontSize];
    }
```

```
        if (STRMATCH(tagText, @"</size>"))
            [string popContext];
    }
    
    return string.string;
}
```

> **Get This Recipe's Code**
>
> To find this recipe's full sample project, point your browser to https://github.com/erica/iOS-6-Advanced-Cookbook and go to the folder for Chapter 3.

Drawing with Core Text

Attributed strings extend beyond those UIKit classes that present them, such as text views, labels, refresh controls, and so forth. UIKit offers string drawing extensions that enable you to create custom UIView classes that support attributed strings. You can accomplish this in iOS 6 with almost no work. Listing 3-2 details a custom view that stores an attributed string and draws its contents through a `drawRect:` method.

Listing 3-2   **Attributed String View**

```
@interface ASView : UIView
@property (nonatomic, strong) NSAttributedString *attributedString;
@end

@implementation ASView
- (id) initWithFrame:(CGRect)frame
{
    if (self = [super initWithFrame:frame])
        self.backgroundColor = [UIColor clearColor];
    return self;
}

- (void) drawRect:(CGRect)rect
{
    [super drawRect: rect];
    [_attributedString drawInRect:self.bounds];
}
@end
```

Compare Listing 3-2 to Listing 3-3, which shows the same custom view class as built prior to iOS 6 using Core Text calls. The simplicity of the new `drawRect:` method demonstrates how much attributed text features have integrated themselves into UIKit, hiding Core Text's C-based API with a robust and usable Objective-C one.

Listing 3-3 **Attributed String Core Text View**

```objc
@interface CTView : UIView
@property (nonatomic, strong) NSAttributedString *attributedString;
@end

@implementation CTView
- (id) initWithFrame:(CGRect)frame
{
    if (self = [super initWithFrame:frame])
        self.backgroundColor = [UIColor clearColor];
    return self;
}

- (void) drawRect:(CGRect)rect
{
    [super drawRect: rect];
    CGContextRef context = UIGraphicsGetCurrentContext();

    // Flip the context
    CGContextSetTextMatrix(context, CGAffineTransformIdentity);
    CGContextTranslateCTM(context, 0, self.bounds.size.height);
    CGContextScaleCTM(context, 1.0, -1.0);

    // Slightly inset from the edges of the view
    CGMutablePathRef path = CGPathCreateMutable();
    CGRect insetRect = CGRectInset(self.frame, 20.0f, 20.0f);
    CGPathAddRect(path, NULL, insetRect);

    // Build the framesetter
    CTFramesetterRef framesetter =
        CTFramesetterCreateWithAttributedString(
            (__bridge CFAttributedStringRef)_attributedString);

    // Draw the text
    CTFrameRef destFrame = CTFramesetterCreateFrame(
        framesetter, CFRangeMake(0, _attributedString.length),
        path, NULL);
    CTFrameDraw(destFrame, context);

    // Clean up
    CFRelease(framesetter);
    CFRelease(path);
```

```
        CFRelease(destFrame);
    }
@end
```

The `drawRect:` method in Listing 3-3 uses raw Core Text to perform all the tasks hidden by the UIKit string drawing call in Listing 3-2. It starts by flipping its context. This enables the text to start drawing from the top-left corner, which is the standard in UIKit but not in raw Core Text. It then creates an inset rectangle using the view bounds as a starting point and moving in slightly from there. Finally, it establishes a frame setter object, which is responsible for managing the elements that break down into individual characters.

Those elements are called *lines*, each of which contains a number of character *runs*. Each run contains a series of text *glyphs*, which are character images like the pictures for "a" or "b." Each glyph in a run shares common attributes such as font size or color. Finally, the frame setter uses all this information to build a *frame*, which can then draw the text to a graphics context.

By changing the path, you can achieve eye-catching visual effects quite cheaply. You'll see some of this in the recipes later in this chapter. For this example, when you adjust the size of the inset rectangle, you find that the text adapts and re-adjusts its wrapping to match its new boundaries. Other ways to change the shape include adding ellipses and arcs to the path. Core Text ably handles nonrectangular paths. Figure 3-11 shows the result of changing Listing 3-3 from using a rectangular frame (`CGPathAddRect()`) to an elliptical one (`CGPathAddEllipseInRect()`).

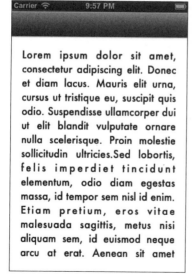 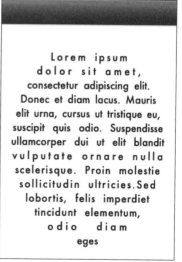

Figure 3-11   Core Text automatically handles the change from a rectangular path to an elliptical one.

Creating Image Cut-Outs

Adding images to your Core Text layout involves two key steps. First, draw those images before flipping your context coordinate system. The `UIImage` drawing routines assume an origin at the top-left corner. Second, modify your text path so that the frame setter does draw text where your images are supposed to go.

Listing 3-4 creates a manual layout that inserts two images. You can see the typesetting created by this code in Figure 3-12. The text wraps around the images, respecting the path cut-outs built in the `drawRect:` method. This example uses a ragged-right, left-aligned paragraph style, which explains the various gaps to the left of each image.

Figure 3-12   Cut out areas of a Core Graphics path to provide room for figures. The Core Text typesetter accommodates those gaps.

Listing 3-4   **Adding Images to Core Text Flow**

```
// Flip a rectangle within an outer coordinate system
CGRect CGRectFlipVertical(CGRect innerRect, CGRect outerRect)
{
    CGRect rect = innerRect;
    rect.origin.y = outerRect.origin.y + outerRect.size.height -
        (rect.origin.y + rect.size.height);
```

```
        return rect;
}

- (void) drawRect: (CGRect) rect
{
    NSAttributedString *string = [self string];

    [super drawRect: rect];
    CGContextRef context = UIGraphicsGetCurrentContext();
    CGRect aRect = CGRectInset(self.bounds, 10.0f, 10.0f);

    // Draw the background
    [[UIColor whiteColor] set];
    CGContextFillRect(context, self.bounds);

    // Perform the image drawing in the normal context geometry
    CGRect imageRect1 =
        CGRectMake(150.0f, 500.0f, 200.0f, 300.0f);
    [[UIImage imageNamed:@"Default.png"]
        drawInRect:CGRectInset(imageRect2, 10.0f, 10.0f)];

    // Draw the Bear (public domain)
    CGRect imageRect2 =
        CGRectMake(500.0f, 100.0f, 187.5f, 150.0f);
    [[UIImage imageNamed:@"Bear.jpg"]
        drawInRect:CGRectInset(imageRect1, 10.0f, 10.0f)];

    // Flip the context for the Core Text layout
    CGContextSetTextMatrix(context, CGAffineTransformIdentity);
    CGContextTranslateCTM(context, 0, self.frame.size.height);
    CGContextScaleCTM(context, 1.0, -1.0);

    // Start the path
    CGMutablePathRef path = CGPathCreateMutable();
    CGPathAddRect(path, NULL, aRect);

    // Cut out the two image areas
    CGPathAddRect(path, NULL,
        CGRectFlipVertical(imageRect1, self.bounds));
    CGPathAddRect(path, NULL,
        CGRectFlipVertical(imageRect2, self.bounds));

    // Create framesetter
    CTFramesetterRef framesetter =
        CTFramesetterCreateWithAttributedString(
            (__bridge CFAttributedStringRef)string);
```

```
    // Draw the text
    CTFrameRef theFrame = CTFramesetterCreateFrame(framesetter,
        CFRangeMake(0, string.length), path, NULL);
    CTFrameDraw(theFrame, context);

    // Clean up
    CFRelease(path);
    CFRelease(theFrame);
    CFRelease(framesetter);
}
```

Recipe: Drawing Core Text onto a Scroll View

Although UIKit support for string drawing simplifies text layout, CoreText is still important to know and use for more complicated but not uncommon cases. Listing 3-4 demonstrated how to lay text around images. Recipe 3-5 shows how to build a Core Text-powered scrolling text view.

Custom Core Text rendering can be extended to scroll views if you work carefully. Recipe 3-5 adapts Listing 3-3's Core Text drawing to draw attributed text onto a custom scroll view. The updates include a method to calculate the scroll view's content size each time the view geometry changes, typically from reorienting the iPhone, and math that takes that large content size into account when drawing.

The `updateContentSize` method works by requesting a suggested frame size. The method passes the source text, the view width, and a dummy height (that is, max float). This lets the frame setter calculate how much vertical height that text will occupy for a fixed width, allowing it to ignore any height constraints. This method uses the returned frame to set up the scroll view's content size.

In the `drawRect:` method, Recipe 3-5's math has to compensate for the fact that the drawing is taking place within the scroll view's large content view, not its frame. Both the context flip and the destination rectangle use the content size.

Finally, in the `initWithFrame:` method, the scroll view's content mode is set to redraw. This forces the view to refresh whenever its bounds change.

Recipe 3-5 **Core Text and Scroll Views**

```
@interface CTView : UIScrollView <UIScrollViewDelegate>
@property (nonatomic, strong) NSAttributedString *attributedString;
@end

@implementation CTView
- (id) initWithFrame:(CGRect)frame
{
    if (self = [super initWithFrame:frame])
    {
```

```
        self.backgroundColor = [UIColor clearColor];
        self.contentMode = UIViewContentModeRedraw;
        self.delegate = self;
    }
    return self;
}

// Calculates the content size
- (void) updateContentSize
{
    CTFramesetterRef framesetter =
        CTFramesetterCreateWithAttributedString(
            (__bridge CFAttributedStringRef)_attributedString);
    CFRange destRange = CFRangeMake(0, 0);
    CFRange sourceRange = CFRangeMake(0, _attributedString.length);
    CGSize frameSize = CTFramesetterSuggestFrameSizeWithConstraints(
        framesetter, sourceRange, NULL,
        CGSizeMake(self.frame.size.width, CGFLOAT_MAX), &destRange);
    self.contentSize = CGSizeMake(self.bounds.size.width, frameSize.height);
    CFRelease(framesetter);
}

// This is a scroll-view specific drawRect
- (void) drawRect:(CGRect)rect
{
    [super drawRect: rect];
    CGContextRef context = UIGraphicsGetCurrentContext();

    // Flip the context
    CGContextSetTextMatrix(context, CGAffineTransformIdentity);
    CGContextTranslateCTM(context, 0, self.contentSize.height);
    CGContextScaleCTM(context, 1.0, -1.0);

    CGMutablePathRef path = CGPathCreateMutable();
    CGRect destRect = (CGRect){.size = self.contentSize};
    CGPathAddRect(path, NULL, destRect);

    // Create framesetter
    CTFramesetterRef framesetter =
        CTFramesetterCreateWithAttributedString(
            (__bridge CFAttributedStringRef)_attributedString);

    // Draw the text
    CTFrameRef theFrame = CTFramesetterCreateFrame(framesetter,
        CFRangeMake(0, _attributedString.length), path, NULL);
    CTFrameDraw(theFrame, context);
```

```
    // Clean up
    CFRelease(path);
    CFRelease(theFrame);
    CFRelease(framesetter);
}
@end
```

Get This Recipe's Code

To find this recipe's full sample project, point your browser to https://github.com/erica/iOS-6-Advanced-Cookbook and go to the folder for Chapter 3

Recipe: Exploring Fonts

These days, iOS ships with a fairly rugged collection of fonts, which you can easily explore to your heart's content. The range of fonts may vary by device and operating system, so it's always best to check on-device to see what fonts are available.

Recipe 3-6 iterates through `UIFont`'s collection of font families and their members to build a report, which it embeds in a text view (see Figure 3-13). Created using the helper class from Recipe 3-3, this recipe builds out examples by transiently applying each font to the string's attributes.

Figure 3-13   Attributed strings help you review your onboard fonts.

Recipe: Exploring Fonts

Use this survey app to discover new font names and preview what they look like.

> **Note**
>
> This recipe disables the call to `fontWithTraits` in the helper class. This ensures each font is used exactly without respect to the helper's `bold` and `italics` properties.

Recipe 3-6 Creating a Font List

```
string = [FancyString string];
string.ignoreTraits = YES;

UIFont *headerFont = [UIFont fontWithName:@"Futura" size:24.0f];
UIFont *familyFont = [UIFont fontWithName:@"Futura" size:18.0f];

for (NSString *familyName in [UIFont familyNames])
{
    string.font = headerFont;
    string.foregroundColor = [UIColor redColor];

    // \u25BC is a downward-pointing triangle
    [string appendFormat:@"\u25BC  %@\n", familyName];

    string.foregroundColor = nil;
    string.font = familyFont;

    for (NSString *fontName in [UIFont fontNamesForFamilyName:familyName])
    {
        // \u25A0 is a square
        [string appendFormat:@"\t\u25A0 %@:  ", fontName];
        [string performTransientAttributeBlock:^(){
            string.font = [UIFont fontWithName:fontName size:18.0f];
            string.foregroundColor = [UIColor darkGrayColor];
            [string appendFormat:quickBrownDogString];
        }];
    }
    [string appendFormat:@"\n"];
}
textView.attributedText = string.string;
```

> **Get This Recipe's Code**
>
> To find this recipe's full sample project, point your browser to https://github.com/erica/iOS-6-Advanced-Cookbook and go to the folder for Chapter 3.

Adding Custom Fonts to Your App

You are never limited to the device's repertoire of onboard fonts now that iOS supports custom font integration into applications. This recipe demonstrates how you can add your own fonts into an application for your use. Figure 3-14 shows a screenshot using the Pirulen font along with some basic Lorem Ipsum text.

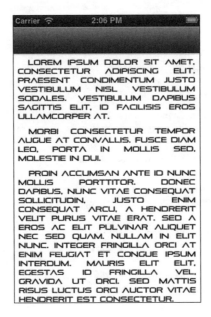

Figure 3-14   Adding custom fonts to your applications enables you to stylize your applications with a unique textual finish.

Add the TrueType font to your Xcode project. Then edit your application's Info.plist file to declare UIAppFonts (aka "Fonts provided by application"). This is an array of filenames that you add to. Figure 3-15 defines a single font entry, namely pirulen.ttf.

Figure 3-15   Edit the Info.plist to add custom fonts.

To use the new font, create fonts as you normally would. Make sure to spell and capitalize the font name exactly:

```
// This call to UIFont uses a custom font name
textView.font = [UIFont fontWithName:@"pirulen"
        size:IS_IPAD ? 28.0f : 12.0f];
```

Recipe: Splitting Core Text into Pages

As you saw in Recipe 3-4, producing attributed strings from markup helps you separate presentation from implementation. It's a flexible approach that lets you edit your source material without affecting your code base. The problem comes when you deal with text that ranges beyond a single page. For those cases, you want to split the text into sections on a page-by-page basis.

Recipe 3-7 shows how you might do that. It uses a Core Text frame setter to return an array of recommended page breaks based on a given point-based page size. You can use this hand-in-hand with pseudo-HTML markup and a page view controller to produce the book shown in Figure 3-16. This book reads in text, converts it from marked-up source into an attributed string, and then breaks it into pages for display in the paged controller.

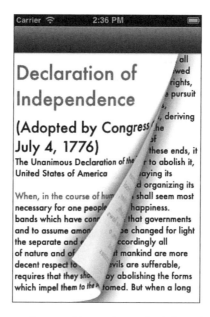

Figure 3-16   Core Text frame setters enable you to break attributed strings into sections that fit a given page size.

Recipe 3-7's page splitter method powers the interface shown in Figure 3-16. This method uses Core Text to iterate through an attributed string, collecting range information on a page-by-page basis. The Core Text `CTFramesetterSuggestFrameSizeWithConstraints` method takes a starting range and a destination size, producing a destination range that fits its attributed string. Although this recipe doesn't use any control attributes, you may want to explore these for your own implementations. Apple's technical Q&A QA1698 (http://developer.apple.com/library/ios/#qa/qa1698) offers some pointers.

Recipe 3-7 **Multipage Core Text**

```objc
- (NSArray *) findPageSplitsForString: (NSAttributedString *)theString
   withPageSize: (CGSize) pageSize
{
    NSInteger stringLength = theString.length;
    NSMutableArray *pages = [NSMutableArray array];

    CTFramesetterRef frameSetter =
        CTFramesetterCreateWithAttributedString(
            (__bridge CFAttributedStringRef) theString);

    CFRange baseRange = {0,0};
    CFRange targetRange = {0,0};
    do {
        CTFramesetterSuggestFrameSizeWithConstraints(
            frameSetter, baseRange, NULL, pageSize, &targetRange);
        NSRange destRange = {baseRange.location, targetRange.length};
        [pages addObject:[NSValue valueWithRange:destRange]];
        baseRange.location += targetRange.length;
    } while(baseRange.location < stringLength);

    CFRelease(frameSetter);
    return pages;
}
```

> **Get This Recipe's Code**
>
> To find this recipe's full sample project, point your browser to https://github.com/erica/iOS-6-Advanced-Cookbook and go to the folder for Chapter 3

Recipe: Drawing Attributed Text into a PDF

The Core Text and attributed listings you've seen so far drew text into views, rendering them in `drawRect` updates. Recipe 3-8 transfers that drawing away from views and into PDFs. It builds a new PDF file, adding content page-by-page, using the same page layout method introduced in Recipe 3-7.

Drawing to a PDF context is similar to drawing to an image context, the difference is in the call to begin a new PDF page. Recipe 3-8 builds its multipage PDF document and stores it to the path passed in the method call.

Recipe 3-8   **Drawing to PDF**

```objc
- (void) dumpToPDFFile: (NSString *) pdfPath
{
    // This is an arbitrary page size. Adjust as desired.
    CGRect theBounds = CGRectMake(0.0f, 0.0f, 480.0f, 640.0f);
    CGRect insetRect = CGRectInset(theBounds, 0.0f, 10.0f);

    NSArray *pageSplits = [self findPageSplitsForString:string
        withPageSize:insetRect.size];
    int offset = 0;

    UIGraphicsBeginPDFContextToFile(pdfPath, theBounds, nil);

    for (NSValue *pageStart in pageSplits)
    {
        UIGraphicsBeginPDFPage();
        NSRange offsetRange = {offset, pageStart.rangeValue.length};
        NSAttributedString *subString =
            [string attributedSubstringFromRange:offsetRange];
        offset += offsetRange.length;
        [subString drawInRect:insetRect];
    }

    UIGraphicsEndPDFContext();
}

- (void) createPDF
{
    NSString *path = [[NSBundle mainBundle]
        pathForResource:@"data" ofType:@"txt"];
    NSString *markup = [NSString stringWithContentsOfFile:path
        encoding:NSUTF8StringEncoding error:nil];
    string = [MarkupHelper stringFromMarkup:markup];

    NSString *destPath = [NSHomeDirectory()
        stringByAppendingPathComponent:@"Documents/results.pdf"];
    [self dumpToPDFFile:destPath];

    // Display the PDF in QuickLook
    QLPreviewController *controller =
        [[QLPreviewController alloc] init];
    controller.dataSource = self;
    [self.navigationController pushViewController:controller animated:YES];
}
```

Get This Recipe's Code

To find this recipe's full sample project, point your browser to https://github.com/erica/iOS-6-Advanced-Cookbook and go to the folder for Chapter 3.

Recipe: Big Phone Text

On the Macintosh, Address Book's big text display is one of my favorite desktop features. It enlarges text to provide an easy-to-read display of short information, information that you might read even from across the room. So why not use the same make-it-big-and-readable philosophy on your iPhone, iPad, or iPod touch, as shown in Figure 3-17? Big text that is unaffected by device rotation with high-contrast display can be a real asset to many applications.

Figure 3-17 High-contrast large text displays can help users show off short snippets of information in a highly readable manner.

That's because big text isn't just about phone numbers. You might display names, level unlock codes, or e-mail addresses, as well as those phone numbers, making it easier to visually transfer information to someone using another iOS device, a physical notepad, a game console, a computer, or even a third-party smartphone. The motivating features here are that the text is short in length, big in size, and easy to read.

Having an algorithmic tool to display otherwise hard-to-read text is a great software feature, and one whose inspiration I proudly stole from Address Book to create this final recipe's reusable Objective-C class. Recipe 3-9's `BigTextView` class is meant to overlay the key device window and will not respond to device orientation, ensuring that text doesn't start moving as the user reorients the device, especially when showing the information to someone else.

The current implementation dismisses the view with a double-tap to prevent any premature dismissal during typical handling.

Recipe 3-9 **Big Text. Really Big Text.**

```
// Center a rectangle inside a parent
CGRect rectCenteredInRect(CGRect rect, CGRect mainRect)
{
    return CGRectOffset(rect,
                        CGRectGetMidX(mainRect)-CGRectGetMidX(rect),
                        CGRectGetMidY(mainRect)-CGRectGetMidY(rect));
}

// Big Text View
@implementation BigTextView
{
    NSString *baseString;
}

- (id) initWithString: (NSString *) theString
{
    if (self = [super initWithFrame:CGRectZero])
    {
        baseString = theString;

        // Require a double-tap to dismiss
        UITapGestureRecognizer *tapRecognizer =
            [[UITapGestureRecognizer alloc] initWithTarget:self
              action:@selector(dismiss)];
        tapRecognizer.numberOfTapsRequired = 2;
        [self addGestureRecognizer:tapRecognizer];
    }
    return self;
}

- (void) dismiss
{
    [self removeFromSuperview];
}

+ (void) bigTextWithString:(NSString *)theString
{
    BigTextView *theView =
        [[BigTextView alloc] initWithString:theString];

    // Create a dark, translucent backdrop
    theView.backgroundColor =
        [[UIColor darkGrayColor] colorWithAlphaComponent:0.5f];
```

```objc
    // Constrain the view to stretch to its parent size
    UIWindow* window = [[UIApplication sharedApplication] keyWindow];
    [window addSubview:theView];
    PREPCONSTRAINTS(theView);
    STRETCH_VIEW(window, theView);

    return;
}

// Draw out the message
- (void) drawRect:(CGRect)rect
{
    [super drawRect:rect];
    CGContextRef context = UIGraphicsGetCurrentContext();

    // Create a geometry with width greater than height
    CGRect orientedRect = self.bounds;
    if (orientedRect.size.height > orientedRect.size.width)
        orientedRect.size = CGSizeMake(
            orientedRect.size.height, orientedRect.size.width);

    // Rotate 90 deg to write text horizontally
    // along window's vertical axis
    CGContextRotateCTM(context, -M_PI_2);
    CGContextTranslateCTM(context, -self.frame.size.height, 0.0f);

    // Draw a light gray rounded-corner backsplash
    [[[UIColor darkGrayColor] colorWithAlphaComponent:0.75f] set];
    CGRect insetRect = CGRectInset(orientedRect,
        orientedRect.size.width * 0.05f,
        orientedRect.size.height * 0.35f);
    [[UIBezierPath bezierPathWithRoundedRect:insetRect
        cornerRadius:32.0f] fill];
    CGContextFillPath(context);

    // Inset again for the text
    insetRect = CGRectInset(insetRect,
        insetRect.size.width * 0.05f,
        insetRect.size.height * 0.05f);

    // Iterate until finding a set of font
    // traits that fits this rectangle
    UIFont *textFont;
    NSString *fontFace = @"HelveticaNeue-Bold";
    CGSize fullSize = CGSizeMake(CGFLOAT_MAX, CGFLOAT_MAX);
```

```objc
    for (CGFloat fontSize = 18; fontSize < 300; fontSize++ )
    {
        // Search until the font size is too big
        textFont = [UIFont fontWithName:fontFace size: fontSize];
        CGSize textSize = [baseString sizeWithFont:textFont
            constrainedToSize:fullSize];
        if (textSize.width > insetRect.size.width)
        {
            // Ease back on font size to prior level
            textFont = [UIFont fontWithName:fontFace
                size: fontSize - 1];
            break;
        }
    }

    // Establish a frame that just encloses the text at the maximum size
    CGSize textSize = [baseString sizeWithFont:textFont
        constrainedToSize:fullSize];
    CGRect textFrame = (CGRect){.size = textSize};
    CGRect centerRect = rectCenteredInRect(textFrame, insetRect);

    // Draw the string in white
    [[UIColor whiteColor] set];
    [baseString drawInRect:centerRect withFont:textFont];
}
@end
```

> **Get This Recipe's Code**
>
> To find this recipe's full sample project, point your browser to https://github.com/erica/iOS-6-Advanced-Cookbook and go to the folder for Chapter 3.

Summary

This chapter introduced many ways to creatively use text in your iOS applications. In this chapter, you read about attributed strings and building solutions using Core Text. Here are a few further thoughts before you move to the next chapter:

- With the proper utility classes, attributed strings are amazingly simple to use. They offer an easy way to create aesthetically pleasing presentations with little coding. Now that UIKit supports attributed strings for so many classes, it's time to jump off the plain `NSString` bandwagon and enter the powerful world of attributes.

- Some typesetting features seem to have missed the iOS 6 cut-off. Although iOS 6 is wonderful for text, iOS 7 should be amazing. This chapter detailed a number of places (such as adding traits to fonts) where iOS 6 didn't *quite* move development completely out from Core Text. Responder features such as bolding and italics offer a tantalizing taste of where iOS is heading.

- Serious work requires serious typesetting. Don't be afraid to use Core Text for laying out complicated pages that mix text with images. You've seen examples that demonstrate how to create flexible paths. Use them to build all kinds of mind-blowing layouts.

- If you plan to use custom typesetting with page view controllers, schedule time to work out the math and layout—especially if you intend to let users adjust font sizes, and particularly when moving from portrait to landscape layouts. There's a good reason Apple has an entire team working on iBooks.

- iOS ships with an excellent and robust set of fonts, but you aren't limited to that set. Feel free to license and bundle the fonts you need, so long as they're iOS-compatible.

- Are you looking for the recipe that lays out text along a Bezier path? It's in the next chapter. Chapter 4, "Geometry," creates a better match to the underlying problem, which is more about math than Core Text. You'll find it there.

4

Geometry

Although UIKit requires less applied math than, say, Core Animation or Open GL, geometry plays an important role when working with Bezier paths and view transforms. Why do you need geometry? It helps you manipulate views in nonstandard ways, including laying out text along custom paths and performing path-follow types of animation. If your eyes glaze over at the mention of Bezier curves, convex hulls, and splines, this chapter helps demystify these terms, enabling you to add some powerful customization options to your toolbox.

Recipe: Retrieving Points from Bezier Paths

In mathematics, Bezier curves refer to a parameterized smoothed curve, created by control points applied to a line segment. In UIKit, Bezier *paths* define shapes built from straight and curved line segments, which may include Bezier curves and circle arcs. Members of the `UIBezierPath` class store sequences of segments, which may be open or closed, continuous or noncontinuous. Paths can include the following:

- Straight lines, created by `moveToPoint:` and `addLineToPoint:` calls.
- Cubic Bezier curve segments created by calling `addCurveToPoint:controlPoint1:controlPoint2:`.
- Quadratic Bezier curve segments built by `addQuadCurveToPoint:controlPoint:` calls.
- Arcs added with calls to `addArcToCenter:radius:startAngle:endAngle:clockwise:`.

Class methods enable you to build rectangles, ovals, rounded rectangles, and arcs, offering single-call access to common path styles. When working with free-form drawing, you often create paths interactively, responding to user touches along the way.

Although Bezier paths offer a lot of utility for working with drawn paths, they do not directly offer a way to retrieve an array of their points. Recipe 4-1 does this by collecting an array of source points. It uses the `CGPathApply` function to iterate a Bezier path's elements, supplying a custom function (`getPointsFromBezier()`) called on each element.

This function retrieves the element type (such as move-to-point, add-line-to-point, add-curve-to-point, add-quad-curve-to-point, or close-subpath) and uses that information to pull out the points that make up the path. This ability to move to and from path and point representation offers a way for you to take advantage of both worlds.

Path objects provide core-level access to drawing features such as filling and stroking. You can set a path's line width and dash style. This makes it easy to incorporate paths into a view's `drawRect:` implementation. A simple call to `[myPath stroke]` performs all the drawing you need to show a path's outline.

Moving from a path to points lets you apply math functions to your custom drawing. You can smooth the path, find its bounding box, apply a point thinning function, and so forth. The points offer concrete access to the data that makes up the path.

Throughout this chapter, you can find routines that move in both directions: from path objects to point arrays, and back. Recipe 4-1 makes this possible by providing a property (`points`) that returns an array of points extracted from a Bezier path, and a class method (`pathwithPoints:`) that establishes a Bezier path from an array of points.

Recipe 4-1 **Extracting Bezier Path Points**

```
#define POINT(_INDEX_) \
    [(NSValue *)[points objectAtIndex:_INDEX_] CGPointValue]
#define VALUE(_INDEX_) \
    [NSValue valueWithCGPoint:points[_INDEX_]]

void getPointsFromBezier(void *info, const CGPathElement *element)
{
    NSMutableArray *bezierPoints = (__bridge NSMutableArray *)info;
    CGPathElementType type = element->type;
    CGPoint *points = element->points;
    if (type != kCGPathElementCloseSubpath)
    {
        if ((type == kCGPathElementAddLineToPoint) ||
            (type == kCGPathElementMoveToPoint))
            [bezierPoints addObject:VALUE(0)];
        else if (type == kCGPathElementAddQuadCurveToPoint)
            [bezierPoints addObject:VALUE(1)];
        else if (type == kCGPathElementAddCurveToPoint)
            [bezierPoints addObject:VALUE(2)];
    }
}

- (NSArray *)points
{
    NSMutableArray *points = [NSMutableArray array];
    CGPathApply(self.CGPath, (__bridge void *)points, getPointsFromBezier);
```

```
    return points;
}

+ (UIBezierPath *) pathWithPoints: (NSArray *) points
{
    UIBezierPath *path = [UIBezierPath bezierPath];
    if (points.count == 0) return path;
    [path moveToPoint:POINT(0)];
    for (int i = 1; i < points.count; i++)
        [path addLineToPoint:POINT(i)];
    return path;
}
```

> **Get This Recipe's Code**
>
> To find this recipe's full sample project, point your browser to https://github.com/erica/iOS-6-Cookbook and go to the folder for Chapter 4.

Recipe: Thinning Points

The points created in Recipe 4-1 are built from touch events. Every time users move their finger, a new point joins the underlying Bezier path. Paths clutter quickly, as you see in Figure 4-1 (left). This screen shot shows the touch points captured during a user interaction. Each small circle represents a touch event as the user strokes the screen.

Many of these points are collinear, representing redundant elements. Figure 4-1 (right) shows a thinned representation of the same path, with numerous items removed. This version eliminates points that fail to inflect the line, so you see several long straight segments without any circles. The math that creates this streamlined version tests for alignment and removes points that aren't playing a role in defining the shape.

Consider Figure 4-2. It shows three points: A, P, and B. The angle formed between the AP and PB line segments is nearly flat, almost but not quite 180 degrees. Eliminating point P in this sequence produces a line segment (AB) that's close to the original (APB) but with one fewer point. This creates a simpler representation with a minimal loss of fidelity. As you can see in Figure 4-1, the before-and-after versions are nearly identical despite a 4–5x reduction in data points.

Recipe 4-2 details the thinning algorithm. It iterates through the path's points, using Recipe 4-1's point retrieval. It examines each point (point P in Figure 4-2), looking one point back (to point A) and one point forward (to point B). It creates the two vectors, AP and BP, and calculates their dot product.

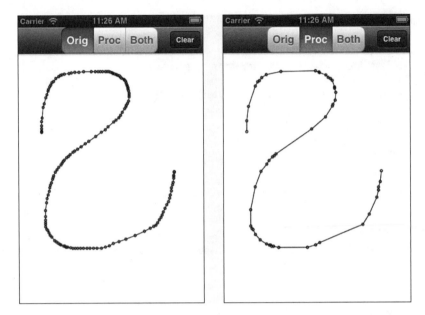

Figure 4-1 Thinning enables you to remove collinear points to reduce the number of items used to build a curve. This is important if you are storing the strokes or supporting undo/redo. It enables you to simultaneously reduce your app's memory overhead and increase its performance and responsiveness.

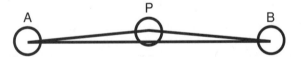

Figure 4-2 The angle formed between the AP and PB line segments is nearly flat.

Dot products are algebraic operations that multiply the elements of two vectors. The result is directly related to the cosine of the angle between the two vectors. Recipe 4-2 compares the cosine value to a test value, which ranges between –1 (cosine of Pi, or 180 degrees) to –0.865 (cosine of 150 degrees). This corresponds to a tolerance, which you supply as a function parameter. It ranges between 0 (no tolerance, the angle must be exactly 180) to 1 (highest tolerance, allowing angles up to 150 degrees).

As the tolerance moves toward 1, the algorithm enables greater divergence from strict collinearity. Using this check, the function eliminates collinear and nearly collinear points, returning a greatly thinned Bezier path.

There's a trade-off here of memory savings versus the accuracy of the depiction. This is similar to JPEG image compression. Small compression gives you a small memory savings, with an almost imperceptible decrease in image quality. Pushed too far, your space savings create an unrecognizable blob of an image. As a developer, you decide what level you can use without degrading the user experience. Use smaller values for more precise representations and larger values for less accurate but more memory-limited representations.

Recipe 4-2   **Thinning Bezier Path Points**

```
#define POINT(_INDEX_) \
    [(NSValue *)[points objectAtIndex:_INDEX_] CGPointValue]

// Return dot product of two vectors normalized
static float dotproduct (CGPoint v1, CGPoint v2)
{
    float dot = (v1.x * v2.x) + (v1.y * v2.y);
    float a = ABS(sqrt(v1.x * v1.x + v1.y * v1.y)); // magnitude a
    float b = ABS(sqrt(v2.x * v2.x + v2.y * v2.y)); // magnitude b
    dot /= (a * b);

    return dot;
}

// Pass a tolerance within 0 to 1.
// 0 tolerance uses the tightest checking for colinearity
// As the values loosen, colinearity will be allowed for angles
// further from 180 degrees, up to 150 degrees at a tolerance of 1

UIBezierPath *thinPath(UIBezierPath *path, CGFloat tolerance)
{
    // Retrieve the points
    NSArray *points = path.points;
    if (points.count < 3) return path;

    // Create a new output path
    UIBezierPath *newPath = [UIBezierPath bezierPath];
    CGPoint p1 = POINT(0);
    [newPath moveToPoint:p1];

    CGPoint mostRecent = p1;
    int count = 1;

    // -1 = 180 degrees, -0.985 = 170 degrees,
    // -0.865 = 150 degrees
    CGFloat checkValue = -1.0f + .135 * tolerance;

    // Add only those points that are inflections
    for (int i = 1; i < (points.count - 1); i++)
    {
        CGPoint p2 = POINT(i);
```

```
        CGPoint p3 = POINT(i+1);

        // Cast vectors around p2 origin
        CGPoint v1 = CGPointMake(p1.x - p2.x, p1.y - p2.y);
        CGPoint v2 = CGPointMake(p3.x - p2.x, p3.y - p2.y);
        float dot = dotproduct(v1, v2);

        // Colinear items need to be as close as possible to 180 degrees
        // That means as close to -1 as possible

        if (dot < checkValue) continue;
        p1 = p2;

        mostRecent = POINT(i);
        [newPath addLineToPoint:mostRecent];
        count++;
    }

    // Add final point
    CGPoint finalPoint = POINT(points.count - 1);
    if (!CGPointEqualToPoint(finalPoint, mostRecent))
        [newPath addLineToPoint:finalPoint];

    return newPath;
}
```

> **Get This Recipe's Code**
>
> To find this recipe's full sample project, point your browser to https://github.com/erica/iOS-6-Cookbook and go to the folder for Chapter 4.

Recipe: Smoothing Drawings

Depending on the device in use and the amount of simultaneous processing involved, capturing user gestures may produce results that are rougher than wanted. Figure 4-3 shows the kind of angularity that derives from granular input. Touch events are often limited by CPU demands and are subject to any system events. Using a real-time smoothing algorithm can offset those limitations by interpolating between points using basic splining.

Catmull-Rom splines offer one of the simplest approaches to create continuous curves between key points. This algorithm ensures that each initial point you provide remains part of the final curve, so the resulting path retains the original path's shape. You choose the number of points to interpolate between each of your reference points. The trade-off lies between processing power and greater smoothing. The more points you add, the more CPU resources you consume.

As you can see when using the sample code that accompanies this chapter, a little smoothing goes a long way. That said, newer iOS devices are so responsive that it's hard to draw a particularly jaggy line in the first place.

Figure 4-3 Catmull-Rom smoothing can be applied in real time to improve arcs between touch events. The images shown here are based on an identical gesture input, with and without smoothing applied.

Catmull-Rom interpolates to create smoothed representations. Recipe 4-3 implements this algorithm using four points at a time to calculate intermediate values between the second and third points in each sequence. It uses a granularity you specify, to create extra items between those points. To create the results shown in Figure 4-3, the sample code used a granularity of 4. Recipe 4-3 provides an example of just one kind of real-time geometric processing you might add to your applications.

Recipe 4-3 **Catmull-Rom Splining**

```
#define POINT(_INDEX_) \
    [(NSValue *)[points objectAtIndex:_INDEX_] CGPointValue]

// Return a smoothed path using the supplied granularity
UIBezierPath *smoothedPath(UIBezierPath *path, NSInteger granularity)
{
    NSMutableArray *points = [path.points mutableCopy];
    if (points.count < 4) return [path copy];
    [points insertObject:[points objectAtIndex:0] atIndex:0];
```

```
    [points addObject:[points lastObject]];
    
    UIBezierPath *smoothedPath = [UIBezierPath bezierPath];
    
    // Copy traits
    smoothedPath.lineWidth = path.lineWidth;
    
    // Draw out the first 3 points (0..2)
    [smoothedPath moveToPoint:POINT(0)];
    
    for (int index = 1; index < 3; index++)
        [smoothedPath addLineToPoint:POINT(index)];
    
    for (int index = 4; index < points.count; index++)
    {
        CGPoint p0 = POINT(index - 3);
        CGPoint p1 = POINT(index - 2);
        CGPoint p2 = POINT(index - 1);
        CGPoint p3 = POINT(index);
        
        // now add n points starting at p1 + dx/dy up
        // until p2 using Catmull-Rom splines
        for (int i = 1; i < granularity; i++)
        {
            float t = (float) i * (1.0f / (float) granularity);
            float tt = t * t;
            float ttt = tt * t;
            
            CGPoint pi; // intermediate point
            pi.x = 0.5 * (2*p1.x+(p2.x-p0.x)*t +
                (2*p0.x-5*p1.x+4*p2.x-p3.x)*tt +
                (3*p1.x-p0.x-3*p2.x+p3.x)*ttt);
            pi.y = 0.5 * (2*p1.y+(p2.y-p0.y)*t +
                (2*p0.y-5*p1.y+4*p2.y-p3.y)*tt +
                (3*p1.y-p0.y-3*p2.y+p3.y)*ttt);
            [smoothedPath addLineToPoint:pi];
        }
        
        // Now add p2
        [smoothedPath addLineToPoint:p2];
    }
    
    // finish by adding the last point
    [smoothedPath addLineToPoint:POINT(points.count - 1)];
    
    return smoothedPath;
}
```

> **Get This Recipe's Code**
>
> To find this recipe's full sample project, point your browser to https://github.com/erica/iOS-6-Cookbook and go to the folder for Chapter 4.

Recipe: Velocity-Based Stroking

Velocity-based thickness can add a sense of realism to interactive drawings by mimicking ink-flow. The faster the pen moves, the less ink it can lay down. Slower movements place more ink, creating thicker areas. Figure 4-4 shows how the same tracing looks with and without velocity-based thickness adjustment. Tying line thickness to user velocity is most commonly used when capturing signatures or creating "pen-and-ink" style interfaces.

Figure 4-4 A user-created tracing with (right) and without (left) velocity-based thickness adjustment.

Recipe 4-4 uses point velocities captured during drawing. A custom class called `FlowPath` store dates along with point positions. The time offset and the distance traveled combine to determine what stroke width to apply.

To dampen change a little bit, this approach, which was inspired by any number of superior solutions around the Internet, uses a weighting factor. This limits the degree acceleration that can affect the current velocity, taking an average of the current and previous velocities before calculating the new stroke width.

This is a recipe you want to tweak and adjust to produce the best possible visual design. It offers all the basics you need to start but leaves the "perfect stroke" as an exercise for the reader. Be aware that this routine uses uniform stroke widths for each path length. A more thorough implementation might smooth the path using splines and adjust that width along each interpolation segment.

Recipe 4-4 **Velocity-Based Stroking**

```
#define POINT(_INDEX_) \
    [(NSValue *)[points objectAtIndex:_INDEX_] CGPointValue]

// Calculate the velocity between two point events
- (CGFloat) velocityFrom:(int) j to:(int) i
{
    CGFloat dPos = distance(POINT(j), POINT(i));
    CGFloat dTime = (DATE(j) - DATE(i));
    return dPos / dTime;
}

// Return a stroke width based on velocity
// Feel free to tweak this all you like
- (CGFloat) strokeWidth: (CGFloat) velocity
{
    CGFloat multiplier = 2.0f;
    CGFloat base = 5.0f;
    CGFloat adjusted = base - (log2f(velocity) / multiplier);
    adjusted = MIN(MAX(adjusted, 0.4), base);
    return multiplier * adjusted * _lineWidth;
}

// Create a Bezier path from p0 to p1 for drawing
UIBezierPath *bPath(CGPoint p0, CGPoint p1)
{
    UIBezierPath *path = [UIBezierPath bezierPath];
    [path moveToPoint:p0];
    [path addLineToPoint:p1];
    return path;
}

// Stroke the custom FlowPath
- (void) stroke
{
    if (points.count < 2) return;

    // Store the most recent velocity
    CGFloat lastVelocity = [self velocityFrom:1 to:0];
```

```
    // Adjustable weighting for slight filtering
    CGFloat weight = 0.5f;

    UIBezierPath *path;
    for (int i = 1; i < points.count; i++)
    {
        // Adjust the velocity so it doesn't change too much
        // at any given time
        CGFloat velocity = [self velocityFrom:i to:i-1];
        velocity = weight*velocity + (1.0f - weight)*lastVelocity;
        lastVelocity = velocity;
        CGFloat strokeWidth = [self strokeWidth:velocity];

        // Stroke each segment
        path = bPath(POINT(i - 1), POINT(i));
        path.lineWidth = strokeWidth;
        [path stroke];

    }
}
```

Get This Recipe's Code

To find this recipe's full sample project, point your browser to https://github.com/erica/iOS-6-Cookbook and go to the folder for Chapter 4.

Recipe: Bounding Bezier Paths

Bezier paths produce irregular shapes. Because of that, you may need to retrieve a path's bounding box (the minimal rectangle that encloses the path) or convex hull (the smallest convex shape that encloses the path) to do intersection testing or provide a visual backdrop. Figure 4-5 shows the bounding box and convex hull for a sketched face.

The Bezier path's built-in bounds property corresponds to the curve's bounding box. Recipe 4-5 shows how to create a path's convex hull. It works by sorting the points geometrically along the X-axis and then along the Y-axis. The recipe calculates the bottom hull and then the top hull, testing each point to determine if it falls within the object (to the left for the lower hull, to the right for the upper hull) or outside (right for lower, left for upper). If the point falls outside, the method extends the hull to accommodate.

After calculating the bounding points, the convex hull method builds a new UIBezierPath instance to store them and returns that object.

138   Chapter 4   Geometry

Figure 4-5 A `UIBezierPath` drawing of a human face. The light, outer rectangle marks the path's bounding box. The inner, heavy outline is the path's convex hull.

Recipe 4-5 **Bounding Boxes and Convex Hulls**

```
#define POINT(_INDEX_) \
    [(NSValue *)[points objectAtIndex:_INDEX_] CGPointValue]

@implementation UIBezierPath (Bounding)

// Create a zero-sized rectangle at a point
static CGRect pointRect(CGPoint point)
{
    return (CGRect){.origin=point};
}

// Return an array of sorted points along X and then Y
- (NSArray *) sortedPoints
{
    NSArray *sorted = [self.points sortedArrayUsingComparator:
        ^NSComparisonResult(id item1, id item2)
    {
        NSValue *v1 = (NSValue *) item1;
        NSValue *v2 = (NSValue *) item2;
```

```objc
        CGPoint p1 = v1.CGPointValue;
        CGPoint p2 = v2.CGPointValue;

        if (p1.x == p2.x)
            return [@(p1.y) compare:@(p2.y)];
        else
            return [@(p1.x) compare:@(p2.x)];
    }];
    return sorted;
}

// Test a point's half-plane
static float halfPlane(CGPoint p1, CGPoint p2, CGPoint testPoint)
{
    return (p2.x-p1.x)*(testPoint.y-p1.y) - (testPoint.x-p1.x)*(p2.y-p1.y);
}

// Return a path's convex hull
- (UIBezierPath *) convexHull
{
    /*
     minmin = top left, min x, min y
     minmax = bottom left, min x, max y
     maxmin = top right, max x, min y
     maxmax = bottom right, max x, max y
     */

    NSMutableArray *output = [NSMutableArray array];
    NSInteger bottom = 0;
    NSInteger top = -1;
    NSInteger i;

    // Pre-sort the points
    NSArray *points = self.sortedPoints;
    NSInteger lastIndex = points.count - 1;

    // Location of top-left corner
    NSInteger minmin = 0;
    CGFloat xmin = POINT(0).x;

    // Locate minmax, bottom left
    for (i = 1; i <= lastIndex; i++)
        if (POINT(i).x != xmin)
            break;
    NSInteger minmax = i - 1;
```

```objc
    // If the bottom left is the final item
    // check whether to add both minmin & minmax
    if (minmax == lastIndex)
    {
        output[++top] = points[minmin];
        if (POINT(minmax).y != POINT(minmin).y)
        {
            // add the second point, and close the path
            output[++top] = points[minmax];
            output[++top] = points[minmin];
        }

        for (int i = top + 1; i < output.count; i++)
            [output removeObjectAtIndex:i];

        return [UIBezierPath pathWithPoints:output];
    }

    // Search for top right, max x, min y by moving
    // back from max x, max y at final index
    NSInteger maxmin = lastIndex;
    CGFloat xmax = POINT(lastIndex).x;
    for (i = lastIndex - 1; i >= 0; i--)
        if (POINT(i).x != xmax)
            break;
    maxmin = i + 1;

    // Compute Lower Hull
    output[++top] = points[minmin]; // top left
    i = minmax; // bottom left

    while (++i < maxmin) // top right
    {
        // Test against TopLeft-TopRight
        if ((halfPlane(POINT(minmin),
            POINT(maxmin), POINT(i)) >= 0) &&
            (i < maxmin))
            continue;

        while (top > 0)
        {
            // Find points that extend the hull and add them
            if (halfPlane([output[top - 1] CGPointValue],
                [output[top] CGPointValue], POINT(i)) > 0)
                break;
```

```
            else
                top--;
        }
        output[++top] = points[i];
    }

    // Ensure the hull is continuous when going from lower to upper
    NSInteger maxmax = lastIndex;
    if (maxmax != maxmin)
        output[++top] = points[maxmax];

    // Compute Upper Hull
    bottom = top;
    i = maxmin;
    while (--i >= minmax)
    {
        if ((halfPlane(POINT(maxmax),
            POINT(minmax), POINT(i)) >= 0) &&
            (i > minmax))
            continue;

        while (top > bottom)
        {
            // Add points that extend the hull
            if (halfPlane([output[top - 1] CGPointValue],
                [output[top] CGPointValue], POINT(i)) > 0)
                break;
            else
                top--;
        }
        output[++top] = points[i];
    }

    // Again ensure continuity at the end
    if (minmax != minmin)
        output[++top] = points[minmin];

    NSMutableArray *results = [NSMutableArray array];
    for (int i = 0; i <= top; i++)
        [results addObject:output[i]];

    return [UIBezierPath pathWithPoints:results];
}
@end
```

Chapter 4 Geometry

> **Get This Recipe's Code**
> To find this recipe's full sample project, point your browser to https://github.com/erica/iOS-6-Cookbook and go to the folder for Chapter 4.

Recipe: Fitting Paths

UIKit Bezier paths provide an easy way to use vector graphics in your application. Their geometric representations enable you to scale and place art without regard to pixels. Recipe 4-6 details the work involved in drawing a path into a custom rectangle, as demonstrated in Figure 4-6. This picture's custom line drawing is projected into a small inset box in real time as it's being drawn.

Figure 4-6 The larger Bezier path drawn by the user is projected into the small darker box.

The recipe works by calculating a path's bounding box and fitting that size into a destination rectangle supplied as a method parameter. This is simply a matter of checking the horizontal and vertical scaling factors. The smaller scale wins, and the results are offset to center the scaled path within the destination rectangle.

This recipe projects each point from the original rectangle to the destination one using simple geometry. It calculates the point's vector from the initial origin, scales it, and then casts that vector from the destination origin. This solution works well for point-based `UIBezierPath` instances but runs up short when faced with path objects that support arcs and lines. Read on to discover a more general solution that handles both point and curve-based paths.

Recipe 4-6 Fitting Paths into Custom Rectangles

```
// Determine the scale that allows a size to fit into
// a destination rectangle
CGFloat AspectScaleFit(CGSize sourceSize, CGRect destRect)
{
    CGSize destSize = destRect.size;
    CGFloat scaleW = destSize.width / sourceSize.width;
    CGFloat scaleH = destSize.height / sourceSize.height;
    return MIN(scaleW, scaleH);
}

// Create a rectangle that will fit an item while preserving
// its original aspect
CGRect AspectFitRect(CGSize sourceSize, CGRect destRect)
{
    CGSize destSize = destRect.size;
    CGFloat destScale = AspectScaleFit(sourceSize, destRect);

    CGFloat newWidth = sourceSize.width * destScale;
    CGFloat newHeight = sourceSize.height * destScale;

    float dWidth = ((destSize.width - newWidth) / 2.0f);
    float dHeight = ((destSize.height - newHeight) / 2.0f);

    CGRect rect = CGRectMake(destRect.origin.x + dWidth,
        destRect.origin.y + dHeight, newWidth, newHeight);
    return rect;
}

// Add two points
CGPoint PointAddPoint(CGPoint p1, CGPoint p2)
{
    return CGPointMake(p1.x + p2.x, p1.y + p2.y);
}

// Subtract a point from a point
CGPoint PointSubtractPoint(CGPoint p1, CGPoint p2)
{
    return CGPointMake(p1.x - p2.x, p1.y - p2.y);
}

// Project a point from a native rectangle into a destination
// rectangle
```

```
NSValue *adjustPoint(CGPoint p, CGRect native, CGRect dest)
{
    CGFloat scaleX = dest.size.width / native.size.width;
    CGFloat scaleY = dest.size.height / native.size.height;

    CGPoint point = PointSubtractPoint(p, native.origin);
    point.x *= scaleX;
    point.y *= scaleY;
    CGPoint destPoint = PointAddPoint(point, dest.origin);

    return [NSValue valueWithCGPoint:destPoint];
}

// Fit a path into a rectangle
- (UIBezierPath *) fitInRect: (CGRect) destRect
{
    // Calculate an aspect-preserving destination rectangle
    NSArray *points = self.points;
    CGRect bounding = self.bounds;
    CGRect fitRect = AspectFitRect(bounding.size, destRect);

    // Project each point from the original to the
    // destination rectangle
    NSMutableArray *adjustedPoints = [NSMutableArray array];
    for (int i = 0; i < points.count; i++)
        [adjustedPoints addObject:adjustPoint(
            POINT(i), bounding, fitRect)];

    return [UIBezierPath pathWithPoints:adjustedPoints];
}
```

> **Get This Recipe's Code**
>
> To find this recipe's full sample project, point your browser to https://github.com/erica/iOS-6-Cookbook and go to the folder for Chapter 4.

Working with Curves

`UIBezierPath` aren't just points and straight lines as you've seen so far. Instances can be constructed from complex elements, including cubic and quadratic curves. Each element has a type, which describes the role the element plays in the path, and from zero to three parameters, which are used to construct the element. A path-closing element uses no parameters. A cubic Bezier curve element has three: an end-point and two control points.

Listing 4-1 expands the concepts from Recipe 4-1 to retrieve path elements instead of points. Each element is represented as an array consisting of an element type followed by its parameters. The bezierElements method returns that array.

It's easy enough to go the other direction as well. The listing's pathWithElements: class method takes this element array to return the original path constructed from these elements. As with Recipe 4-1, this pair of methods enables you to retrieve path elements as points, evaluate and adjust them, and then reconstruct a proper path with the updated items.

Listing 4-1   **Bezier Elements**

```
// Construct an array of Bezier Elements
void getBezierElements(void *info, const CGPathElement *element)
{
    NSMutableArray *bezierElements = (__bridge NSMutableArray *)info;
    CGPathElementType type = element->type;
    CGPoint *points = element->points;

    switch (type)
    {
        case kCGPathElementCloseSubpath:
            [bezierElements addObject:@[@(type)]];
            break;
        case kCGPathElementMoveToPoint:
        case kCGPathElementAddLineToPoint:
            [bezierElements addObject:@[@(type), VALUE(0)]];
            break;
        case kCGPathElementAddQuadCurveToPoint:
            [bezierElements addObject:
                @[@(type), VALUE(0), VALUE(1)]];
            break;
        case kCGPathElementAddCurveToPoint:
            [bezierElements addObject:
                @[@(type), VALUE(0), VALUE(1), VALUE(2)]];
            break;
    }
}

// Retrieve the element array
- (NSArray *) bezierElements
{
    NSMutableArray *elements = [NSMutableArray array];
    CGPathApply(self.CGPath,
        (__bridge void *)elements, getBezierElements);
    return elements;
}
```

```objc
// Construct a path from its elements
+ (UIBezierPath *) pathWithElements: (NSArray *) elements
{
    UIBezierPath *path = [UIBezierPath bezierPath];
    if (elements.count == 0) return path;

    for (NSArray *points in elements)
    {
        if (!points.count) continue;
        CGPathElementType elementType = [points[0] integerValue];
        switch (elementType)
        {
            case kCGPathElementCloseSubpath:
                [path closePath];
                break;
            case kCGPathElementMoveToPoint:
                if (points.count == 2)
                    [path moveToPoint:POINT(1)];
                break;
            case kCGPathElementAddLineToPoint:
                if (points.count == 2)
                    [path addLineToPoint:POINT(1)];
                break;
            case kCGPathElementAddQuadCurveToPoint:
                if (points.count == 3)
                    [path addQuadCurveToPoint:POINT(2)
                        controlPoint:POINT(1)];
                break;
            case kCGPathElementAddCurveToPoint:
                if (points.count == 4)
                    [path addCurveToPoint:POINT(3)
                        controlPoint1:POINT(1) controlPoint2:POINT(2)];
                break;
        }
    }

    return path;
}
```

Fitting Elements

After you have access to elements, you can adjust system-supplied paths such as rounded rectangles, ellipses, and so forth, just as easily as you would point-based paths. Listing 4-2 updates the concepts from Recipe 4-2 to enable projection of the points and control points for a general

Bezier path case. Figure 4-7 shows how a path constructed of quadratic and cubic Bezier paths can be moved and resized to a destination rectangle.

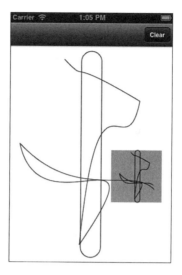

Figure 4-7 UIBezierPaths can be constructed of quadratic and cubic curves and lines. This path, which consists of a rounded rectangle and several random cubic Bezier curves, is projected into a custom rectangle using element recovery from the original path.

The listing works by iterating through the element arrays. It adjusts each point (destination and control points) from its original geometry, projecting it into the destination rectangle. The adjusted elements are then passed to the reconstruction class method from Listing 4-1 to return an updated Bezier path.

Listing 4-2 **Fitting Element-Based Bezier Paths**

```
// Project a Bezier Path into a custom rectangle
- (UIBezierPath *) fitElementsInRect: (CGRect) destRect
{
    CGRect bounding = self.bounds;
    CGRect fitRect = AspectFitRect(bounding.size, destRect);

    NSArray *elements = self.bezierElements;
    NSMutableArray *adjustedElements = [NSMutableArray array];
    for (NSArray *points in elements)
    {
        if (!points.count) continue;
        NSMutableArray *outArray = [NSMutableArray array];
        [outArray addObject:points[0]]; // NSNumber, type
```

```
        for (int i = 1; i < points.count; i++)
            [outArray addObject:adjustPoint(
                POINT(i), bounding, fitRect)];
        [adjustedElements addObject:outArray];
    }

    return [UIBezierPath pathWithElements:adjustedElements];
}
```

Recipe: Moving Items Along a Bezier Path

Animation represents a common use for Bezier paths. For example, you might move a view along a custom path drawn by the user. The rectangle in Figure 4-8 moves along its path, orienting itself to the path as it goes. Recipe 4-7 supports this behavior by returning a point and a slope when given an offset.

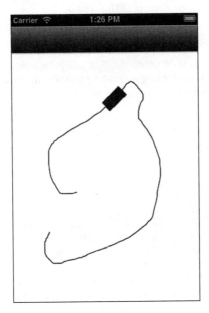

Figure 4-8 The user's tracing provides a path for the rectangle to move along. It adjusts its rotation to match that of the path.

The offset may vary between 0 and 1 and represents the percentage of progress along the path. A 0% progress returns the start point; 100% returns the end point. To determine the points between, this recipe calculates the total distance along the path (length) and builds an array from its component points, precalculating their progress in the pointPercentArray method.

A simple search and interpolation enables the `pointAtPercent:slope:` method to return a point that is exactly, for example, 67.25% along the path. The method identifies the points just before and just after that item. This allows it to return a slope (dy/dx) defined by those points, providing a basis for calculating the rotation angle for the animated object. To calculate the angle, provide the slope to an arctan function. Apply the rotation using affine transforms.

Recipe 4-7   **Retrieving Points and Slopes from Bezier Paths**

```
// Return distance between two points
static float distance (CGPoint p1, CGPoint p2)
{
    float dx = p2.x - p1.x;
    float dy = p2.y - p1.y;

    return sqrt(dx*dx + dy*dy);
}

// Return the length of a Bezier path
- (CGFloat) length
{
    NSArray *points = self.points;
    float totalPointLength = 0.0f;
    for (int i = 1; i < points.count; i++)
        totalPointLength += distance(POINT(i), POINT(i-1));
    return totalPointLength;
}

- (NSArray *) pointPercentArray
{
    // Use total length to calculate the percent of path
    // consumed at each control point
    NSArray *points = self.points;
    int pointCount = points.count;

    float totalPointLength = self.length;
    float distanceTravelled = 0.0f;

    NSMutableArray *pointPercentArray = [NSMutableArray array];
    [pointPercentArray addObject:@(0.0)];

    for (int i = 1; i < pointCount; i++)
    {
        distanceTravelled += distance(POINT(i), POINT(i-1));
        [pointPercentArray addObject:
            @(distanceTravelled / totalPointLength)];
    }
```

Chapter 4 Geometry

```objc
    // Add a final item just to stop with. Probably not needed.
    [pointPercentArray addObject:[NSNumber numberWithFloat:1.1f]]; // 110%

    return pointPercentArray;
}

// Return a point and its slope at a given offset
- (CGPoint) pointAtPercent: (CGFloat) percent withSlope: (CGPoint *) slope
{
    NSArray *points = self.points;
    NSArray *percentArray = self.pointPercentArray;
    CFIndex lastPointIndex = points.count - 1;

    if (!points.count)
        return CGPointZero;

    // Check for 0% and 100%
    if (percent <= 0.0f) return POINT(0);
    if (percent >= 1.0f) return POINT(lastPointIndex);

    // Find a corresponding pair of points in the path
    CFIndex index = 1;
    while ((index < percentArray.count) &&
           (percent >
                ((NSNumber *)percentArray[index]).floatValue))
        index++;

    // Calculate the intermediate distance between the two points
    CGPoint point1 = POINT(index -1);
    CGPoint point2 = POINT(index);

    float percent1 =
        [[percentArray objectAtIndex:index - 1] floatValue];
    float percent2 =
        [[percentArray objectAtIndex:index] floatValue];
    float percentOffset =
        (percent - percent1) / (percent2 - percent1);

    float dx = point2.x - point1.x;
    float dy = point2.y - point1.y;

    // Store dy, dx for retrieving arctan
    if (slope) *slope = CGPointMake(dx, dy);

    // Calculate new point
    CGFloat newX = point1.x + (percentOffset * dx);
    CGFloat newY = point1.y + (percentOffset * dy);
```

```
    CGPoint targetPoint = CGPointMake(newX, newY);

    return targetPoint;
}
```

> **Get This Recipe's Code**
> To find this recipe's full sample project, point your browser to https://github.com/erica/iOS-6-Cookbook and go to the folder for Chapter 4.

Recipe: Drawing Attributed Text Along a Bezier Path

The interpolation and slope math returned by Recipe 4-7 is not limited to animation. These methods can set text along a Bezier path (see Figure 4–9). Recipe 4-8's text layout routine respects string attributes, so you can mix and match fonts, colors, and sizes while adhering to the underlying path geometry.

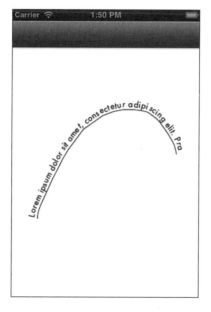

Figure 4-9   Recipe 4-8 carefully typesets its path along a user-drawn path, wobbles and all.

The recipe calculates the rendered size of each character in the attributed string, determining the bounding height and width. Knowing this size enables the recipe to determine how much of the path each character (or glyph) consumes:

```objc
- (CGSize) renderedSize
{
    CGRect bounding = [self boundingRectWithSize:CGSizeMake(
        CGFLOAT_MAX, CGFLOAT_MAX) options:0 context:nil];
    return bounding.size;
}
```

It calculates where the glyph center would appear, if laid out on a line and uses that distance as a percentage of the path's length. Recipe 4-7's point and slope function returns the position and angle for placement.

The method performs this placement by translating and rotating the context. This enables the string to render using `NSAttributedString`'s `drawAtPoint:` method. A graphics stack pops the context back to its original state after drawing each glyph.

After consuming the entire path, the routine stops adding text, clipping any remaining characters from view. If you want to ensure that the entire string appears, use the bounding rect routines (`boundingRectWithSize:` and `sizeWithFont:`) to calculate which fonts fit a given size.

Recipe 4-8 **Laying Out Text Along a Bezier Path**

```objc
- (void) drawAttributedString: (NSAttributedString *) string
    withOptions: (StringRenderingOptions) renderingOptions
{
    if (!string) return;

    NSArray *points = self.points;
    int pointCount = points.count;
    if (pointCount < 2) return;

    // Please do not send over anything with a new line
    NSAttributedString *baseString = string.versionWithoutNewLines;

    // Keep a running tab of how far the glyphs have travelled to
    // be able to calculate the percent along the point path
    float glyphDistance = 0.0f;

    // Should the renderer squeeze/stretch the text to fit?
    BOOL fitText = (renderingOptions & RenderStringToFit) != 0;
    float lineLength = fitText ? baseString.renderedWidth : self.length;

    // Optionally force close path
    BOOL closePath = (renderingOptions & RenderStringClosePath) != 0;
    if (closePath) [self addLineToPoint:POINT(0)];
```

```objc
// Establish the context
CGContextRef context = UIGraphicsGetCurrentContext();
CGContextSaveGState(context);

// Set the initial positions -- skip?
CGPoint textPosition = CGPointMake(0.0f, 0.0f);
CGContextSetTextPosition(context, textPosition.x, textPosition.y);

for (int loc = 0; loc < baseString.length; loc++)
{
    // Retrieve item
    NSRange range = NSMakeRange(loc, 1);
    NSAttributedString *item = [baseString attributedSubstringFromRange:range];

    // Calculate the percent travel
    CGFloat glyphWidth = item.renderedWidth;
    glyphDistance += glyphWidth;
    CGFloat percentConsumed = glyphDistance / lineLength;
    if (percentConsumed > 1.0f) break; // stop when all consumed

    // Find a corresponding pair of points in the path
    CGPoint slope;
    CGPoint targetPoint = [self pointAtPercent:percentConsumed withSlope:&slope];

    // Set the x and y offset
    CGContextTranslateCTM(context, targetPoint.x, targetPoint.y);
    CGPoint positionForThisGlyph = CGPointMake(textPosition.x, textPosition.y);

    // Rotate
    float angle = atan(slope.y / slope.x);
    if (slope.x < 0) angle += M_PI; // going left, update the angle
    CGContextRotateCTM(context, angle);

    // Place the glyph
    positionForThisGlyph.x -= glyphWidth;

    if ((renderingOptions & RenderStringOutsidePath) != 0)
    {
            positionForThisGlyph.y -= item.renderedHeight;
    }
    else if ((renderingOptions & RenderStringInsidePath) != 0)
    {
        // no op
    }
    else // over path or default
```

```
        {
                positionForThisGlyph.y -= item.renderedHeight / 2.0f;
        }
        
        // Draw the glyph
        [item drawAtPoint:positionForThisGlyph]; // was textPosition
        
        // Reset context transforms
        CGContextRotateCTM(context, -angle);
        CGContextTranslateCTM(context, -targetPoint.x, -targetPoint.y);
    }
    
    CGContextRestoreGState(context);
}
```

> **Get This Recipe's Code**
> To find this recipe's full sample project, point your browser to https://github.com/erica/iOS-6-Cookbook and go to the folder for Chapter 4.

Recipe: View Transforms

Affine transforms represent one of the most-used and most-feared features in UIKit. Tied into direct interaction, you often run across them when working with gesture recognizers, animation, and any kind of view scaling and rotation. Much of the transform's frustration factor ties into the opaqueness of the underlying structure and the lack of easy human-relatable methods. Simple tweaks can transform (if you pardon the pun) the `CGAffineTransform` structure into friendlier Objective-C-based properties and methods.

Basic Transforms

Affine transforms enable you scale, rotate, and translate UIView objects in your apps. You generally create a transform and apply it to your view using one of the following patterns. You either create and apply a new transform using one of the "make" functions

```
float angle = theta * (PI / 100);
CGAffineTransform transform = CGAffineTransformMakeRotation(angle);
myView.transform = transform;
```

or you layer a new change onto an existing transform using one of the "action" functions: rotate, scale, or translate.

```
CGAffineTransform transform = CGAffineTransformRotate(myView.transform, angle);
myView.transform = transform;
```

Creating a new transform resets whatever changes have already been applied to a view. If your view was already scaled larger through a transform, for example, the first of these two samples would override that scaling and replace it with rotation. Figure 4-10 shows before and after for this scenario. The outer, larger, scaled view is replaced by the unscaled, smaller, rotated view. (The red circle marks the top-right corner of the view.)

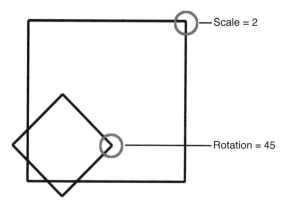

Figure 4-10 Setting a transform replaces whatever view transform had previously been set.

In the second example (see Figure 4-11), the transform is layered on so it adds to, rather than replaces, the scaling. In this case, the scaled view rotates but the scaling remains unaffected.

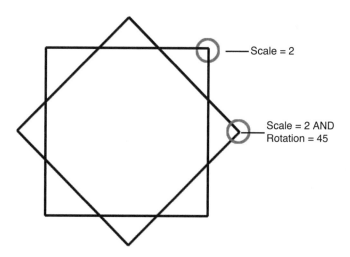

Figure 4-11 Appending transforms preserve previous settings.

Under the Hood

Every transform is represented by an underlying transformation matrix, which is set up as follows.

This matrix corresponds to a simple C structure:

```
struct CGAffineTransform {
   CGFloat a;
   CGFloat b;
   CGFloat c;
   CGFloat d;
   CGFloat tx;
   CGFloat ty;
};
typedef struct CGAffineTransform CGAffineTransform;
```

The UIKit framework defines a variety of helper functions specific to graphics and drawing operations. These include several affine-specific utilities. You can print out a view's transform via UIKit's `NSStringFromCGAffineTransform()` function. Its inverse is `CGAffineTransformFromString()`. Here's what the transform values look like for the scaled (by a factor of 1.5) and rotated (by a factor of Pi/4) view previously discussed:

```
2012-08-31 09:43:20.837 HelloWorld[41450:c07]
    [1.06066, 1.06066, -1.06066, 1.06066, 0, 0]
```

These raw numbers aren't especially helpful. Specifically, this representation does not tell you exactly how much the view has been scaled or rotated. Fortunately, there's a way around that.

Retrieving Transform Values

You can easily calculate specific transform values from the affine structure's a, b, c, d, tx, and ty entries. Here are five methods that return a view's scale (in X and Y), rotation, and translation (in X and Y). The last two of these methods are admittedly trivial but are included for completeness:

```
- (CGFloat) xscale
{
    CGAffineTransform t = self.transform;
    return sqrt(t.a * t.a + t.c * t.c);
}

- (CGFloat) yscale
{
    CGAffineTransform t = self.transform;
    return sqrt(t.b * t.b + t.d * t.d);
}
```

```
- (CGFloat) rotation
{
    CGAffineTransform t = self.transform;
    return atan2f(t.b, t.a);
}

- (CGFloat) tx
{
    CGAffineTransform t = self.transform;
    return t.tx;
}

- (CGFloat) ty
{
    CGAffineTransform t = self.transform;
    return t.ty;
}
```

Using these methods, you can determine exactly how much your view's been rotated or scaled. That's particularly helpful when you have been using gesture recognizers to interactively stretch, shrink, and rotate views with combined transforms. These methods can help when set bounds on scaling. For example, you might want to limit a view's scale to just twice its normal size or keep it from shrinking below one-half the original size. Checking the current scale lets you do that.

Setting Transform Values

With the right math, it's just as easy to set transform values such as rotation and x-translation as it is to retrieve them. Every transform can be calculated from its components:

```
CGAffineTransform makeTransform(CGFloat xScale, CGFloat yScale,
    CGFloat theta, CGFloat tx, CGFloat ty)
{
    CGAffineTransform transform = CGAffineTransformIdentity;

    transform.a = xScale * cos(theta);
    transform.b = yScale * sin(theta);
    transform.c = xScale * -sin(theta);
    transform.d = yScale * cos(theta);
    transform.tx = tx;
    transform.ty = ty;

    return transform;
}
```

Say you want to set a view's y-scale independently. Here's how you might do that using the view properties defined earlier in this write-up:

```
- (void) setYscale: (CGFloat) yScale
{
    self.transform = makeTransform(self.xscale, yScale,
        self.rotation, self.tx, self.ty);
}
```

Keep in mind that rotating a view that's scaled in just one direction may produce distortion. Figure 4-12 shows two images. The first represents a view with a natural 1:1.5 aspect, (The view is 100 points in width, 150 points in height.) It's been rotated about 45 degrees or so to the right. The second image is a view with a natural 1:1 aspect (100 by 100 points). It's been scaled in Y by 1.5 and then rotated the same 45-or-so-degrees. Notice the distortion. The top-to-bottom Y scaling remains 1.5 (along the top-left corner to right-bottom-corner axis) so the view skews to accommodate.

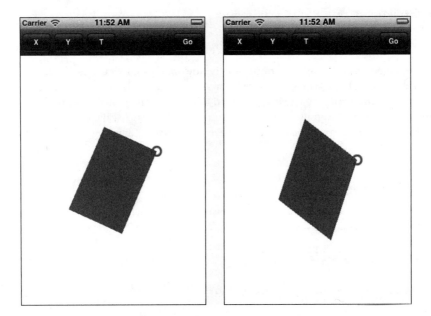

Figure 4-12   Unbalanced scaling may produce skewed views.

Retrieving View Point Locations

In addition to asking, "What is the view's current rotation?" and "By how much is it scaled?," developers perform math that relates a view's post-transform geometry. To do this, you need to specify where frame elements appear onscreen.

A view's center remains meaningful during the transition from pretransform to post-transform without incident. The value may change, especially after scaling, but the property is valid regardless of whatever transform has been applied. This center property always refers to the geometric center of the view's frame within the parent's coordinate system.

The frame is not so resilient. After rotation, a view's origin may be completely decoupled from the view. Look at Figure 4-13. It shows a rotated view on top of its original frame (the smallest of the outlines) and the updated frame (the largest gray outline). The circles indicate the view's top-right corner before and after rotation.

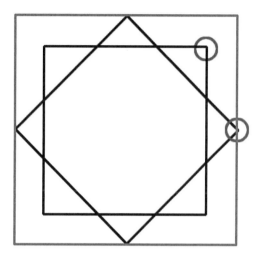

Figure 4-13 A view's frame changes during rotation and may no longer be meaningful. The gray outer rectangle corresponds to the rotated view's frame.

After the transform is applied, the frame updates to the minimum bounding box that encloses the view. Its new origin (the top-left corner of the outside box) has essentially nothing to do with the updated view origin (the top-left corner of the inner box). iOS does not provide a way to retrieve that adjusted point.

Recipe 4-9 defines view methods that perform that math for you. They return a transformed view's corners: top left, top right, bottom left, and bottom right. These coordinates are defined in the parent view; so if you want to add a new view on top of the top circle, you place its center at theView.transformedTopRight.

Recipe 4-9 Transformed View Access

```
// Coordinate utilities
- (CGPoint) offsetPointToParentCoordinates: (CGPoint) aPoint
{
    return CGPointMake(aPoint.x + self.center.x,
        aPoint.y + self.center.y);
}
```

```objc
- (CGPoint) pointInViewCenterTerms: (CGPoint) aPoint
{
    return CGPointMake(aPoint.x - self.center.x,
        aPoint.y - self.center.y);
}

- (CGPoint) pointInTransformedView: (CGPoint) aPoint
{
    CGPoint offsetItem = [self pointInViewCenterTerms:aPoint];
    CGPoint updatedItem = CGPointApplyAffineTransform(
        offsetItem, self.transform);
    CGPoint finalItem =
        [self offsetPointToParentCoordinates:updatedItem];
    return finalItem;
}

- (CGRect) originalFrame
{
    CGAffineTransform currentTransform = self.transform;
    self.transform = CGAffineTransformIdentity;
    CGRect originalFrame = self.frame;
    self.transform = currentTransform;

    return originalFrame;
}

// These four methods return the positions of view elements
// with respect to the current transform

- (CGPoint) transformedTopLeft
{
    CGRect frame = self.originalFrame;
    CGPoint point = frame.origin;
    return [self pointInTransformedView:point];
}

- (CGPoint) transformedTopRight
{
    CGRect frame = self.originalFrame;
    CGPoint point = frame.origin;
    point.x += frame.size.width;
    return [self pointInTransformedView:point];
}

- (CGPoint) transformedBottomRight
```

```
{
    CGRect frame = self.originalFrame;
    CGPoint point = frame.origin;
    point.x += frame.size.width;
    point.y += frame.size.height;
    return [self pointInTransformedView:point];
}

- (CGPoint) transformedBottomLeft
{
    CGRect frame = self.originalFrame;
    CGPoint point = frame.origin;
    point.y += frame.size.height;
    return [self pointInTransformedView:point];
}
```

> **Get This Recipe's Code**
>
> To find this recipe's full sample project, point your browser to https://github.com/erica/iOS-6-Cookbook and go to the folder for Chapter 4.

Recipe: Testing for View Intersection

Recipe 4-10 checks whether two transformed views intersect. The code also works with views that have not been transformed so that you can use it with any two views; although, it's a bit pointless to do so. (You can use the `CGRectIntersectsRect()` function for simple untransformed frames.) This custom intersection method works best for views whose frames do not represent their underlying geometry, like the one shown in Figure 4-13.

The `intersectsView:` method applies an axis separation algorithm for convex polygons. For each edge of each view, it tests whether all the points in one view fall on one side of the edge, and whether all the points of the other view fall on the other side. This test is based on the half plane function, which returns a value indicating whether a point is on the left or right side of an edge.

As soon as it finds an edge that satisfies this condition, the `intersectsView:` method returns NO. The views cannot geometrically intersect if there's a line that separates all the points in one object from all the points in the other.

If all eight tests fail (four edges on the first view, four edges on the second), the method concludes that the two views do intersect. It returns YES.

Recipe 4-10 Retrieving Transform Values

```
// The following three methods move points into and out of the
// transform coordinate system whose origin is at the view center

- (CGPoint) offsetPointToParentCoordinates: (CGPoint) aPoint
{
    return CGPointMake(aPoint.x + self.center.x,
        aPoint.y + self.center.y);
}

- (CGPoint) pointInViewCenterTerms: (CGPoint) aPoint
{
    return CGPointMake(aPoint.x - self.center.x, aPoint.y - self.center.y);
}

- (CGPoint) pointInTransformedView: (CGPoint) aPoint
{
    CGPoint offsetItem = [self pointInViewCenterTerms:aPoint];
    CGPoint updatedItem = CGPointApplyAffineTransform(
        offsetItem, self.transform);
    CGPoint finalItem =
        [self offsetPointToParentCoordinates:updatedItem];
    return finalItem;
}

// Return the original frame without transform
- (CGRect) originalFrame
{
    CGAffineTransform currentTransform = self.transform;
    self.transform = CGAffineTransformIdentity;
    CGRect originalFrame = self.frame;
    self.transform = currentTransform;

    return originalFrame;
}

// These four methods return the positions of view elements
// with respect to the current transform

- (CGPoint) transformedTopLeft
{
    CGRect frame = self.originalFrame;
    CGPoint point = frame.origin;
    return [self pointInTransformedView:point];
}
```

```objc
- (CGPoint) transformedTopRight
{
    CGRect frame = self.originalFrame;
    CGPoint point = frame.origin;
    point.x += frame.size.width;
    return [self pointInTransformedView:point];
}

- (CGPoint) transformedBottomRight
{
    CGRect frame = self.originalFrame;
    CGPoint point = frame.origin;
    point.x += frame.size.width;
    point.y += frame.size.height;
    return [self pointInTransformedView:point];
}

- (CGPoint) transformedBottomLeft
{
    CGRect frame = self.originalFrame;
    CGPoint point = frame.origin;
    point.y += frame.size.height;
    return [self pointInTransformedView:point];
}

// Determine if two views intersect, with respect to any
// active transforms

// After extending a line, determine which side of the half
// plane defined by that line, a point will appear
BOOL halfPlane(CGPoint p1, CGPoint p2, CGPoint testPoint)
{
    CGPoint base = CGPointMake(p2.x - p1.x, p2.y - p1.y);
    CGPoint orthog = CGPointMake(-base.y, base.x);
    return (((orthog.x * (testPoint.x - p1.x)) +
        (orthog.y * (testPoint.y - p1.y))) >= 0);
}

// Utility test for testing view points against a proposed line
BOOL intersectionTest(CGPoint p1, CGPoint p2, UIView *aView)
{
    BOOL tlTest = halfPlane(p1, p2, aView.transformedTopLeft);
    BOOL trTest = halfPlane(p1, p2, aView.transformedTopRight);
    if (tlTest != trTest) return YES;
```

```objc
    BOOL brTest = halfPlane(p1, p2, aView.transformedBottomRight);
    if (tlTest != brTest) return YES;

    BOOL blTest = halfPlane(p1, p2, aView.transformedBottomLeft);
    if (tlTest != blTest) return YES;

    return NO;
}

// Determine whether the view intersects a second view
// with respect to their transforms
- (BOOL) intersectsView: (UIView *) aView
{
    if (!CGRectIntersectsRect(self.frame, aView.frame)) return NO;

    CGPoint A = self.transformedTopLeft;
    CGPoint B = self.transformedTopRight;
    CGPoint C = self.transformedBottomRight;
    CGPoint D = self.transformedBottomLeft;

    if (!intersectionTest(A, B, aView))
    {
        BOOL test = halfPlane(A, B, aView.transformedTopLeft);
        BOOL t1 = halfPlane(A, B, C);
        BOOL t2 = halfPlane(A, B, D);
        if ((t1 != test) && (t2 != test)) return NO;
    }
    if (!intersectionTest(B, C, aView))
    {
        BOOL test = halfPlane(B, C, aView.transformedTopLeft);
        BOOL t1 = halfPlane(B, C, A);
        BOOL t2 = halfPlane(B, C, D);
        if ((t1 != test) && (t2 != test)) return NO;
    }
    if (!intersectionTest(C, D, aView))
    {
        BOOL test = halfPlane(C, D, aView.transformedTopLeft);
        BOOL t1 = halfPlane(C, D, A);
        BOOL t2 = halfPlane(C, D, B);
        if ((t1 != test) && (t2 != test)) return NO;
    }
    if (!intersectionTest(D, A, aView))
    {
        BOOL test = halfPlane(D, A, aView.transformedTopLeft);
        BOOL t1 = halfPlane(D, A, B);
        BOOL t2 = halfPlane(D, A, C);
```

```
            if ((t1 != test) && (t2 != test)) return NO;
    }

    A = aView.transformedTopLeft;
    B = aView.transformedTopRight;
    C = aView.transformedBottomRight;
    D = aView.transformedBottomLeft;

    if (!intersectionTest(A, B, self))
    {
        BOOL test = halfPlane(A, B, self.transformedTopLeft);
        BOOL t1 = halfPlane(A, B, C);
        BOOL t2 = halfPlane(A, B, D);
        if ((t1 != test) && (t2 != test)) return NO;
    }
    if (!intersectionTest(B, C, self))
    {
        BOOL test = halfPlane(B, C, self.transformedTopLeft);
        BOOL t1 = halfPlane(B, C, A);
        BOOL t2 = halfPlane(B, C, D);
        if ((t1 != test) && (t2 != test)) return NO;
    }
    if (!intersectionTest(C, D, self))
    {
        BOOL test = halfPlane(C, D, self.transformedTopLeft);
        BOOL t1 = halfPlane(C, D, A);
        BOOL t2 = halfPlane(C, D, B);
        if ((t1 != test) && (t2 != test)) return NO;
    }
    if (!intersectionTest(D, A, self))
    {
        BOOL test = halfPlane(D, A, self.transformedTopLeft);
        BOOL t1 = halfPlane(D, A, B);
        BOOL t2 = halfPlane(D, A, C);
        if ((t1 != test) && (t2 != test)) return NO;
    }

    return YES;
}
```

Get This Recipe's Code

To find this recipe's full sample project, point your browser to https://github.com/erica/iOS-6-Cookbook and go to the folder for Chapter 4.

Summary

This chapter surveyed a variety of practical geometric recipes related to paths and views. There was, admittedly, a lot of math mixed in with those solutions. Before moving to the next chapter, here are a few final thoughts to take with you:

- A great deal of this chapter is concerned with exposing properties normally hidden from you by the UIKit API. Just because a class expresses itself in a way that makes sense to its developers (transform, A, B, C, D, tx, ty, and so on) does not mean that it is semantically the best fit to its users (rotation, scale, translation, and points). If you can expose the underlying math, feel free to express those classes through categories using terminology that better fits the way people use them.

- Many of the Bezier path examples used in this chapter are based on user tracings on the screen. These recipes aren't limited to that. You can just as easily use these routines with programmatically created paths or ones drawn using third-party tools such as PixelCut's PaintCode ($99, http://paintcodeapp.com).

5
Networking

As an Internet-connected device, the iPhone and its other iOS family members are particularly well suited to retrieving remote data and accessing web-based services. Apple has lavished the platform with a solid grounding in all kinds of network computing and its supporting technologies. The Networking chapter in the *Core Cookbook* introduced network status checks, synchronous and asynchronous downloads, JSON, and XML parsing. This chapter continues that theme by introducing more advanced techniques. These include authentication challenges, using the system keychain, working with OAuth, and so forth. Here are handy approaches that should help with your development.

Recipe: Secure Credential Storage

Before diving into actual networking, consider the role of user credentials. They form a fundamental part of any modern network access scenario. You use credentials to log in to social services, to upload to file-sharing sites, or to retrieve information from a user's account. There are hardly any network development tasks these days that don't involve some sort of username and password situation. For that reason, the first two recipes in this chapter introduce secure credential storage.

iOS's credential system offers a secure mechanism for storing and retrieving usernames and passwords. Credentials are specific to a server or "realm." You don't use the same credentials with Facebook, for example, that you would on Twitter or imgur, or other Web sites. Each server therefore has its own protection "space", and you have access to only credentials that are specific to that host.

You specify the scope of the credentials, establish how long the credentials will last, and allow the system-supplied storage to manage that information. You can use three kinds of persistence with credential storage:

- `NSURLCredentialPersistenceNone` means the credential must be used immediately and cannot be stored for future use.
- `NSURLCredentialPersistenceForSession` enables the credential to be stored for just this application session.

- `NSURLCredentialPersistencePermanent` permits the credential to be stored to the user's keychain and be shared with other applications as well.

Here is how you might create a credential. You create a key/value pair (in the form of username and password) and specify the persistence:

```
NSURLCredential *credential = [NSURLCredential credentialWithUser:key
    password:value persistence: NSURLCredentialPersistencePermanent];
```

Credentials also enable you to store certificates associated with a signing identity and to establish server trust authentication.

Add credentials to storage by assigning them to a protection space. Protection spaces store the authentication details for a single host. Each protection space defines a scope for which the credential is valid. For example, you might create a protection space specific to ericasadun.com as follows:

```
[[NSURLProtectionSpace alloc] initWithHost:@"ericasadun.com" port:0
    protocol:@"http" realm:nil authenticationMethod:nil]
```

Setting the port to 0 means Use the Default Port. For http, this defaults to port 80. Pass a different value if needed. The protocol defines the scheme used to access resources: via http, https, ftp, and so forth.

You save credentials to a central credential storage, which is a system-supplied singleton. Here's how you might store a credential for a particular protection space. The "setting" method saves that credential for you:

```
[[NSURLCredentialStorage sharedCredentialStorage]
    setCredential:credential forProtectionSpace:self.protectionSpace];
```

Recipe 5-1 builds a helper class that stores and retrieves secure information. Each instance is primed with a specific host, and its protection space is tied to that host. With the helper, you can add new credentials, set new values for exist credentials, remove credentials, and basically treat the credential storage system as a dictionary.

This dictionary support is provided because the helper class implements keyed subscript methods. This means you can use the class as follows, taking advantage of Objective-C literal shortcuts:

```
CredentialHelper *helper =
    [CredentialHelper helperWithHost:@"ericasadun.com"];
helper[@"John Doe"] = @"FooBarGaz";
NSLog(@"%@", helper[@"John Doe"]);
```

The protection space details are encapsulated in the helper, making this a simple solution for most secure solution needs.

Recipe 5-1   **Credential Helper**

```
@implementation CredentialHelper

// Create a helper specific to a supplied host
+ (id) helperWithHost: (NSString *) host
{
    CredentialHelper *helper = [[CredentialHelper alloc] init];
    helper.host = host;
    return helper;
}

// Return the current protection space for the host
- (NSURLProtectionSpace *) protectionSpace
{
    if (!_host) return nil;
    return [[NSURLProtectionSpace alloc] initWithHost:_host
        port:0 protocol:@"http" realm:nil authenticationMethod:nil];
}

// Return all the credentials for the supplied protection space
- (NSDictionary *) credentials
{
    if (!_host) return nil;

    NSDictionary *credentials =
        [[NSURLCredentialStorage sharedCredentialStorage]
            credentialsForProtectionSpace:self.protectionSpace];
    return credentials;
}

// Store a credential
- (void) storeCredential: (NSString *) value forKey: (NSString *) key
{
    if (!_host)
    {
        NSLog(@"Error: Cannot store credential for nil host");
        return;
    }

    NSURLCredential *credential =
        [NSURLCredential credentialWithUser:key password:value
            persistence: NSURLCredentialPersistencePermanent];
    [[NSURLCredentialStorage sharedCredentialStorage]
        setCredential:credential
            forProtectionSpace:self.protectionSpace];
}
```

```objc
// Store a credential but make it act as the default
- (void) storeDefaultCredential: (NSString *) value
    forKey: (NSString *) key
{
    if (!_host)
    {
        NSLog(@"Error: Cannot store credential for nil host");
        return;
    }

    NSURLCredential *credential =
        [NSURLCredential credentialWithUser:key password:value
            persistence: NSURLCredentialPersistencePermanent];
    [[NSURLCredentialStorage sharedCredentialStorage]
        setDefaultCredential:credential
        forProtectionSpace:self.protectionSpace];
}

// Return the default credential
- (NSURLCredential *) defaultCredential
{
    return [[NSURLCredentialStorage sharedCredentialStorage]
        defaultCredentialForProtectionSpace:self.protectionSpace];
}

// Remove a credential associated with a username (key)
- (void) removeCredential: (NSString *) key
{
    NSArray *keys = self.credentials.allKeys;
    if (![keys containsObject:key])
    {
        NSLog(@"Key %@ not found in credentials. \
            Skipping remove request.", key);
        return;
    }
    [[NSURLCredentialStorage sharedCredentialStorage]
        removeCredential:self.credentials[key]
        forProtectionSpace:self.protectionSpace];
}

// Remove all credentials associated with the protection space
- (void) removeAllCredentials
{
    NSArray *keys = self.credentials.allKeys;
    for (NSString *key in keys)
        [self removeCredential:key];
}
```

```objc
// Return a credential for a given user (key)
- (NSURLCredential *) credentialForKey: (NSString *) key
{
    if (!_host) return nil;
    return self.credentials[key];
}

// Return the password (value) for a given user (key)
- (NSString *) valueForKey:(NSString *)key
{
    NSURLCredential *credential = [self credentialForKey:key];
    if (!credential) return nil;
    return credential.password;
}

// Provide indexed access to the helper, e.g. helper[username]
// and helper[username] = value
- (id) objectForKeyedSubscript: (NSString *) key
{
    return [self valueForKey:key];
}
- (void) setObject: (NSString *) newValue
    forKeyedSubscript: (NSString *) aKey
{
    [self storeCredential: newValue forKey: aKey];
}
@end
```

> **Get This Recipe's Code**
>
> To find this recipe's full sample project, point your browser to https://github.com/erica/iOS-6-Advanced-Cookbook and go to the folder for Chapter 5.

Recipe: Entering Credentials

There's a lot you can do to simplify your user's life when it comes to entering and retrieving credentials. Proactively auto-filling passwords is part of that. Recipe 5-2 demonstrates how.

Recipe 5-2 uses the credential storage mechanism introduced in Recipe 5-1 to create a secure password entry form. As its view appears, the recipe loads any default credentials if available. It then sets those in the username and password fields. Upon being dismissed with *Done*, the recipe stores any updates. If the user taps Cancel rather than Done, the dialog is dismissed without saving.

The `storeCredentials` method always sets the default credential. The most recently updated username/password pair is saved (or resaved) to the keychain and marked as the default.

The recipe implements several text field delegate methods as well. If users tap Done on the keyboard, it's treated as if they had tapped Done on the navigation bar. The text field resigns first responder, the data is stored, and the password-entry view controller is dismissed.

Another text field delegate method looks for the start of editing. Whenever the user edits the name field, the password is invalidated and cleared. At the same time, the should-change-characters method examines any text in the username field. When it finds a credential that matches the new username, it automatically updates the password field. This allows users to switch between several usernames for a given service without retyping in the password each time they do so.

You can easily expand this recipe to list known usernames to offer one-tap access to stored credentials.

Recipe 5-2 Password Entry View Controller

```
@implementation PasswordBaseController
{
    UITextField *userField;
    UITextField *passField;
    CredentialHelper *helper;
}

- (id) initWithHost: (NSString *) hostName
{
    if (self = [super init])
    {
        _host = hostName;
        helper = [CredentialHelper helperWithHost:hostName];
    }
    return self;
}

- (void) listCredentials
{
    // Never log passwords in production code
    NSLog(@"Protection space for %@ has %d credentials:",
        _host, helper.credentialCount);
    for (NSString *userName in helper.credentials.allKeys)
        NSLog(@"%@: %@", userName, helper[userName]);
}

// Save the current credentials as default
- (void) storeCredentials
```

```objc
{
    if (!userField.text.length) return;
    [helper storeDefaultCredential:passField.text
        forKey:userField.text];
    [self listCredentials];
}

// Remove the currently-displayed credential
// from the credential storage
- (void) remove
{
    [helper removeCredential:userField.text];
    [self listCredentials];

    // Update GUI
    userField.text = @"";
    passField.text = @"";
    UIBarButtonItem *removeButton = self.navigationItem.leftBarButtonItems[1];
    removeButton.enabled = NO;
}

// Finish and store credentials
- (void) done
{
    [self storeCredentials];
    [self dismissViewControllerAnimated:YES completion:nil];
}

// Finish without updating credentials
- (void) cancel
{
    [self dismissViewControllerAnimated:YES completion:nil];
}

#pragma mark - Text Edits

// User tapping Done confirms changes. Store credentials
- (BOOL)textFieldShouldReturn:(UITextField *)textField
{
    [textField resignFirstResponder];
    [self done];
    return YES;
}

// Only enable Cancel on edits
- (void)textFieldDidBeginEditing:(UITextField *)textField
```

```objc
{
    UIBarButtonItem *cancelButton = self.navigationItem.leftBarButtonItems[0];
    cancelButton.enabled = YES;
}

- (BOOL)textFieldShouldClear:(UITextField *)textField
{
    // Empty the passfield upon a username clear
    if (textField == userField)
        passField.text = @"";
    return YES;
}

// Watch for known usernames during text edits
- (BOOL)textField:(UITextField *)textField
    shouldChangeCharactersInRange:(NSRange)range
    replacementString:(NSString *)string
{
    if (textField != userField) return YES;

    // Initially disable remove until there's a credential match
    UIBarButtonItem *removeButton =
        self.navigationItem.leftBarButtonItems[1];
    removeButton.enabled = NO;

    // Preemptively clear password field until there's a value for it
    passField.text = @"";

    // Calculate the target string that will occupy the username field
    NSString *username = [textField.text
        stringByReplacingCharactersInRange:range withString:string];
    if (!username) return YES;
    if (!username.length) return YES;

    // Always check if there's a matching password on file
    NSURLCredential *credential = [helper credentialForKey:username];
    if (!credential)
        return YES;

    // Match!
    passField.text = credential.password;
    removeButton.enabled = YES;

    // Never log passwords in production code!
    NSLog(@"Found match: %@: %@", username, passField.text);

    return YES;
```

```objc
}

#pragma mark - Load Defaults

- (void) viewWillAppear:(BOOL)animated
{
    // Disable the cancel button, there are no edits to cancel
    UIBarButtonItem *cancelButton =
        self.navigationItem.leftBarButtonItems[0];
    cancelButton.enabled = NO;

    // Disable the remove button, until a credential has been matched
    UIBarButtonItem *removeButton =
        self.navigationItem.leftBarButtonItems[1];
    removeButton.enabled = NO;

    NSURLCredential *credential = helper.defaultCredential;
    if (credential)
    {
        // Populate the fields
        userField.text = credential.user;
        passField.text = credential.password;

        // Enable credential removal
        removeButton.enabled = YES;
    }
}

- (void) loadView
{
    [super loadView];
    self.navigationItem.rightBarButtonItem =
        BARBUTTON(@"Done", @selector(done));
    self.navigationItem.leftBarButtonItems =  @[
    BARBUTTON(@"Cancel", @selector(cancel)),
    BARBUTTON(@"Remove", @selector(remove)),
    BARBUTTON(@"List", @selector(listCredentials)),
    ];

    userField = [self textField];
    userField.placeholder = @"User Name";

    passField = [self textField];
    passField.secureTextEntry = YES;
    passField.placeholder = @"Password";
}
@end
```

Get This Recipe's Code

To find this recipe's full sample project, point your browser to https://github.com/erica/iOS-6-Advanced-Cookbook and go to the folder for Chapter 5.

Recipe: Handling Authentication Challenges

Some Web sites are protected with usernames and passwords. `NSURLConnection` helps you access pages by responding to a site's authentication challenges. To test authentication, Recipe 5-3 connects to http://ericasadun.com/Private, which was set up for use with this authentication discussion. This test folder uses the username `PrivateAccess` and password `tuR7!mZ#eh`.

To test an unauthorized connection—that is, you will be refused—set the password to `nil` or to a nonsense string. When the password is set to `nil`, the challenge will be sent a `nil` credential, producing an immediate failure. With nonsense strings, the challenge will fail after the server rejects the credentials.

Each time the challenge fails, the connection's `previousFailureCount` property increases by one. You can easily check this count to see if you've already failed a challenge and, if so, implement some kind of can't-connect handler. You might, for example, want to request a different password to try.

This recipe does two things to demonstrate failed and successful authentications that you will not want to do in your own code:

- It clears the shared URL cache after each connection to ensure that your request can fail after a success. If you do not do so, the cache remembers the authentication and continues connecting even if you send garbage down the line. Don't put this into production code because users expect authentication to persist.

- The credentials used here are created on-demand and use a persistence of None. Again, this is not an entirely realistic use unless you want your users to enter their credentials for every web page access. That would not be a successful or satisfying browsing experience.

Recipe 5-3   Authentication with NSURLCredential Instances

```
- (void)connection:(NSURLConnection *)connection
    didReceiveData:(NSData *)data
{
    // Load the data into the view
    NSString *source = DATASTR(data);
    [webView loadHTMLString:source baseURL:_url];

    // Force clean the cache -- so you can "fail" after success
    [[NSURLCache sharedURLCache] removeAllCachedResponses];
}
```

```objc
- (void)connection:(NSURLConnection *) connection
    willSendRequestForAuthenticationChallenge:
        (NSURLAuthenticationChallenge *)challenge
{
    if (!challenge.previousFailureCount)
    {
        // Build a one-time use credential
        NSURLCredential *credential =
            [NSURLCredential credentialWithUser:@"PrivateAccess"
                password:_shouldFail ? @"foo" : @"tuR7!mZ#eh"
                persistence:NSURLCredentialPersistenceNone];
        [challenge.sender useCredential:credential
            forAuthenticationChallenge:challenge];
    }
    else
    {
        // Stop challenge after first failure
        [challenge.sender cancelAuthenticationChallenge:challenge];
        [webView loadHTMLString:@"<h1>Failed</h1>" baseURL:nil];
    }
}


- (BOOL)connectionShouldUseCredentialStorage:(NSURLConnection *)connection;
{
    NSLog(@"Being queried about credential storage. Saying no.");
    return NO;
}
```

> **Get This Recipe's Code**
>
> To find this recipe's full sample project, point your browser to https://github.com/erica/iOS-6-Advanced-Cookbook and go to the folder for Chapter 5.

Recipe: Uploading Data

Recipe 5-4 uses POST to create a multipart form data submission. It uploads images to the imgur.com service. Imgur offers anonymous image uploading without credentials, so it offers a particularly good example for simple posting. "Anonymous" refers to your users and not to you as a developer. An anonymous upload enables you to upload images to imgur without those images being tied to a specific user account. As a developer, you still need to register for the API and use it to place your requests. Request your API key at https://imgur.com/register/api_anon. Make sure to add your key to the sample code for Recipe 5-4.

The challenge for Recipe 5-4 is to create a properly formatted body that can be used by the imgur service. It implements a method that generates form data from a dictionary of keys and values. For the purposes of this example, the objects in that dictionary are limited to strings and images. You can extend this approach for other data types by changing the content type string with different MIME types. It's obviously not an ideal way to create this data, but it's been getting the job done for many years now.

This recipe uses a synchronous request to perform the upload, which can take up to a minute or so to process. To avoid blocking GUI updates, the entire submission process is embedded into an `NSOperation` subclass. Operations encapsulate code and data for a single task, enabling you to run that task asynchronously.

Using `NSOperation` objects lets you submit them to an asynchronous `NSOperationQueue`. Operation queues manage the execution of individual operations. Each operation is prioritized and placed into the queue, where it is executed in that priority order.

Whenever you subclass `NSOperation`, make sure to implement a `main` method. This method is called when the operation executes. When `main` returns, the operation finishes.

Recipe 5-4   **Uploading Images to imgur**

```
#define NOTIFY_AND_LEAVE(MESSAGE) {[self bail:MESSAGE]; return;}
#define STRDATA(STRING) \
    ([STRING dataUsingEncoding:NSUTF8StringEncoding])
#define SAFE_PERFORM_WITH_ARG(THE_OBJECT, THE_SELECTOR, THE_ARG)\
    (([THE_OBJECT respondsToSelector:THE_SELECTOR]) ? \
        [THE_OBJECT performSelectorOnMainThread:THE_SELECTOR \
            withObject:THE_ARG waitUntilDone:NO] : nil)

// Form data constants
#define IMAGE_CONTENT(_FILENAME_) \
    @"Content-Disposition: form-data; name=\"%@\";\
filename=\"_FILENAME_\"\r\nContent-Type: image/jpeg\r\n\r\n"
#define STRING_CONTENT \
    @"Content-Disposition: form-data; name=\"%@\"\r\n\r\n"
#define MULTIPART \
    @"multipart/form-data; boundary=------------0x0x0x0x0x0x0x0x"


@implementation ImgurUploadOperation

// Operation failure. Send an error off to the delegate.
- (void) bail: (NSString *) message
{
    SAFE_PERFORM_WITH_ARG(_delegate,
        @selector(handleImgurOperationError:), message);
}
```

```objc
// Create multipart data from a dictionary
- (NSData*)generateFormDataFromPOSTDictionary:(NSDictionary*)dict
{
    NSString *boundary = @"------------0x0x0x0x0x0x0x0x";
    NSArray *keys = [dict allKeys];
    NSMutableData *result = [NSMutableData data];

    for (int i = 0; i < keys.count; i++)
    {
        // Start part
        id value = dict[keys[i]];
        NSString *start =
            [NSString stringWithFormat:@"--%@\r\n", boundary];
        [result appendData:STRDATA(start)];

        if ([value isKindOfClass:[NSData class]])
        {
            // handle image data
            NSString *formstring =
                [NSString stringWithFormat:IMAGE_CONTENT(@"Cookbook.jpg"),
                    [keys objectAtIndex:i]];
            [result appendData:STRDATA(formstring)];
            [result appendData:value];
        }
        else
        {
            // all non-image fields assumed to be strings
            NSString *formstring =
                [NSString stringWithFormat:STRING_CONTENT,
                    [keys objectAtIndex:i]];
            [result appendData: STRDATA(formstring)];
            [result appendData:STRDATA(value)];
        }

        // End of part
        NSString *formstring = @"\r\n";
        [result appendData:STRDATA(formstring)];
    }

    // End of form
    NSString *formstring =
        [NSString stringWithFormat:@"--%@--\r\n", boundary];
    [result appendData:STRDATA(formstring)];
    return result;
}
```

```objc
- (void) main
{
    if (!_image)
        NOTIFY_AND_LEAVE(@"ERROR: Please set image before uploading.");
    
    // Establish the post dictionary contents
    NSMutableDictionary *postDictionary =
        [NSMutableDictionary dictionary];
    postDictionary[@"key"] = IMGUR_API_KEY;
    postDictionary[@"title"] = @"Random Image";
    postDictionary[@"caption"] =
        @"Created by the iOS Developer's Cookbook";
    postDictionary[@"type"] = @"base64";
    postDictionary[@"image"] =
        [UIImageJPEGRepresentation(_image, 0.65)
            base64EncodedString];
    
    // Create the post data from the post dictionary
    NSData *postData =
        [self generateFormDataFromPOSTDictionary:postDictionary];
    
    // Establish the API request.
    NSString *baseurl = @"http://api.imgur.com/2/upload.json";
    NSURL *url = [NSURL URLWithString:baseurl];
    NSMutableURLRequest *urlRequest =
        [NSMutableURLRequest requestWithURL:url];
    if (!urlRequest)
        NOTIFY_AND_LEAVE(@"ERROR: Error creating the URL Request");
    
    [urlRequest setHTTPMethod: @"POST"];
    [urlRequest setValue:MULTIPART
        forHTTPHeaderField: @"Content-Type"];
    [urlRequest setHTTPBody:postData];
    
    // Submit & retrieve results
    NSError *error;
    NSURLResponse *response;
    NSData* result = [NSURLConnection
        sendSynchronousRequest:urlRequest
        returningResponse:&response error:&error];
    if (!result)
    {
        [self bail:[NSString stringWithFormat:
            @"Submission error: %@", error.localizedFailureReason]];
        return;
    }
```

```objc
    // Success. Return results
    SAFE_PERFORM_WITH_ARG(_delegate,
        @selector(finishedImgurOperationWithData:), result);
}

// Helper method to return a pre-populated operation
+ (id) operationWithDelegate: (id <ImgurUploadOperationDelegate>) delegate andImage:
(UIImage *) image
{
    ImgurUploadOperation *op = [[ImgurUploadOperation alloc] init];
    op.delegate = delegate;
    op.image= image;
    return op;
}
@end
```

> **Get This Recipe's Code**
>
> To find this recipe's full sample project, point your browser to https://github.com/erica/iOS-6-Advanced-Cookbook and go to the folder for Chapter 5.

Recipe: Building a Simple Web Server

A web server provides one of the cleanest ways to serve data off your phone to another computer on the same network. You don't need special client software. Any browser can list and access web-based files. Best of all, a web server requires just a few key routines. You must establish the service, creating a loop that listens for a request (`startServer`), and then pass those requests onto a handler (`handleWebRequest:`) that responds with the requested data. Recipe 5-5 shows a `WebHelper` class that handles establishing and controlling a basic web service that serves the same image currently shown on the iOS device screen.

The loop routine uses low-level socket programming to establish a listening port and catch client requests. When the client issues a GET command, the server intercepts that request and passes it to the web request handler. The handler could decompose it, typically to find the name of the desired data file, but in this example, it serves back an image, regardless of the request specifics.

Recipe 5-5 Serving iPhone Files Through a Web Service

```objc
#define SAFE_PERFORM_WITH_ARG(THE_OBJECT, THE_SELECTOR, THE_ARG)\
    (([THE_OBJECT respondsToSelector:THE_SELECTOR]) ? \
    [THE_OBJECT performSelectorOnMainThread:THE_SELECTOR \
        withObject:THE_ARG waitUntilDone:NO] : nil)
```

```
@implementation WebHelper
// Process the external request by sending an image
// (Customize this to do something more interesting.)
- (void) handleWebRequest: (int) fd
{
    // Request an image from the delegate
    if (!_delegate) return;
    if (![_delegate respondsToSelector:@selector(image)]) return;
    UIImage *image = (UIImage *)[_delegate performSelector:@selector(image)];
    if (!image) return;

    // Produce a jpeg header
    NSString *outcontent = [NSString stringWithFormat:
        @"HTTP/1.0 200 OK\r\nContent-Type: image/jpeg\r\n\r\n"];
    write (fd, [outcontent UTF8String], outcontent.length);

    // Send the data and close
    NSData *data = UIImageJPEGRepresentation(image, 0.75f);
    write (fd, data.bytes, data.length);
    close(fd);
}

// Listen for external requests
- (void) listenForRequests
{
    @autoreleasepool {
        static struct sockaddr_in cli_addr;
        socklen_t length = sizeof(cli_addr);

        while (1) {
            if (!isServing) return;

            if ((socketfd = accept(listenfd,
                (struct sockaddr *)&cli_addr, &length)) < 0)
            {
                isServing = NO;
                [[NSOperationQueue mainQueue]
                    addOperationWithBlock:^(){
                        SAFE_PERFORM_WITH_ARG(delegate,
                        @selector(serviceWasLost), nil);
                }];
                return;
            }
            [self handleWebRequest:socketfd];
        }
    }
}
```

```
// Begin serving data
- (void) startServer
{
    static struct sockaddr_in serv_addr;

    // Set up socket
    if((listenfd = socket(AF_INET, SOCK_STREAM,0)) < 0)
    {
        isServing = NO;
        SAFE_PERFORM_WITH_ARG(delegate,
            @selector(serviceCouldNotBeEstablished), nil);
        return;
    }

    // Serve to a random port
    serv_addr.sin_family = AF_INET;
    serv_addr.sin_addr.s_addr = htonl(INADDR_ANY);
    serv_addr.sin_port = 0;

    // Bind
    if (bind(listenfd, (struct sockaddr *)&serv_addr,
        sizeof(serv_addr)) <0)
    {
        isServing = NO;
        SAFE_PERFORM_WITH_ARG(delegate,
            @selector(serviceCouldNotBeEstablished), nil);
        return;
    }

    // Find out what port number was chosen.
    int namelen = sizeof(serv_addr);
    if (getsockname(listenfd, (struct sockaddr *)&serv_addr,
        (void *) &namelen) < 0) {
        close(listenfd);
        isServing = NO;
        SAFE_PERFORM_WITH_ARG(delegate,
            @selector(serviceCouldNotBeEstablished), nil);
        return;
    }

    chosenPort = ntohs(serv_addr.sin_port);

    // Listen
    if(listen(listenfd, 64) < 0)
    {
        isServing = NO;
        SAFE_PERFORM_WITH_ARG(delegate,
```

```
                @selector(serviceCouldNotBeEstablished), nil);
        return;
    }

    isServing = YES;
    [NSThread
        detachNewThreadSelector:@selector(listenForRequests)
        toTarget:self withObject:NULL];
    SAFE_PERFORM_WITH_ARG(delegate,
        @selector(serviceWasEstablished:), self);
}

- (void) stopService
{
    printf("Shutting down service\n");
    _isServing = NO;
    close(listenfd);
    SAFE_PERFORM_WITH_ARG(_delegate, @selector(serviceDidEnd), nil);
}

+ (id) serviceWithDelegate:(id)delegate
{
    if (![[UIDevice currentDevice] networkAvailable])
    {
        NSLog(@"Not connected to network");
        return nil;
    }

    WebHelper *helper = [[WebHelper alloc] init];
    helper.delegate = delegate ;
    [helper startServer];
    return helper;
}
@end
```

> **Get This Recipe's Code**
>
> To find this recipe's full sample project, point your browser to https://github.com/erica/iOS-6-Advanced-Cookbook and go to the folder for Chapter 5.

Recipe: OAuth Utilities

OAuth is the bane of many an iOS developer. Developed as an open standard for protecting user credentials, it provides a secure way to grant sensitive access on a limited basis. Many

Recipe: OAuth Utilities

popular APIs are based around OAuth and its controversial successor OAuth 2.0, which remains under development. For most developers, OAuth is fussy, detail-oriented, and a pain to use.

Apple has simplified some of the OAuth situation with its new Accounts framework, but at the time of this writing, the framework is limited to Facebook, Sina Weibo, and Twitter. In addition, a private OAuth support framework remains out of bounds for developer API calls.

Recipe 5-6 introduces a number of utilities that may help you when working with OAuth. Its code base goes back some number of years and is built around the CommonCrypto library. It builds an OAuthRequestSigner class, whose job it is to build, sign, and encode requests. This implementation is hard-coded to use HMAC-SHA1 signing.

Recipe 5-6 forms the basis for all the signing requests discussed in Recipe 5-7, which introduces the OAuth token exchange process.

> **Note**
>
> The sample code for Recipe 5-6 uses imgur OAuth credentials. Register for your keys at https://imgur.com/register/api_oauth. You will need these to compile and run the sample code.

Recipe 5-6 Basic OAuth Signing Utilities

```
#import <CommonCrypto/CommonHMAC.h>

@implementation OAuthRequestSigner
// Sign the clear text with the secret key
+ (NSString *) signClearText: (NSString *)text
        withKey: (NSString *) secret
{
    NSData *secretData = STRDATA(secret);
    NSData *clearTextData = STRDATA(text);

    //HMAC-SHA1
    CCHmacContext hmacContext;
    uint8_t digest[CC_SHA1_DIGEST_LENGTH] = {0};
    CCHmacInit(&hmacContext, kCCHmacAlgSHA1,
        secretData.bytes, secretData.length);
    CCHmacUpdate(&hmacContext,
        clearTextData.bytes, clearTextData.length);
    CCHmacFinal(&hmacContext, digest);

    // Convert to a base64-encoded result,
    // Thanks to Matt Gallagher's NSData category
    NSData *out = [NSData dataWithBytes:digest
        length:CC_SHA1_DIGEST_LENGTH];
```

```objc
        return [out base64EncodedString];
}

// RFC 3986
+ (NSString *) urlEncodedString: (NSString *) string
{
    NSString *result = (__bridge_transfer NSString *)
        CFURLCreateStringByAddingPercentEscapes(
            kCFAllocatorDefault, (__bridge CFStringRef)string,
            NULL,  CFSTR(":/?#[]@!$&'()*+,;="),
            kCFStringEncodingUTF8);
    return result;
}

// Return url-encoded signed request
+ (NSString *) signRequest: (NSString *)
    baseRequest withKey: (NSString *) secret
{
    NSString *signedRequest = [OAuthRequestSigner
        signClearText:baseRequest withKey:secret];
    NSString *encodedRequest = [OAuthRequestSigner
        urlEncodedString:signedRequest];
    return encodedRequest;
}

// Return a nonce (á random value)
+ (NSString *) oauthNonce;
{
    CFUUIDRef theUUID = CFUUIDCreate(NULL);
    NSString *nonceString =
        (__bridge_transfer NSString *)CFUUIDCreateString(
            NULL, theUUID);
    CFRelease(theUUID);
    return nonceString;
}

// Build a token dictionary from a
// key=value&key=value&key=value string
+ (NSDictionary *) dictionaryFromParameterString:
    (NSString *) resultString
{
    if (!resultString) return nil;
    NSMutableDictionary *tokens = [NSMutableDictionary dictionary];
    NSArray *pairs = [resultString componentsSeparatedByString:@"&"];
    for (NSString *pairString in pairs)
```

```objc
    {
        NSArray *pair =
            [pairString componentsSeparatedByString:@"="];
        if (pair.count != 2) continue;
        tokens[pair[0]] = pair[1];
    }
    return tokens;
}

// Build a string from an oauth dictionary
+ (NSString *) parameterStringFromDictionary: (NSDictionary *) dict
{
    NSMutableString *outString = [NSMutableString string];

    // Sort keys
    NSMutableArray *keys =
        [NSMutableArray arrayWithArray:[dict allKeys]];
    [keys sortUsingSelector:@selector(caseInsensitiveCompare:)];

    // Add sorted items to parameter string
    for (int i = 0; i < keys.count; i++)
    {
        NSString *key = keys[i];
        [outString appendFormat:@"%@=%@", key, dict[key]];
        if (i < (keys.count - 1))
            [outString appendString:@"&"];
    }

    return outString;
}

// Create a base oauth (header) dictionary
+ (NSMutableDictionary *) oauthBaseDictionary: (NSString *) consumerKey;
{
    NSMutableDictionary *dict = [NSMutableDictionary dictionary];
    dict[@"oauth_consumer_key"] = consumerKey;
    dict[@"oauth_nonce"] = [OAuthRequestSigner oauthNonce];
    dict[@"oauth_signature_method"] = @"HMAC-SHA1";
    dict[@"oauth_timestamp"] =
        [NSString stringWithFormat:@"%d", (int)time(0)];
    dict[@"oauth_version"] = @"1.0";
    return dict;
}
```

```objc
+ (NSMutableString *) baseRequestWithEndpoint: (NSString *) endPoint
    dictionary: (NSDictionary *)dict
    andRequestMethod: (NSString *) method
{
    NSMutableString *baseRequest = [NSMutableString string];
    NSString *encodedEndpoint =
        [OAuthRequestSigner urlEncodedString:endPoint];
    [baseRequest appendString:
        [NSString stringWithFormat:@"%@&%@&",
            method, encodedEndpoint]];
    NSString *baseParameterString =
        [OAuthRequestSigner parameterStringFromDictionary:dict];
    NSString *encodedParamString =
        [OAuthRequestSigner urlEncodedString:baseParameterString];
    [baseRequest appendString:encodedParamString];
    return baseRequest;
}
@end
```

> **Get This Recipe's Code**
>
> To find this recipe's full sample project, point your browser to https://github.com/erica/iOS-6-Advanced-Cookbook and go to the folder for Chapter 5.

Recipe: The OAuth Process

Working with OAuth generally requires five steps before you can access a service. These steps enable you to authenticate your application with the provider and ensure that the user has properly granted permission to the application. The following sections detail how these steps proceed in a standard OAuth application.

Step 1: Request Tokens from the API Provider

To start, your application needs base OAuth tokens to work with. The application posts a request to a server's token endpoint, for example, https://api.imgur.com/oauth/request_token.

This request includes the application's unique consumer key ("oauth_consumer_key") supplied by the API provider, and the standard elements of an OAuth header dictionary. These are a nonce (a unique nonsense string, "oauth_nonce"), a timestamp ("oauth_timestamp"), a signature method ("oauth_signature_method" such as HMAC-SHA1), and a version ("oauth_version").

You construct a parameter string from these items and sign it with the application's secret key. Pro tip: Make sure you append "&" to the key. If the key is "XYZZYPLUGH", you sign with "XYZZYPLUGH&". That's because the signing key is always composed of a combination of the application (consumer) secret key and the end-user (token) secret key, and at this stage of the process, the token secret key remains as-yet undefined.

You add the signature to the request and send it up to the API provider. The `requestTokens:` method in Recipe 5-7 demonstrates this process.

Step 2: Retrieve and Store Tokens

Upon receiving an authenticated token request, the service sends back a number of items in its response. These include an important pair, specifically a user token ("oauth_token") and a user secret token ("oauth_token_secret"). You extract these values by decomposing the parameter-encoded string returned by the service to its component keys and values. Recipe 5-7's `processTokens:` method shows how this is done.

Because the user has not yet granted access for your application to use his credentials, these are temporary items. They enable your app to move to the next stage of the process but no further. Your app cannot use them to make general requests. Store these items securely in the user keychain, and not in user defaults.

Step 3: Request User Access

Users must now visit a web page and authorize your application. You construct the URL from an API endpoint and the OAuth user token returned in step 2. The URL will look something like this, but the specific URL will vary by provider:

https://api.imgur.com/oauth/authorize?oauth_token=*token*

This URL includes a base authorization endpoint that uses the token returned by step 2 of this process. Figure 5-1 shows the imgur version of this authorization screen. Control passes to a web page, and your app is responsible for monitoring and retrieving a verifier from this process.

In the simplest possible case, you can open the URL in Safari or direct your user to visit a Web site and enter the URL. Most iOS developers would prefer to keep control within the app, however, and should supply an in-app `UIWebView` to display the authentication screen, as shown in Figure 5-1.

Step 4: Retrieve an OAuth Verifier Token

The API authorization web interface concludes its interaction by providing a verification code, as shown in Figure 5-2. The authentication process is conducted entirely in HTML and lies outside of standard Objective-C interaction.

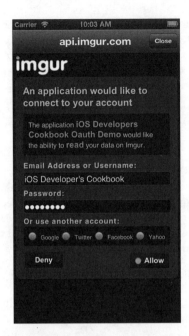

Figure 5-1 Users must explicitly grant access to OAuth-based applications to allow API calls.

Figure 5-2 Users must explicitly grant access to OAuth-based applications to allow API calls.

Your app should either instruct users to copy this code so that they can type it in or should retrieve it directly from the page source using JavaScript, for example:

```
// Look for the "Your verification code is" pattern that imgur provides
- (void)webViewDidFinishLoad: (UIWebView *)aWebView
{
    // Retrieve HTML
    NSString *string = [webView
        stringByEvaluatingJavaScriptFromString:
            @"document.getElementById('content').innerHTML;"];

    // Check for code
    NSString *searchString = @"Your verification code is";
    if ([string rangeOfString:searchString].location == NSNotFound)
        return;

    // Split into lines
    NSArray *lines = [string componentsSeparatedByCharactersInSet:
        [NSCharacterSet newlineCharacterSet]];
    for (NSString *line in lines)
    {
        // Search each line for code
        if ([line rangeOfString:searchString].location == NSNotFound)
            continue;
        NSString *code = [self scanForCode:line];
        if (code)
        {
            // Store the credential
            CredentialHelper *helper =
                [CredentialHelper helperWithHost:HOST];
            [helper storeCredential:code forKey:@"oauth_verifier"];
        }
    }

    // Automatically close the window
    [self close];
}
```

As you can tell, this HTML/Objective-C hybridization is ugly, hard-to-maintain, and subject to failure should the API-provider change any wording. That's why some providers now offer xAuth access instead. xAuth enables apps to directly supply the same credentials that users normally type into an OAuth authentication web page. That's handy for iOS apps because you don't have to display a web page and then use JavaScript to scrape for results. Sadly, many providers are moving away from xAuth.

You supply the username ("x_auth_username"), password ("x_auth_password"), the OAuth token, and set an authentication mode ("x_auth_mode") such as "client_auth", for client authentication. This enables you to bypass the web page authentication phase entirely and provide a seamless experience for the application user.

Not all API providers offer xAuth options, and those that do tend to limit the scope of that access to guard against abuse. Check with your API provider to see if xAuth is an option for your application.

Step 5: Authenticate with the Access Token

Use the verification code returned in step 4 to complete the authentication process, as shown in Recipe 5-7's `authenticate:` method. Submit a final token request to your API provider, supplying the verifier as part of your request. This request uses a different endpoint from step 1, for example:

> https://api.imgur.com/oauth/access_token

Your request contains both the code ("oauth_verifier") and the OAuth token returned in step 2.

The way you sign your requests now changes. The first token request in step 2 was signed only with the application's consumer secret key. Starting with step 5, and continuing from there, all requests must be signed with both the consumer secret key and the OAuth secret token (consumer_secret&user_token_secret). What's more, this step returns a new OAuth secret token.

That means the secret token returned in step 2 is used exactly once, to request a new pair of keys. This step returns a new OAuth token and a new OAuth secret token. Discard your previous values; replace them with these new ones. Make sure they're stored to the secure keychain and not to user defaults. Your application is now authorized and ready to submit requests.

> **Note**
> Always use the keychain to save your OAuth tokens. Never store OAuth credentials to end user-readable places such as plain-text files, data base entries, or user defaults. Several well-publicized hacks have allowed access to these tokens, opening up user accounts on popular social networking services to mischievous exploits. Protect your users by safeguarding their tokens, as demonstrated in Recipe 5-1.

Recipe 5-7 **The OAuth Process**

```
// Request initial tokens
- (BOOL) requestTokens: (NSString *) tokenEndpoint
{
    NSURL *endpointURL = [NSURL URLWithString:tokenEndpoint];
    
    // Create the preliminary (no token) dictionary
    NSMutableDictionary *dict = [self baseDictionary];
    if (!dict) return NO;
```

```objc
    // Create signature
    NSMutableString *baseRequest =
        [OAuthRequestSigner baseRequestWithEndpoint:tokenEndpoint
            dictionary:dict andRequestMethod:@"POST"];
    NSString *secretKey =
        [_consumerSecret stringByAppendingString:@"&"];
    dict[@"oauth_signature"] =
        [OAuthRequestSigner signRequest:baseRequest withKey:secretKey];

    // Produce the token request
    NSString *bodyString = [OAuthRequestSigner
        parameterStringFromDictionary:dict];
    NSMutableURLRequest *request =
        [NSMutableURLRequest requestWithURL:endpointURL];
    request.HTTPMethod = @"POST";
    request.HTTPBody = STRDATA(bodyString);
    [request setValue:@"application/x-www-form-urlencoded"
        forHTTPHeaderField:@"Content-Type"];

    // Request the tokens
    NSError *error;
    NSURLResponse *response;
    NSData *tokenData =
        [NSURLConnection sendSynchronousRequest:request
            returningResponse:&response error:&error];
    if (!tokenData)
    {
        NSLog(@"Failed to retrieve tokens: %@",
            error.localizedFailureReason);
        return NO;
    }

    return [self processTokenData:tokenData];
}

// Process and store tokens
- (BOOL) processTokens: (NSData *) tokenData
{
    NSString *tokenResultString = DATASTR(tokenData);

    // Check that we've received the right data
    NSRange range = [tokenResultString
        rangeOfString:@"oauth_token_secret"];
    if (range.location == NSNotFound)
    {
        NSLog(@"Failed to retrieve tokens: %@", tokenResultString);
        return NO;
    }
```

```objc
    // Convert the tokens
    NSDictionary *tokens = [OAuthRequestSigner
        dictionaryFromParameterString:tokenResultString];
    if (!tokens)
    {
        NSLog(@"Unable to process tokens: %@", tokenResultString);
        return NO;
    }

    // Store the tokens
    for (NSString *key in tokens.allKeys)
        credentialHelper[key] = tokens[key];

    return YES;
}

// Finish the end-user authentication step
- (BOOL) authenticate: (NSString *) accessEndpoint
{
    NSURL *endpointURL = [NSURL URLWithString:accessEndpoint];

    // This verifier invalidates after use
    NSString *access_verifier = credentialHelper[@"oauth_verifier"];
    if (!access_verifier)
    {
        NSLog(@"Error: Expected but did not find verifier");
        return NO;
    }

    // Add the token and verifier
    NSMutableDictionary *dict = [self baseDictionary];
    if (!dict) return NO;
    dict[@"oauth_token"] = credentialHelper[@"oauth_token"];
    dict[@"oauth_verifier"] = credentialHelper[@"oauth_verifier"];

    // Create signature
    NSMutableString *baseRequest =
        [OAuthRequestSigner baseRequestWithEndpoint:accessEndpoint
            dictionary:dict andRequestMethod:@"POST"];
    NSString *compositeKey = [NSString stringWithFormat:@"%@&%@",
        _consumerSecret, credentialHelper[@"oauth_token_secret"]];
    dict[@"oauth_signature"] =
        [OAuthRequestSigner signRequest:baseRequest
            withKey:compositeKey];

    // Build the request
    NSString *bodyString =
```

```objc
        [OAuthRequestSigner parameterStringFromDictionary:dict];
    NSMutableURLRequest *request =
        [NSMutableURLRequest requestWithURL:endpointURL];
    request.HTTPMethod = @"POST";
    request.HTTPBody = STRDATA(bodyString);
    [request setValue:@"application/x-www-form-urlencoded"
        forHTTPHeaderField:@"Content-Type"];

    // Place the request
    NSError *error;
    NSURLResponse *response;
    NSData *resultData =
        [NSURLConnection sendSynchronousRequest:request
            returningResponse:&response error:&error];

    // Check for common issues
    if (!resultData)
    {
        NSLog(@"Failed to retrieve tokens: %@",
            error.localizedFailureReason);
        return NO;
    }

    // Convert results to a string
    NSString *resultString = DATASTR(resultData);
    if (!resultString)
    {
        NSLog(@"Expected but did not get result string with tokens");
        return NO;
    }

    // Process the tokens
    NSDictionary *tokens = [OAuthRequestSigner
        dictionaryFromParameterString:resultString];
    if ([tokens.allKeys containsObject:@"oauth_token_secret"])
    {
        NSLog(@"Success. App is verified.");
        for (NSString *key in tokens.allKeys)
            credentialHelper[key] = tokens[key];

        // Clean up
        [credentialHelper removeCredential:@"oauth_verifier"];
        credentialHelper[@"authenticated"] = @"YES";
    }
    return YES;
}
```

> **Get This Recipe's Code**
>
> To find this recipe's full sample project, point your browser to https://github.com/erica/iOS-6-Advanced-Cookbook and go to the folder for Chapter 5.

Summary

This chapter introduced advanced network-supporting technologies. You saw how to work with the keychain, handle secure authentication challenges, upload data, use basic OAuth, and more. Here are a few thoughts to take away before leaving this chapter:

- Most of Apple's networking support is provided through low-level C-based routines. If you can find a friendly Objective-C wrapper to simplify your programming work, consider using it. The only drawback occurs when you specifically need tight networking control at the most basic level of your application, which is rare. There are superb resources out there. Just Google for them.

- iOS's system keychain has become so incredibly simple to use, there's no reason not to take advantage of it—especially when working with secure credentials. Protect your user and safeguard your application's reputation.

- If you can get access to xAuth rather than OAuth, use it. You'll generally apply with a specific application and some providers require screen shots, descriptions, and so on. In return, you gain a cleaner interface without having to scrape web pages.

6
Images

Images are abstract representations, storing data that makes up pictures. This chapter introduces Cocoa Touch images, specifically the `UIImage` class, and teaches you all the basic know-how you need for working with image data on iOS. In this chapter, you learn how to load, store, and modify image data in your applications. You discover how to process image data to create special effects, how to access images on a byte-by-byte basis, and more.

Image Sources

iOS images are generally stored in one of several well-defined places on the device. These sources include the photo album, the application bundle, the sandbox, iCloud, and the Internet, among others. Here's a review of the most common image sources you'll use in your applications:

- **Photo album:** The iOS's photo album contains a camera roll (for camera-ready units), a saved pictures roll, and photos synced from a user's Photo Stream, computer, or transferred from another digital device using a camera connection kit. Users can request images from this album using the interactive dialog supplied by the `UIImagePickerController` class. The dialog enables users to browse through stored photos and select the image they want to work with on an album-by-album basis.

- **Application bundle:** Your application bundle includes static images you add to an Xcode project. These resources appear within the application bundle (`[NSBundle mainBundle]`) along with your application executable, Info.plist file, and other material included in the project. Examine the included files by inspecting the Target > Build Phases > Copy Bundle Resources pane for your project.

    Your application cannot modify these files, but it can load and display them within the application using `UIImage`'s handy `imageNamed:` method. Images accessed in this manner can be selected for device-specific resolutions using standard file naming conventions such as "@2x" for Retina display.

- **Sandbox:** Your application can also write image files into its sandbox and read them back as needed. The sandbox enables you to store files to the Documents, Library, and tmp folders.

    The top-level Documents directory can be populated by and accessed from iTunes when the application enables `UIFileSharingEnabled` in the Info.plist file. Although other parts of iOS outside the sandbox are technically readable, Apple has made it clear that these areas are off limits for App Store applications.

    > **Note**
    >
    > When you use the Document file-sharing option, make sure you store any application-specific files that should not be shared directly with your users in your Library folder.

- **iCloud:** Apple's iCloud service enables you to store documents in a shared central location (folders stored in /private/var/mobile/Library/Mobile Documents/) and access them from all your users' computers and iOS devices. On iOS, you might use the `UIDocument` class to load iCloud images into your apps.

- **Internet:** Your application can download images using URL resources to point to web-based files. To make this work, iOS needs an active Internet connection, but when established the data from a remote image is just as accessible as data stored locally.

- **Pasteboard:** Applications can use image data stored in the system pasteboard. Image data is indistinguishable from any other kind of pastable data. You can recover that data by querying the `pasteboardTypes`. This returns an array of uniform type identifiers that specify what kind of data is currently available on the pasteboard.

    Image UTIs are typically of the form public.png, public.tiff, or public.jpg, or similar. Each application can choose whether it can handle the contents of the pasteboard based on those UTIs.

    You can read more about the system pasteboard and its uses in Chapter 2, "Documents and Data Sharing."

- **Shared Data:** iOS applications can open and display image files sent by other applications. Applications that declare support for image file types (by defining a `CFBundleDocumentTypes` array in their Info.plist file) can be called upon to open those files. Shared data is also discussed in detail in Chapter 2.

- **Generated Images:** You are not restricted to using prebuilt images in your applications. You can generate images on-the-fly as needed. UIKit's graphics calls enable you to create images from code, building `UIImage` objects as needed. For example, you might create a color swatch as follows:

```
// Return a swatch with the given color
- (UIImage *) swatchWithColor:(UIColor *) color
    andSize: (CGFloat) side
{
    UIGraphicsBeginImageContext(CGSizeMake(side, side));
    CGContextRef context = UIGraphicsGetCurrentContext();
```

```
[color setFill];
CGContextFillRect(context, CGRectMake(0.0f, 0.0f, side, side));
UIImage *image = UIGraphicsGetImageFromCurrentImageContext();
UIGraphicsEndImageContext();
return image;
}
```

Place Quartz drawing calls between the begin and end context statements. Retrieve the current image by calling `UIGraphicsGetImageFromCurrentImageContext()`. Generated images are like any other `UIImage` instance. You can display them in image views, pass them to table cells to display, place them into buttons, and so forth. Learn more about Quartz drawing specifics by consulting Apple's *Quartz 2D Programming Guide*.

Reading Image Data

An image's file location controls how you read its data. You'd imagine that you could just use a method such as `UIImage`'s `imageWithContentsOfFile:` to load your images, regardless of their source. In reality, you cannot.

For example, photo album pictures and their paths are hidden from direct application access unless you have retrieved an `ALAsset` URL. Apple's user-granted privileges ensure a level of privacy that's breeched only when users explicitly grant permission to applications for access. Even then, you must use Asset Library calls to read those files and not `UIImage` calls.

Images stored in the application bundle are accessed in a different way than those stored on a Web site or in the sandbox. Each kind of image data requires a nuanced approach for access and display.

What follows is a summary of common image reading methods, and how you can use these to read in image data to your application.

`UIImage` Convenience Methods

The `UIImage` class offers a simple method that loads any image stored in the application bundle. Call `imageNamed:` with a filename, including its extension, for example:

`myImage = [UIImage imageNamed:@"icon.png"];`

This method looks for an image with the supplied name in the top-level folder of the application bundle. If found, the image loads and is cached by iOS. That means the image is memory managed by that cache.

If your data is stored in another location in your sandbox, use `imageWithContentsOfFile:` instead. Pass it a string that points to the image file's path. This method will not cache the object the way `imageNamed:` does.

As a rule, `imageNamed:` does more work upfront. It non-lazily decodes the file into raw pixels and adds it to the cache, but it gives you memory management for free. Use it for images that you use over and over, especially for small ones. Caching large images may cause more re-reading overhead as you exhaust memory. The method has an undeserved bad reputation due to early iOS bugs and severe memory limitations on early iOS devices, but those issues have long since been resolved and newer devices ship with far more RAM.

Both the `named` and `contents` methods offer a huge advantage over other image-loading approaches. For higher-scale devices, namely the iPhone 4 and later, the third-generation iPad and later, and the fifth-generation iPod touch and later, with their higher-pixel density screens, you can use these methods to automatically select between low- and high-resolution images. For screen scales of 2.0, the method first searches for filenames using a `@2x` naming hint.

Naming Hints

A call to `[UIImage imageNamed:@"basicArt.png"]` loads the file basicArt@2x.png, if available, on higher-scale devices, and basicArt.png on 1.0 scale devices. The "@2x" hint indicates higher resolution art for Retina equipment. This naming scheme is tied into both general image loading and more nuanced application launching. The overall scheme looks like this:

```
<basename> <usage_specific_modifiers> <scale_modifier> <device_modifier>.png
```

Here's how the components of the naming break down:

- The basename refers to any filename, such as basicArt, Default, icon, and so forth.

- Usage-specific modifiers are used for launching and system-required items, namely the Default image, icon image, Spotlight image, Settings image, and so forth. These modifiers consist of items such as orientation hints (-Landscape, -Portrait, -PortraitUpsideDown, and so on), size tips (-small, -72, -144, -50, -100, -568h, and so on), and URL schema. Say your app supports a custom scheme named xyz. It can be launched from Mobile Safari via xyz:*whatever*. You can add –xyz as a usage-specific modifier to your Default images, for example, Default-xyz@2x.png. This enables your app to launch with that resource when accessed from that scheme.

- The scale modifier is the optional @2x. Should Apple ever introduce additional scales, developers might add new modifiers here.

- Device modifiers refer to the various device families, specifically ~iphone and ~ipad. The iPhone 5 and fifth-generation iPod touch belong to the ~iphone family, even if their 4" screens are taller than the 3.5" iPhone 4, iPhone 4S, and fourth-generation iPod. The iPad mini belongs to the ~ipad family.

These naming hints and multi-item resources reflect how cluttered the iOS family has grown over time. A typical application now ships with more than a dozen default and icon images. For this reason, this is a likely area for Apple to evolve, hopefully moving to vector graphics (or even better solutions) and away from multiple files, especially as updated screen sizes and resolutions ship on new devices.

> **Note**
>
> You can find some initial work on PDF-UIImage integration on Github, created by third-party developers. Repositories such as mindbrix/UIImage-PDF attempt to move images into scalable vector assets. Hopefully, Apple will eventually re-engineer image representations, removing the need to use these component modifiers for much longer.

> **Note**
>
> iOS supports the following image types: PNG, JPG, JPEG, TIF, TIFF, GIF, BMP, BMPF, ICO, CUR, XBM, and PDF. The `UIImage` class does not read or display PDF files; use `UIWebView` instead.

Finding Images in the Sandbox

By default, each sandbox contains three folders: Documents, Library, and tmp. User-generated data files, including images, normally reside in the Documents folder. This folder does exactly what the name suggests. You store documents to and access them from this directory. Keep document file data here that is created by or browsed from your program. You can also use the Documents folder and iTunes to allow desktop-based users direct access to data.

The Library folder contains user defaults and other state information for your program. Use the Library folder for any application support files that must be retained between successive runs and that are not meant for general end-user access.

The tmp folder provides a place to create transient (short-lived) files on-the-fly. Unlike in tmp, files in Documents and Library are not transient. iTunes backs up all Documents and most Library files (items in Library/Caches are not backed up, just like on the Mac) whenever iOS syncs. In contrast, iOS discards any tmp files when it reboots or when it encounters low storage conditions on the device.

This is the standard way to locate the Documents folder:

```
NSArray *paths = [NSSearchPathForDirectoriesInDomains(
    NSDocumentDirectory, NSUserDomainMask, YES);
return [paths lastObject];
```

You can also reliably locate the top sandbox folder by calling `NSHomeDirectory()`. This lets you navigate down one level to Documents with full assurance of reaching the proper destination. The following function provides a handy way to return paths to the Documents folders:

```
NSString *documentsFolder()
{
    return [NSHomeDirectory()
        stringByAppendingPathComponent:@"Documents"];
}
```

To load an image, append its filename to the returned path and tell UIImage to create a new image with those contents. This code loads a file named image.png from the top level of the documents folder and returns a UIImage instance initialized with that data:

```
path = [documentsFolder() stringByAppendingPathComponent:@"image.png"];
return [UIImage imageWithContentsOfFile:path];
```

This call returns `nil` if the path is not valid or points to a nonimage resource.

Loading Images from URLs

The `UIImage` class can load images from `NSData` instances, but it cannot do so directly from URL strings or `NSURL` objects. So, supply `UIImage` with data already downloaded from a URL. This snippet asynchronously downloads the latest United States weather map from weather.com and then creates a new image using the weather data:

```
NSString *map =
    @"http://maps.weather.com/images/maps/current/curwx_720x486.jpg";
NSOperationQueue *queue = [[NSOperationQueue alloc] init];
[queue addOperationWithBlock: ^{
    // Load the weather data
    NSURL *weatherURL = [NSURL URLWithString:map];
    NSData *imageData = [NSData dataWithContentsOfURL:weatherURL];

    // Update the image on the main thread using the main queue
    [[NSOperationQueue mainQueue] addOperationWithBlock:^{
        UIImage *weatherImage = [UIImage imageWithData:imageData];
        imageView.image = weatherImage;
    }];
}];
```

First, it constructs an `NSURL` object, and then it creates an `NSData` instance initialized with the contents of that URL. The data returned helps build the `UIImage` instance, which is added to an image view, taking care to perform that update on the main thread.

A more thorough implementation would return a placeholder image, cache the retrieved data locally, and update the main thread when the placeholder could be replaced with the downloaded asset.

Reading from the Assets Library

An Asset URL provides direct access to media stored outside your application's sandbox. This access persists indefinitely and does not limit you to images returned by an image picker. A typical assets library URL looks something like this:

```
assets-library://asset/asset.JPG?id=553F6592-43C9-45A0-B851-28A726727436&ext=JPG
```

The following snippet demonstrates how to access image assets. It creates a pair of blocks that handle asset access conditions for both success and failure. Upon retrieving an asset, the code creates a data representation, stores it to a `CGImageRef`, and then converts that into a standard `UIImage`:

```
// Retrieve an image from an asset URL
- (void) imageFromAssetURL: (NSURL *) assetURL into: (UIImage **) image
{
    ALAssetsLibrary *library = [[ALAssetsLibrary alloc] init];
    ALAssetsLibraryAssetForURLResultBlock resultsBlock = ^(ALAsset *asset)
    {
        ALAssetRepresentation *assetRepresentation =
            [asset defaultRepresentation];
        CGImageRef cgImage =
            [assetRepresentation CGImageWithOptions:nil];
        CFRetain(cgImage); // Thanks Oliver Drobnik. ARC weirdness
        if (image) *image = [UIImage imageWithCGImage:cgImage];
        CFRelease(cgImage);
    };
    ALAssetsLibraryAccessFailureBlock failure = ^(NSError *__strong error)
    {
        NSLog(@"Error retrieving asset from url: %@",
            error.localizedFailureReason);
    };

    [library assetForURL:assetURL
        resultBlock:resultsBlock failureBlock:failure];
}
```

> **Note**
>
> Recipes for loading data from the photo album and from asset URLs are discussed in Chapter 8, "Common Controllers," of *The Core iOS Developer's Cookbook*.

Recipe: Fitting and Filling Images

Often you need to resize an image to fit into a smaller space. You can resize an image view, but its contents may not update as you expect. UIKit offers view content modes that specify exactly how a view adjusts content as its size updates. You set these by adjusting the value of the `contentMode` property.

The most commonly used content modes are as follows. These detail the ways views, including image views, adjust their content to match their current size:

- **Fitting (`UIViewContentModeScaleAspectFit`):** Resizes the image while retaining its proportions, fitting it so every part of the image remains visible. Depending on the image's aspect ratio, the results are either letterboxed or pillarboxed, with some extra area needed to matte the image (see Figure 6-1, top left).

204   Chapter 6   Images

Figure 6-1   These screen shots represent ways to manipulate content to fill available space. Fitting (top left) preserves original aspect ratios, padding as needed with extra space. Filling (top middle) ensures that every available pixel is filled, cropping only those portions that fall outside the frame Centering (top right) uses the original image pixels at a 1:1 scale of pixels to points, cropping from the center out. Squeezing (bottom left) fits the original pixels to the destination aspect ratio, scaling as needed without preserving aspect. The original image appears at the bottom right.

- **Filling (`UIViewContentModeScaleAspectFill`):** Punches out part of the image to match the available space. This approach crops any elements that fall outside the pixel display area, either to the top and bottom, or to the left and right. The image is usually centered, with only the smaller dimension being fully displayed (see Figure 6-1, top middle).

- **Centering (`UIViewContentModeCenter`):** Places the image at its natural scale directly in the center of the image. Images larger than the pixel display area are cropped. Those smaller are matted as needed (see Figure 6-1, top right). Similar modes allow placement at the top, bottom, left, right, top left, top right, bottom left, and bottom right of the view.
- **Squeezing (`UIViewContentModeScaleToFill`):** Adjusts the image's aspect to fit it into the available space. All pixels are filled.

At times, you'll want to respect content rules while redrawing an image into fewer bytes. This allows you to mimic this behavior while reducing memory overhead.

Recipe 6-1 shows how to create mode-adjusted images that respect a destination size. These methods are implemented using a `UIImage` category and return new image instances. You pass a target size and a wanted content mode. The recipe returns a new image based on that size.

The context background is not initialized in this recipe. Pillarboxed or letterboxed pixels remain transparent. You can easily update this code to provide a backing fill color before drawing the image into its destination rectangle. This is left as an exercise for you.

Recipe 6-1 **Applying Image Aspect**

```
// Calculate the destination scale for filling
CGFloat CGAspectScaleFill(CGSize sourceSize, CGRect destRect)
{
    CGSize destSize = destRect.size;
    CGFloat scaleW = destSize.width / sourceSize.width;
    CGFloat scaleH = destSize.height / sourceSize.height;
    return MAX(scaleW, scaleH);
}

// Calculate the destination scale for fitting
CGFloat CGAspectScaleFit(CGSize sourceSize, CGRect destRect)
{
    CGSize destSize = destRect.size;
    CGFloat scaleW = destSize.width / sourceSize.width;
    CGFloat scaleH = destSize.height / sourceSize.height;
    return MIN(scaleW, scaleH);
}

// Fit a size into another size
CGSize CGSizeFitInSize(CGSize sourceSize, CGSize destSize)
{
    CGFloat destScale;
    CGSize newSize = sourceSize;
    
    if (newSize.height && (newSize.height > destSize.height))
    {
        destScale = destSize.height / newSize.height;
```

```
            newSize.width *= destScale;
            newSize.height *= destScale;
        }
        
        if (newSize.width && (newSize.width >= destSize.width))
        {
            destScale = destSize.width / newSize.width;
            newSize.width *= destScale;
            newSize.height *= destScale;
        }
        
        return newSize;
}

// Destination rect by fitting
CGRect CGRectAspectFitRect(CGSize sourceSize, CGRect destRect)
{
    CGSize destSize = destRect.size;
    CGFloat destScale = CGAspectScaleFit(sourceSize, destRect);
    
    CGFloat newWidth = sourceSize.width * destScale;
    CGFloat newHeight = sourceSize.height * destScale;
    
    CGFloat dWidth = ((destSize.width - newWidth) / 2.0f);
    GFloat dHeight = ((destSize.height - newHeight) / 2.0f);
    
    CGRect rect = CGRectMake(dWidth, dHeight, newWidth, newHeight);
    return rect;
}

// Destination rect by filling
CGRect CGRectAspectFillRect(CGSize sourceSize, CGRect destRect)
{
    CGSize destSize = destRect.size;
    CGFloat destScale = CGAspectScaleFill(sourceSize, destRect);
    
    CGFloat newWidth = sourceSize.width * destScale;
    CGFloat newHeight = sourceSize.height * destScale;
    
    CGFloat dWidth = ((destSize.width - newWidth) / 2.0f);
    CGFloat dHeight = ((destSize.height - newHeight) / 2.0f);
    
    CGRect rect = CGRectMake(dWidth, dHeight, newWidth, newHeight);
    return rect;
}
```

```objc
// Return a new image created with a content mode
- (UIImage *) applyAspect: (UIViewContentMode) mode
    inRect: (CGRect) bounds
{
    CGRect destRect;

    UIGraphicsBeginImageContext(bounds.size);
    switch (mode)
    {
        case UIViewContentModeScaleToFill:
        {
            destRect = bounds;
            break;
        }
        case UIViewContentModeScaleAspectFill:
        {
            CGRect rect = CGRectAspectFillRect(self.size, bounds);
            destRect = CGRectCenteredInRect(rect, bounds);
            break;
        }
        case UIViewContentModeScaleAspectFit:
        {
            CGRect rect = CGRectAspectFitRect(self.size, bounds);
            destRect = CGRectCenteredInRect(rect, bounds);
            break;
        }
        case UIViewContentModeCenter:
        {
            CGRect rect = (CGRect){.size = self.size};
            destRect = CGRectCenteredInRect(rect, bounds);
            break;
        }
        default:
            break;
    }

    [self drawInRect:destRect];
    UIImage *newImage = UIGraphicsGetImageFromCurrentImageContext();
    UIGraphicsEndImageContext();
    return newImage;
}
```

Get This Recipe's Code

To find this recipe's full sample project, point your browser to https://github.com/erica/iOS-6-Advanced-Cookbook and go to the folder for Chapter 6.

Recipe: Rotating Images

The easiest way to rotate images involves applying a rotation transform to an image view. That approach does not affect the image data, just its presentation onscreen. Plus transforms give the added bonus of managing the view's touches for free. When you want to rotate the data itself, you can draw data into a rotated context to create an altered image.

There are many reasons to rotate. For example, you may need to adjust a snapshot that was captured at an angle due to the way a user was holding the device. Or you may want to create a collage of rotated and scaled subimages that you merge into a new picture.

Recipe 6-2 establishes a UIImage category that produces images with rotated content. You call it with an angular offset, and it returns a rotated version. In this recipe, the size of the generated image depends on both the original image geometry (its size) and the degree of rotation applied to it.

For example, a square image rotated by 45 degrees is approximately 40% larger in each dimension (see Figure 6-2), occupying a far larger space than the original. Recipe 6-1 calculates this new size by applying an affine transform. Affine transforms are geometric operations that map points in one coordinate system to another coordinate system. iOS uses CGAffineTransform functions, in this case CGRectApplyAffineTransform, to convert rectangles by applying transforms.

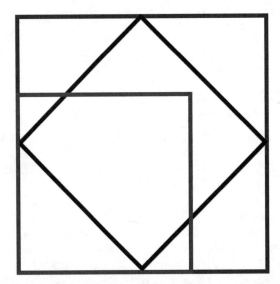

Figure 6-2 Upon rotation, an image may require a far larger space than its original size. The smaller square in the bottom-left corner represents an initial image size. After rotation (to the diamond in the center), the required size to fit within a single image increases to the larger outer square. You choose between expanding the final size, shrinking the results, and cropping parts of the image to fit the available space.

You can, of course, crop the rotated image, cutting off the edges. Alternatively, you may want to expand to encompass the entire picture. This recipe preserves the updated image. It manipulates the image context it draws into, which is sized to fit the entire rotated version. Recipes that follow later in this chapter take the opposite approach, returning an image sized to match the original.

Finally, you could shrink the results, fitting them to the available size. This is usually the least wanted approach because the changed scale will degrade the image contents and may produce user confusion when you present an image that suddenly shrinks onscreen.

Speaking of degradation, should you need to keep rotating the image, make sure you always start with the original data. Each rotation forces pixels to merge, leaving aside any issue of cropping. Multiple rotations applied to the same data produce corrupt results. Rotating the original to each degree required produces cleaner results.

As with Recipe 6-1, unaddressed pixels remain transparent. You could, alternatively, fill the background of the context with whatever color and alpha value you find appropriate to your needs.

Recipe 6-2 Rotating an Image

```
UIImage *rotatedImage(
    UIImage *image, CGFloat rotation)
{
    // Rotation is specified in radians
    // Calculate Destination Size
    CGAffineTransform t = CGAffineTransformMakeRotation(rotation);
    CGRect sizeRect = (CGRect) {.size = image.size};
    CGRect destRect = CGRectApplyAffineTransform(sizeRect, t);
    CGSize destinationSize = destRect.size;

    // Draw image
    UIGraphicsBeginImageContext(destinationSize);
    CGContextRef context = UIGraphicsGetCurrentContext();

    // Translate to set the origin before rotation
    // to the center of the image
    CGContextTranslateCTM(context,
        destinationSize.width / 2.0f,
        destinationSize.height / 2.0f);

    // Rotate around the center of the image
    CGContextRotateCTM(context, rotation);

    // Draw the image into the rotated context
    [image drawInRect:CGRectMake(
        -image.size.width / 2.0f,
```

```
            -image.size.height / 2.0f,
        image.size.width, image.size.height)];
    
    // Save image
    UIImage *newImage =
        UIGraphicsGetImageFromCurrentImageContext();
    UIGraphicsEndImageContext();
    return newImage;
}

@implementation UIImage (Rotation)
- (UIImage *) rotateBy: (CGFloat) theta
{
    return rotatedImage(self, theta);
}

+ (UIImage *) image: (UIImage *) image rotatedBy: (CGFloat) theta
{
    return rotatedImage(image, theta);
}
@end
```

Get This Recipe's Code

To find this recipe's full sample project, point your browser to https://github.com/erica/iOS-6-Advanced-Cookbook and go to the folder for Chapter 6.

Recipe: Working with Bitmap Representations

Although Cocoa Touch provides excellent resolution-independent tools for working with many images, there are times you need to reach down to the bits that underlie a picture and access that data on a bit-by-bit basis. For example, you might apply edge-detection or blurring routines that calculate their results by convolving matrices against actual byte values.

The sample code that supports Recipe 6-3 retrieves bytes that correspond to a point where the user has touched the screen and updates the navigation bar with that color. Accessing pixel-by-pixel color allows the project to look up data from screen coordinates.

Converting Between Coordinate Systems

To access bytes, the touch point must be transformed from the image view's coordinates to the image's. That's made slightly more difficult by setting the view's content mode to `UIViewContentModeScaleAspectFill`. As the previous recipe explains, aspect-fill means that every pixel in the view is filled, extending an image out from its center and clipping excess material either along its sides or its top and bottom.

The following method handles conversion by applying an affine transform to the view's touch point. This scales it from the screen's coordinate system to the image's coordinate system. This takes several steps. The point must first be recentered, then transformed, and then reset back from the image's center. Affine transforms operate with respect to a view's center, so the touch points coordinates must accommodate that change. In the end, the center-transform-recenter process produces a point that falls onto the image data array that respects the image's top-left corner as a (0,0) origin:

```
- (void) handle: (CGPoint) aPoint
{
    // Set origin to the center of the view
    aPoint.x -= self.center.x;
    aPoint.y -= self.center.y;

    // Calculate the point in the image's view coordinate system
    CGFloat imageWidth = self.image.size.width;
    CGFloat imageHeight = self.image.size.height;
    CGFloat xScale = imageWidth / self.frame.size.width;
    CGFloat yScale = imageHeight / self.frame.size.height;
    CGFloat scale = MIN(xScale, yScale);
    CGAffineTransform t = CGAffineTransformMakeScale(scale, scale);
    CGPoint adjustedPoint = CGPointApplyAffineTransform(aPoint, t);

    // Reset the origin to the top-left corner of the image
    adjustedPoint.x += imageWidth / 2.0f;
    adjustedPoint.y += imageHeight / 2.0f;

    // Refresh the image data if needed (it shouldn't be needed)
    if (!_imageData)
        _imageData = self.image.bytes;

    // Retrieve the byte values at the given point —
    Byte *bytes = (Byte *)_imageData.bytes;
    CGFloat red = bytes[redOffset(adjustedPoint.x,
        adjustedPoint.y, imageWidth)] / 255.0f;
    CGFloat green = bytes[greenOffset(adjustedPoint.x,
        adjustedPoint.y, imageWidth)] / 255.0f;
    CGFloat blue = bytes[blueOffset(adjustedPoint.x,
        adjustedPoint.y, imageWidth)] / 255.0f;
    UIColor *color = [UIColor colorWithRed:red
        green:green blue:blue alpha:1.0f];

    // Update the nav bar to match the color at the user's touch
    _controller.navigationController.navigationBar.tintColor = color;
}
```

Looking Up Data

Bitmaps have no natural dimensions. They are just blocks of data. Thus, if you want to access byte data at the point (x,y), you must supply the size of each row. The bytes for each pixel (there are 4 bytes for ARGB data) start at an offset of (y * rowWidth + x) * 4. So, the first pixel in the image starts at 0, and goes through byte 3. The second pixel starts at 4, continuing to byte 7, and so forth.

Each channel occupies one byte. Each 4-byte sequence contains a level for alpha, red, green, and then blue. The first byte holds the alpha (opacity) information. The red channel is offset by 1 byte, the green by 2, and the blue by 3. Four helper functions, allow this method to retrieve a calculated offset into the byte array by supplying x, y, and the row-width. The following functions return offsets for any point (x,y) inside an ARGB bitmap using width w:

```
NSUInteger alphaOffset(NSUInteger x, NSUInteger y, NSUInteger w)
    {return y * w * 4 + x * 4 + 0;}
NSUInteger redOffset(NSUInteger x, NSUInteger y, NSUInteger w)
    {return y * w * 4 + x * 4 + 1;}
NSUInteger greenOffset(NSUInteger x, NSUInteger y, NSUInteger w)
    {return y * w * 4 + x * 4 + 2;}
NSUInteger blueOffset(NSUInteger x, NSUInteger y, NSUInteger w)
    {return y * w * 4 + x * 4 + 3;}
```

The height is not needed for these calculations; the width of each row enables you to determine a two-dimensional point in what is actually a one-dimensional buffer. Each byte ranges from 0 (0%, or 0x00) to 255 (100% or 0xFF). Convert this to a float and divide by 255.0 to retrieve the ARGB value.

Converting Image Data to and from Bitmap Data

To build a bitmap, you draw an image into a bitmap context and then retrieve bytes as a `Byte *` (aka `uint8 *`) buffer, which Recipe 6-3 stores via an `NSData` object. The `bytesFromImage:` class method creates an RGB color space (alternatives are Gray and CMYK), allocates the bitmap data, and creates a bitmap context, which it draws the image into. The bytes are retrieved from the context, placed into an `NSData` object, and all the rest of the CG objects are cleaned up. The Data object takes over responsibility for the bytes in question.

The `imageWithBytes:withSize:` method goes the opposite way. You supply a byte array and the wanted image dimensions. The method creates a new `CGImageRef` with that data, converts it to a `UIImage` instance, and returns that image.

As you'll see in the next recipe, moving back and forth from bitmap representations allows you to introduce image processing to your pictures.

Always make sure to run Xcode's static analyzer on your bitmap routines. It's easy to overlook a release or a free call, and the analyzer can help find errors. The Recipe 6-3's first routine (the `imageWithBytes:` method) releases the bytes passed to it.

Recipe 6-3   **Converting to and from Image Bitmaps**

```objc
@implementation UIImage (Utils)
// Build image from bytes
+ (UIImage *) imageWithBytes: (Byte *) bytes withSize: (CGSize) size
{
    // Create a color space
    CGColorSpaceRef colorSpace = CGColorSpaceCreateDeviceRGB();
    if (colorSpace == NULL)
    {
        fprintf(stderr, "Error allocating color space\n");
        free(bytes);
        return nil;
    }

    // Create the bitmap context
    CGContextRef context = CGBitmapContextCreate(
        bytes, size.width, size.height, 8, size.width * 4,
        colorSpace, kCGImageAlphaPremultipliedFirst);
    if (context == NULL)
    {
        fprintf (stderr, "Error: Context not created!");
        free (bytes);
        CGColorSpaceRelease(colorSpace );
        return nil;
    }

    // Convert to image
    CGImageRef imageRef = CGBitmapContextCreateImage(context);
    UIImage *image = [UIImage imageWithCGImage:imageRef];

    // Clean up
    CGColorSpaceRelease(colorSpace );
    free(CGBitmapContextGetData(context)); // frees bytes
    CGContextRelease(context);
    CFRelease(imageRef);

    return image;
}

// Covert UIImage to byte array
+ (NSData *) bytesFromImage: (UIImage *) image
{
    CGSize size = image.size;

    CGColorSpaceRef colorSpace = CGColorSpaceCreateDeviceRGB();
    if (colorSpace == NULL)
```

```
    {
        fprintf(stderr, "Error allocating color space\n");
        return NULL;
    }

    void *bitmapData = malloc(size.width * size.height * 4);
    if (bitmapData == NULL)
    {
        fprintf (stderr, "Error: Memory not allocated!");
        CGColorSpaceRelease(colorSpace);
        return NULL;
    }

    CGContextRef context = CGBitmapContextCreate(
        bitmapData, size.width, size.height, 8, size.width * 4,
        colorSpace, kCGImageAlphaPremultipliedFirst);
    CGColorSpaceRelease(colorSpace );
    if (context == NULL)
    {
        fprintf (stderr, "Error: Context not created!");
        free (bitmapData);
        return NULL;
    }

    CGRect rect = (CGRect){.size = size};
    CGContextDrawImage(context, rect, image.CGImage);
    Byte *byteData = CGBitmapContextGetData (context);
    CGContextRelease(context);

    NSData *data = [NSData dataWithBytes:byteData
        length:(size.width * size.height * 4)];
    free(bitmapData);

    return data;
}

- (NSData *) bytes
{
    return [UIImage bytesFromImage:self];
}
@end
```

> **Get This Recipe's Code**
>
> To find this recipe's full sample project, point your browser to https://github.com/erica/iOS-6-Advanced-Cookbook and go to the folder for Chapter 6.

Recipe: Basic Image Processing

A somewhat recent addition to iOS, the Accelerate framework offers C-based APIs that allow applications to perform digital signal processing (DSP) using vector and matrix-based arithmetic. According to Apple, its mathematical functions support "speech, sound, audio, and video processing, diagnostic medical imaging, radar signal processing, seismic analysis, and scientific data processing." These routines are engineered for high performance, and are reusable across most Apple platforms. Accelerate offers portable, reliable, highly-optimized code.

Consider Recipe 6-4, which implements image rotation similar to Recipe 6-2. It uses the Accelerate framework instead of Core Graphics context calls. This implementation crops the image to the original size rather than preserving the rotated dimensions.

What stands out about this recipe is its simplicity (just a call to the ARGB rotation routine) and the bookkeeping. The data is packaged as a `vImage_Buffer`. Image buffers consist of a pointer to the data in question and fields that store the image's height, width, and bytes per row. In this case, a row consists of 4 bytes of ARGB data per pixel.

You allocate the output buffer and are responsible for deallocating it after use. In this recipe, the call to Recipe 6-3's `imageWithBytes:withSize:` automatically takes care of that deallocation and returns a `UIImage` instance.

With Accelerate, you can perform most standard image processing routines including scaling, shearing, histograms, gamma adjustments, convolution (see Recipe 6-5), and so forth. Accelerate makes it easy to apply standard routines to image buffers. Recipes 6-3 and 6-4 demonstrate how to make the jump between the UIKit world of `UIImage` instances and the bitmap word of Accelerate's C-based functions.

Recipe 6-4 **Rotating Images with the Accelerate Framework**

```
@implementation UIImage (vImage)

// Return a base buffer without bytes
- (vImage_Buffer) baseBuffer
{
    vImage_Buffer buf;
    buf.height = self.size.height;
    buf.width = self.size.width;
    buf.rowBytes = sizeof(Byte) * self.size.width * 4; // ARGB
    return buf;
}

// Return a buffer for the current image, with its bytes populated
- (vImage_Buffer) buffer
{
    vImage_Buffer buf = [self baseBuffer];
    buf.data = (void *)self.bytes.bytes;
    return buf;
}
```

```objc
// Perform rotation
- (UIImage *) vImageRotate: (CGFloat) theta
{
    vImage_Buffer inBuffer = [self buffer];
    vImage_Buffer outBuffer = [self baseBuffer];
    Byte *outData = (Byte *)malloc(
        outBuffer.rowBytes * outBuffer.height);
    outBuffer.data = (void *) outData;
    uint8_t backColor[4] = {0xFF, 0, 0, 0};

    vImage_Error error = vImageRotate_ARGB8888(
        &inBuffer, &outBuffer, NULL, theta, backColor, 0);
    if (error)
    {
        NSLog(@"Error rotating image: %ld", error);
        free(outData);
        return self;
    }

    // This frees the outData buffer
    return [UIImage imageWithBytes:outData withSize:self.size];
}
@end
```

> **Get This Recipe's Code**
>
> To find this recipe's full sample project, point your browser to https://github.com/erica/iOS-6-Advanced-Cookbook and go to the folder for Chapter 6.

Recipe: Image Convolution

Many common image effects depend on convolving matrices with image data. You might want to sharpen an image, smooth it, or add an "embossed" effect. Creating a two-dimensional array of values with specific characteristics produces these effects. The following matrix defines pixel operations that output the "emboss" effect shown in Figure 6-3:

```objc
- (UIImage *) emboss
{
    static const int16_t kernel[] = {
        -2, -1, 0,
        -1, 1, 1,
         0, 1, 2};
    return [self convolve:kernel side:3];
}
```

Recipe: Image Convolution

Figure 6-3 Embossing an image produces a pseudo-3D effect, where items appear to push forward out of the screen. The embossed image (bottom) is produced by convolving a simple matrix against the original image (top).

To *convolve* an image means to iteratively apply a matrix across all pixels. This particular matrix produces the strongest results when nearby pixels to the top right of a given position are dimmer and pixels to the bottom right are brighter. Because this naturally happens at the right and bottom edges of objects within a scene, the output looks embossed, with items appearing to pop out from the rest of the image. The bottom-right portions of the processed sheep in Figure 6-3 are more pronounced than the original.

Recipe 6-5 demonstrates how to perform a convolution using the Accelerate framework. In steps, it's practically identical to Recipe 6-4. What's notably different is that you provide a matrix to convolve with and a divisor for normalizing results. A couple common operations are added (sharpen and a 5x5 Gaussian blur) to demonstrate how the convolution method might be called. This is a helpful approach for performing blurs on pre-iOS 6 installs.

These routines are fairly basic. You can apply far more complex image enhancements using Core Image, which is discussed in the next recipe.

Recipe 6-5 Convolving Images with the Accelerate Framework

```
// Calculate the matrix sum for non-edge convolution
int32_t getDivisor(NSData *kernel)
{
    const int16_t *matrix = (const int16_t *)kernel.bytes;
    int count = kernel.length / sizeof(const int16_t);

    // Sum up the kernel elements
    int32_t sum = 0;
    for (CFIndex i = 0; i < count; i++)
        sum += matrix[i];
    if (sum != 0) return sum;
    return 1;
}
```

```objc
// Convolve an image against a kernel
- (UIImage *) vImageConvolve: (NSData *) kernel
{
    vImage_Buffer inBuffer = [self buffer];
    vImage_Buffer outBuffer = [self baseBuffer];
    Byte *outData = (Byte *)malloc(
        outBuffer.rowBytes * outBuffer.height);
    outBuffer.data = (void *) outData;
    uint8_t backColor[4] = {0xFF, 0, 0, 0};

    const int16_t *matrix = (const int16_t *)kernel.bytes;
    uint32_t matrixSide = sqrt(kernel.length / sizeof(int16_t));
    int32_t divisor = getDivisor(kernel);

    vImage_Error error = vImageConvolve_ARGB8888(
        &inBuffer, &outBuffer, NULL, 0, 0, matrix,
        matrixSide, matrixSide, divisor,
        backColor, kvImageBackgroundColorFill);
    if (error)
    {
        NSLog(@"Error convolving image: %ld", error);
        free(outData);
        return self;
    }

    return [UIImage imageWithBytes:outData withSize:self.size];
}

- (UIImage *) sharpen
{
    static const int16_t kernel[] = {
        0, -1,  0,
       -1,  8, -1,
        0, -1,  0
    };
    return [self convolve:kernel side:3];
}

- (UIImage *) gauss5
{
    static const int16_t kernel[] = {
        1,  4,  6,  4, 1,
        4, 16, 24, 16, 4,
        6, 24, 36, 24, 6,
        4, 16, 24, 16, 4,
        1,  4,  6,  4, 1
    };
    return [self convolve:kernel side:5];
}
```

Get This Recipe's Code

To find this recipe's full sample project, point your browser to https://github.com/erica/iOS-6-Advanced-Cookbook and go to the folder for Chapter 6.

Recipe: Basic Core Image Processing

Core Image filters enable you to process images quickly. Recipe 6-6 introduces basic CI filtering by applying a simple sepia filter to images. iOS 6 offers a wide range of CI filters that have grown over time. You can query these by looking at the built-in set of filter names:

NSLog(@"%@", [CIFilter filterNamesInCategory: kCICategoryBuiltIn]);

Filters include color adjustments, geometric transforms, cropping, and compositing. Often you feed the results of one filter into the next. For example, to create a picture-in-picture effect, you can scale and translate one image and then composite it over another.

Each filter uses a number of inputs, which include both source data and parameters. These vary by filter. Recipe 6-6 uses a sepia tone filter whose inputs are inputImage and input Intensity. A second perspective filter adds inputs that specify positions for image corners. Figure 6-4 shows the results of this perspective filter implementation. Adjusting the four-corner inputs changes the "direction" of the filter.

Figure 6-4 A perspective filter distorts the image as if seen from another viewpoint.

Look up input specifics in the latest "Core Image Filter Reference" document and *Core Image Programming Guide* from Apple. You can also query inputs directly from the filter by calling `inputKeys`. This returns an array of input parameter names used by the filter.

Nearly all filters offer default settings that provide a basis for you to start playing with them. Calling `setDefaults` loads these values, allowing you to start testing. Pay special attention to Apple's guide, which details what each default means and how the filter uses it. It also offers image examples, so you can see what the filter is meant to do.

As a side note, Recipe 6-6 implements a `coreImageRepresentation` property, which is provided because CI/`UIImage` integration remains a bit wonky. When the method cannot retrieve a `CIImage` instance directly from an image, it builds one from the image's `CGImage` representation. This kind of workaround helps remind you that Core Image remains an evolving technology under iOS.

Recipe 6-6 **Core Image Basics**

```
- (CIImage *) coreImageRepresentation
{
    if (self.CIImage)
        return self.CIImage;
    return [CIImage imageWithCGImage:self.CGImage];
}

- (UIImage *) sepiaVersion: (CGFloat) intensity
{
    CIFilter *filter = [CIFilter
        filterWithName:@"CISepiaTone"
        keysAndValues: @"inputImage",
        self.coreImageRepresentation, nil];
    [filter setDefaults];
    [filter setValue:@(intensity) forKey:@"inputIntensity"];

    CIImage *output = [filter valueForKey:kCIOutputImageKey];
    if (!output)
    {
        NSLog(@"Core Image processing error");
        return nil;
    }

    UIImage *results = [UIImage imageWithCIImage:output];
    return results;
}

- (UIImage *) perspectiveExample
{
    CIFilter *filter = [CIFilter
```

```
            filterWithName:@"CIPerspectiveTransform"
            keysAndValues: @"inputImage",
            self.coreImageRepresentation,
            nil];
    [filter setDefaults];
    [filter setValue:[CIVector vectorWithX:180 Y:600]
        forKey:@"inputTopLeft"];
    [filter setValue:[CIVector vectorWithX:102 Y:20]
        forKey:@"inputBottomLeft"];
    
    CIImage *output = [filter valueForKey:kCIOutputImageKey];
    if (!output)
    {
        NSLog(@"Core Image processing error");
        return nil;
    }
    
    UIImage *results = [UIImage imageWithCIImage:output];
    return results;
}
```

Get This Recipe's Code

To find this recipe's full sample project, point your browser to https://github.com/erica/iOS-6-Advanced-Cookbook and go to the folder for Chapter 6.

Capturing View-Based Screen Shots

At times you need to take a shot of a view or window in its current state or otherwise render a view into an image. Listing 6-1 details how you can draw views into image contexts and retrieve `UIImage` instances. This code works by using Quartz Core's `renderInContext` call for `CALayer` instances. It produces a screen shot not only of the view, but also of all the subviews that view owns.

There are, of course, limits. You cannot make a screen shot of the entire window (the status bar will be missing in action) and, using this code in particular, you cannot make a screen shot of videos or the camera previews. OpenGL ES views also may not be captured.

This kind of view-to-image utility is particularly handy when working with Core Animation as you create transitions from one view hierarchy to another.

Listing 6-1   **Making a Screen Shot of a View**

```
UIImage *imageFromView(UIView *theView)
{
        UIGraphicsBeginImageContext(theView.frame.size);
        CGContextRef context = UIGraphicsGetCurrentContext();

        [theView.layer renderInContext:context];
        UIImage *theImage =
            UIGraphicsGetImageFromCurrentImageContext();

        UIGraphicsEndImageContext();
        return theImage;
}

UIImage *screenShot()
{
        UIWindow *window =
            [[UIApplication sharedApplication] keyWindow];
        return imageFromView(window);
}
```

Drawing into PDF Files

You can draw directly into PDF documents, just as if you were drawing into an image context. As you can see in Listing 6-2, you can use Quartz drawing commands such as `UIImage`'s `drawInRect:` and `CALayer`'s `renderInContext:` methods, along with your full arsenal of other Quartz functions.

You typically draw each image into a new PDF page. You must create at least one page in your PDF document, as shown in this listing. In addition to simple rendering, you can add live links to your PDFs using functions such as `UIGraphicsSetPDFContextURLForRect()`, which links to an external URL from the rectangle you define, and `UIGraphicsSetPDFContextDestinationForRect()`, which links internally within your document. Create your destination points using `UIGraphicsAddPDFContextDestinationAtPoint()`.

Each PDF document is built with a custom dictionary. Pass `nil` if you want to skip the dictionary, or assign values for Author, Creator, Title, Password, AllowsPrinting, and so forth. The keys for this dictionary are listed in Apple's `CGPDFContext` reference documentation.

Listing 6-2   **Drawing an Image into a PDF File**

```
+ (void) saveImage: (UIImage *) image
    toPDFFile: (NSString *) path
{
    CGRect theBounds = (CGRect){.size=image.size};
```

```
UIGraphicsBeginPDFContextToFile(path, theBounds, nil);
{
    UIGraphicsBeginPDFPage();
    [image drawInRect:theBounds];
}
UIGraphicsEndPDFContext();
}
```

Recipe: Reflection

The final two recipes for this chapter add a bit of razzle dazzle to your images. They offer a natural fit to the special effects introduced by Core Image and Accelerate. They are both part of the QuartzCore framework, used for creating, manipulating, and drawing images.

Reflections enhance the reality of onscreen objects. They provide a little extra visual spice beyond the views-floating-over-a-backsplash, which prevails as the norm. Reflections have become especially easy to implement because of the `CAReplicatorLayer` class. This class does what its name implies; it replicates a view's layer and enables you to apply transforms to that replication.

Recipe 6-7 shows how you can use replicator layers to build a view controller that automatically reflects content. A custom reflector view class overrides `layerClass` to ensure that the view's layer defaults to a replicator. That replicator is given the `instanceCount` of 2, so it duplicates the original view to a second instance. Its `instanceTransform` specifies how the second instance is manipulated and placed onscreen: flipped, squeezed, and then moved vertically below the original view. This creates a small reflection at the foot of the original view.

Due to the replicator, that reflection is completely "live." Any changes to the main view immediately update in the reflection layer, which you can test yourself using the sample code that accompanies this chapter. You can add scrolling text views, web views, switches, and so forth. Any changes to the original view are replicated, creating a completely passive reflection system.

Good reflections use a natural attrition as the "reflection" moves further away from the original view. This listing adds an optional gradient overlay (using the `usesGradientOverlay` property) that creates a visual drop-off from the bottom of the view. This code adds the gradient as a `CAGradientLayer`.

This code introduces an arbitrary gap between the bottom of the view and its reflection. Here, it is hardwired to 10 points, and demonstrated in Figure 6-5, but you can adjust this as you like.

To make this recipe work within the context of a view controller, its child classes must do three things. First, add custom views to the controller's backsplash instead of its view, for example, `[self.backsplash addSubview:cannedTable]`. Second, be sure to call the super implementation for `viewDidAppear`. Finally, make sure to update the reflection each time the orientation changes by calling `setupReflection` for the backsplash.

Chapter 6 Images

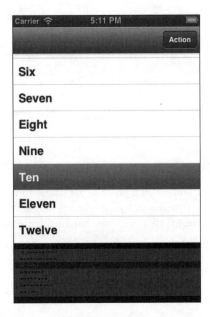

Figure 6-5 This layer-based reflection updates in real time to match the view it reflects. When using a ReflectionViewController like this, you need to add subviews to the controller's reflecting backdrop, not its view.

Recipe 6-7 Creating Reflections

```
@implementation ReflectingView
// Always use a replicator as the base layer
+ (Class) layerClass
{
    return [CAReplicatorLayer class];
}

- (void) setupGradient
{
    // Add a new gradient layer to the parent
    UIView *parent = self.superview;
    if (!gradient)
    {
        gradient = [CAGradientLayer layer];
        CGColorRef c1 = [[UIColor blackColor]
            colorWithAlphaComponent:0.5f].CGColor;
        CGColorRef c2 = [[UIColor blackColor]
            colorWithAlphaComponent:0.9f].CGColor;
        [gradient setColors:[NSArray arrayWithObjects:
            (__bridge id)c1, (__bridge id)c2, nil]];
```

```objc
        [parent.layer addSublayer:gradient];
    }
    
    // Place the gradient just below the view using the
    // reflection's geometry
    float desiredGap = 10.0f; // gap between view and its reflection
    CGFloat shrinkFactor = 0.25f; // reflection size
    CGFloat height = self.bounds.size.height;
    CGFloat width = self.bounds.size.width;
    CGFloat y = self.frame.origin.y;
    
    [gradient setAnchorPoint:CGPointMake(0.0f,0.0f)];
    [gradient setFrame:CGRectMake(0.0f, y + height + desiredGap,
        width, height * shrinkFactor)];
    [gradient removeAllAnimations];
    
    [gradient setAnchorPoint:CGPointMake(0.0f,0.0f)];
    [gradient setFrame:CGRectMake(0.0f, y + height + desiredGap,
        maxDimension, height * shrinkFactor)];
    [gradient removeAllAnimations];
}

- (void) setupReflection
{
    CGFloat height = self.bounds.size.height;
    CGFloat shrinkFactor = 0.25f;
    
    CATransform3D t = CATransform3DMakeScale(1.0, -shrinkFactor, 1.0);
    
    // Scaling centers the shadow in the view.
    // Translate the results in shrunken terms
    float offsetFromBottom = height * ((1.0f - shrinkFactor) / 2.0f);
    float inverse = 1.0 / shrinkFactor;
    float desiredGap = 10.0f;
    t = CATransform3DTranslate(t, 0.0, -offsetFromBottom * inverse
        - height - inverse * desiredGap, 0.0f);
    
    CAReplicatorLayer *replicatorLayer =
        (CAReplicatorLayer*)self.layer;
    replicatorLayer.instanceTransform = t;
    replicatorLayer.instanceCount = 2;
    
    // Gradient use must be explicitly set
    if (usesGradientOverlay)
        [self setupGradient];
    else
    {
```

```
        // Darken the reflection when not using a gradient
        replicatorLayer.instanceRedOffset = -0.75;
        replicatorLayer.instanceGreenOffset = -0.75;
        replicatorLayer.instanceBlueOffset = -0.75;
    }
}
@end
```

> **Get This Recipe's Code**
>
> To find this recipe's full sample project, point your browser to https://github.com/erica/iOS-6-Advanced-Cookbook and go to the folder for Chapter 6.

Recipe: Emitters

Add a bit of sparkle to your applications by introducing emitters. Emitters render particles in real time using a suite of custom properties that you adjust. Recipe 6-8 creates a violent purple cloud of particles that moves with your touch on the screen. As with Recipe 6-7's reflections, emitters enable you to add interactive visual interest to your applications.

Creating an emitter is based on establishing a layer and adding it to a view. You establish the emitter by adding a particle emitter cell and tweaking its characteristics. Here, each particle is based on an image, which in this case is loaded from the spark.png file included in the sample code.

Beyond that, it's up to you to decide how quickly particles are created, how long they live, their speed, color, and so forth. This kind of recipe is a tweaker's delight. Feel free to explore how each parameter affects the particle display created as you trace your finger across the screen.

Recipe 6-8 Adding Emitters

```
@implementation SparkleTouchView
{
    CAEmitterLayer *emitter;
}

- (id) initWithFrame: (CGRect) aFrame
{
    if (!(self = [super initWithFrame:aFrame])) return self;

    return self;
}

- (void) touchesBegan:(NSSet *)touches withEvent:(UIEvent *)event
{
```

```
    float multiplier = 0.25f;

    CGPoint pt = [[touches anyObject] locationInView:self];

    //Create the emitter layer
    emitter = [CAEmitterLayer layer];
    emitter.emitterPosition = pt;
    emitter.emitterMode = kCAEmitterLayerOutline;
    emitter.emitterShape = kCAEmitterLayerCircle;
    emitter.renderMode = kCAEmitterLayerAdditive;
    emitter.emitterSize = CGSizeMake(100 * multiplier, 0);

    //Create the emitter cell
    CAEmitterCell* particle = [CAEmitterCell emitterCell];
    particle.emissionLongitude = M_PI;
    particle.birthRate = multiplier * 1000.0;
    particle.lifetime = multiplier;
    particle.lifetimeRange = multiplier * 0.35;
    particle.velocity = 180;
    particle.velocityRange = 130;
    particle.emissionRange = 1.1;
    particle.scaleSpeed = 1.0; // was 0.3
    particle.color = [[COOKBOOK_PURPLE_COLOR
        colorWithAlphaComponent:0.5f] CGColor];
    particle.contents =
        (__bridge id)([UIImage imageNamed:@"spark.png"].CGImage);
    particle.name = @"particle";

    emitter.emitterCells = [NSArray arrayWithObject:particle];
    [self.layer addSublayer:emitter];
}

- (void) touchesMoved:(NSSet *)touches withEvent:(UIEvent *)event
{
    CGPoint pt = [[touches anyObject] locationInView:self];

    // Disable implicit animations
    [CATransaction begin];
    [CATransaction setValue:(id)kCFBooleanTrue
        forKey:kCATransactionDisableActions];
    emitter.emitterPosition = pt;
    [CATransaction commit];
}

- (void) touchesEnded:(NSSet *)touches withEvent:(UIEvent *)event
{
    [emitter removeFromSuperlayer];
```

```
    emitter = nil;
}

- (void) touchesCancelled:(NSSet *)touches withEvent:(UIEvent *)event
{
    [self touchesEnded:touches withEvent:event];
}
@end
```

> **Get This Recipe's Code**
> To find this recipe's full sample project, point your browser to https://github.com/erica/iOS-6-Advanced-Cookbook and go to the folder for Chapter 6.

Summary

This chapter introduced many ways to handle images. You read about generating images, resizing them, rotating them, and processing them using both the Accelerate and Core Image frameworks. Before leaving this chapter, here are some thoughts about the recipes you saw here:

- Moving between `UIImage` and bitmap representations enables you to access images on a byte-by-byte level. This enables you to perform image processing and test touches against content. Exposing images at a byte level means you always have full control over an image content, not just an opaque image object.

- The way you load and manipulate objects varies by source. Items in the application bundle are imported using method calls that cannot be used for those found in the sandbox, on the Internet, or in the system assets library. When loaded, the `UIImage` class provides a unified workhorse for displaying that data.

- The Accelerate framework may be ugly, but it's super-reliable. The Core Image framework remains an evolving beast on iOS and may not provide as favorable an experience, although it keeps getting better over time.

- iOS's QuartzCore library enables you work with graphics at a low level. A lightweight 2-D rendering system, you can use these calls for any kind of drawing operations you work with, whether rendering views in `drawRect:` or creating images on-the-fly.

7
Cameras

Cameras kick images up to the next level. They enable you to integrate live feeds and user-directed snapshots into your applications and provide raw data sourced from the real world. In this chapter, you read about image capture. You discover how to take pictures using Apple-sourced classes and how to roll your own from scratch. You learn about controlling image metadata and how to integrate live feeds with advanced filtering. This chapter focuses on image capture from a hardware point of view. Whether you're switching on the camera flash LED or detecting faces, this chapter introduces the ins and outs of iOS image capture technology.

Recipe: Snapping Photos

iOS offers a variety of system-supplied view controllers that enable developers to offer system features with a minimum of programming. The iOS image picker controller is one of these. It enables users to snap photos with a device's built-in camera. Because cameras are not available on all iOS units (specifically, older iPod touch and iPad devices), begin by checking whether the system running the application supports camera usage:

```
if ([UIImagePickerController isSourceTypeAvailable:
    UIImagePickerControllerSourceTypeCamera]) ...
```

The rule is this: Never offer camera-based features for devices without cameras. Although iOS 6 was deployed only to camera-ready devices, Apple has not committed to this as policy. As unlikely as it sounds, it could introduce new models without cameras. Until Apple says otherwise, assume the possibility exists for a noncamera system, even under modern iOS releases. Further, assume this method can accurately report state for camera-enabled devices whose source has been disabled through some future system setting.

Setting Up the Picker

You instantiate a camera version of the image picker the way you create a picture selection one. Just change the source type to from library or camera roll to camera. As with other modes, you can allow or disallow image editing as part of the photo-capture process by setting the `allowsEditing` property.

Although the setup is the same, the user experience differs slightly (see Figure 7-1). The camera picker offers a preview that displays after the user taps the camera icon to snap a photo. This preview enables users to Retake the photo or Use the photo as is. Once they tap Use, control passes to the next phase. If you've enabled image editing, the user does so next. If not, control moves to the standard "did finish picking" method in the delegate.

Figure 7-1 The camera version of the image picker controller offers a distinct user experience for snapping photos.

Most modern devices offer more than one camera. The iPhone 3GS, the last remaining iOS 6 dinosaur, does not. Assign the `cameraDevice` property to select which camera you want to use. The rear camera is always the default.

The `isCameraDeviceAvailable:` class method queries whether a camera device is available. This snippet checks to see if the front camera is available, and if so selects it:

```
if ([UIImagePickerController isCameraDeviceAvailable:
    UIImagePickerControllerCameraDeviceFront])
    picker.cameraDevice = UIImagePickerControllerCameraDeviceFront;
```

Here are a few more points about the camera or cameras that you can access through the `UIImagePickerController` class:

- You can query the device's capability to use flash using the `isFlashAvailableForCameraDevice:` class method. Supply either the front or back device constant. This method returns `YES` for available flash, or otherwise `NO`.

- When a camera supports flash, you can set the `cameraFlashMode` property directly to auto (`UIImagePickerControllerCameraFlashModeAuto`, which is the default), to always used (`UIImagePickerControllerCameraFlashModeOn`), or always off (`UIImagePickerControllerCameraFlashModeOff`). Selecting "off" disables the flash regardless of ambient light conditions.

- Choose between photo and video capture by setting the `cameraCaptureMode` property. The picker defaults to photo capture mode. You can test what modes are available for a device using `availableCaptureModesForCameraDevice:`. This returns an array of `NSNumber` objects, each of which encodes a valid capture mode, either photo (`UIImagePickerControllerCameraCaptureModePhoto`) or video (`UIImagePickerControllerCameraCaptureModeVideo`).

Displaying Images

When working with photos, keep image size in mind. Snapped pictures, especially those from high-resolution cameras, can be quite large compared to screen sizes, even in the age of Retina displays. Those captured from front-facing video cameras use lower-quality sensors and are smaller.

Content modes provide a view solution for displaying large images. They allow image views to scale their embedded images to available screen space. Consider using one of the following modes:

- The `UIViewContentModeScaleAspectFit` mode ensures that the entire image is shown with the aspect ratio retained. The image may be padded with empty rectangles on the sides or the top and bottom to preserve that aspect.

- The `UIViewContentModeScaleAspectFill` mode displays as much of the image as possible, while filling the entire view. Some content may be clipped so that the entire view's bounds are filled.

Saving Images to the Photo Album

Save a snapped image (or any `UIImage` instance, actually) to the photo album by calling `UIImageWriteToSavedPhotosAlbum()`. This function takes four arguments. The first is the image to save. The second and third arguments specify a callback target and selector, typically your primary view controller and `image:didFinishSavingWithError:contextInfo:`. The fourth argument is an optional context pointer. Whatever selector you use, it must take three arguments: an image, an error, and a pointer to the passed context information.

Recipe 7-1 uses this function to demonstrate how to snap a new image, allow user edits, and then save it to the photo album. Users will be asked to permit access to their photo library on this call. Should they deny access, the attempt will fail.

You can customize the iOS access alert by adding an `NSPhotoLibraryUsageDescription` key to your project's Info.plist file. This entry describes the reason your app wants to access the photo library and is displayed in the dialog box that requests access.

Recipe 7-1 **Snapping Pictures**

```objc
// "Finished saving" callback method
- (void)image:(UIImage *)image
    didFinishSavingWithError: (NSError *)error
    contextInfo:(void *)contextInfo;
{
    // Handle the end of the image write process
    if (!error)
        NSLog(@"Image written to photo album");
    else
        NSLog(@"Error writing to photo album: %@", error.localizedFailureReason);
}

// Save the returned image
- (void)imagePickerController:(UIImagePickerController *)picker
    didFinishPickingMediaWithInfo:(NSDictionary *)info
{
    // Use the edited image if available
    UIImage __autoreleasing *image =
        info[UIImagePickerControllerEditedImage];

    // If not, grab the original image
    if (!image) image = info[UIImagePickerControllerOriginalImage];

    NSURL *assetURL = info[UIImagePickerControllerReferenceURL];
    if (!image && !assetURL)
    {
        NSLog(@"Cannot retrieve an image from selected item. Giving up.");
    }
    else if (!image)
    {
        NSLog(@"Retrieving from Assets Library");
        [self loadImageFromAssetURL:assetURL into:&image];
    }

    if (image)
    {
        // Save the image
        UIImageWriteToSavedPhotosAlbum(image, self,
            @selector(image:didFinishSavingWithError:contextInfo:), NULL);
        imageView.image = image;
    }

    [self performDismiss];
}
```

```
- (void) loadView
{
    self.view = [[UIView alloc] init];

    // Only present the "Snap" option for camera-ready devices
    if ([UIImagePickerController isSourceTypeAvailable:
        UIImagePickerControllerSourceTypeCamera])
            self.navigationItem.rightBarButtonItem =
                SYSBARBUTTON(UIBarButtonSystemItemCamera, @selector(snapImage)));
}
```

> **Get This Recipe's Code**
>
> To find this recipe's full sample project, point your browser to https://github.com/erica/iOS-6-Cookbook and go to the folder for Chapter 7.

Recipe: Enabling a Flashlight

The iPhone 4 was the first iOS device to offer a built-in LED camera flash. Since then the flash has migrated to several newer units. It's simple to control that LED from your application using the torch mode. Recipe 7-2 demonstrates this functionality, enabling users to toggle the light on and off. As you would expect, powering the "torch" is not especially battery-friendly for long use but isn't particularly harmful for short durations.

This recipe begins by surveying the available onboard camera devices. This number may range from 0 (no onboard cameras) through 2 (for current generation devices). Each camera may or may not support onboard flash. The fifth-generation iPod touch does; the third-generation iPad does not. To discover whether a unit provides a torch-ready LED flash, turn to AVFoundation. This framework, which powers many recipes in this chapter, offers programmatic interfaces for testing and interacting with audio-visual capture.

Recipe 7-2 queries the current unit to recover a list of all onboard capture devices (that is, cameras). It checks to see if each capture device offers a torch (namely, an LED flash that can provide general illumination) and whether that torch supports being switched on.

Upon finding any compliant capture device, it stores that information to a local instance variable and the app reveals a simple switch, which calls back to the `toggleLightSwitch` method. If it cannot find a torch-ready device, the recipe's `supportsTorchMode` method returns NO and the app basically shuts down.

The `toggleLightSwitch` method handles the work of switching the light on and off. This method demonstrates the basic flow for this process. Before setting capture device properties—these include focus, exposure, or the flash's on/off state—you must acquire an exclusive lock.

The code locks the camera device, updates its torch mode to either on or off, and then unlocks it.

As a rule, always release the device lock as soon as you're done making your updates. Holding onto a lock can affect other applications, which may be trying to access the same hardware. Use this lock/modify/unlock process to apply updates to capture device settings.

In addition to torch mode, you can check whether a unit supports basic camera flash (hasFlash), white balance adjustment (isWhiteBalanceModeSupported:), exposure adjustment (isExposureModeSupported:), and autofocus modes (isFocusModeSupported:). Tests and updates for these other features follow the same flow you see in Recipe 7-2.

Recipe 7-2 **Controlling Torch Mode**

```
@implementation TestBedViewController
{
    UISwitch *lightSwitch;
    AVCaptureDevice *device;
}

- (void) toggleLightSwitch
{
    // Lock the device
    if ([device lockForConfiguration:nil])
    {
        // Toggle the light on or off
        device.torchMode = (lightSwitch.isOn) ?
            AVCaptureTorchModeOn : AVCaptureTorchModeOff;

        // Unlock and proceed
        [device unlockForConfiguration];
    }
}

- (BOOL) supportsTorchMode
{
    // Survey all the onboard capture devices
    NSArray *devices = [AVCaptureDevice devices];
    for (AVCaptureDevice *aDevice in devices)
    {
        if ((aDevice.hasTorch) &&
            [aDevice isTorchModeSupported:AVCaptureTorchModeOn])
        {
            device = aDevice;
             return YES;
        }
    }
```

```
        // No torch-ready camera found. Return NO.
        return NO;
}

- (void) loadView
{
    self.view = [[UIView alloc] init];
    self.view.backgroundColor = [UIColor whiteColor];

    // Initially hide the light switch
    lightSwitch = [[UISwitch alloc] init];
    [self.view addSubview:lightSwitch];
    lightSwitch.alpha = 0.0f; // Alternatively, setHidden:

    // Only reveal the switch for torch-ready units
    if ([self supportsTorchMode])
        lightSwitch.alpha = 1.0f;
    else
        self.title = @"Flash/Torch not available";
}
@end
```

Get This Recipe's Code

To find this recipe's full sample project, point your browser to https://github.com/erica/iOS-6-Cookbook and go to the folder for Chapter 7.

Recipe: Accessing the AVFoundation Camera

AVFoundation enables you to access to camera buffer without using the cumbersome image picker. It's faster and more responsive than the picker but does not offer that friendly built-in interface. For example, you might want to build an application that evaluates the contents of the camera for augmented reality or that allows users to play with a video feed.

AVFoundation enables you to retrieve both a live preview as well as raw image buffer data from the camera, providing valuable tools in your development arsenal. To start with AVFoundation, you need to build an Xcode project that uses quite a number of frameworks.

Here are frameworks you should consider adding and the roles they play in your application. Only the first three are required. You may want to add all of them. The extra frameworks are commonly used with AVFoundation video feeds and image capture:

- **AVFoundation**—Manage and play audio-visual media in your iOS applications: <AVFoundation/AVFoundation.h>.

- **CoreVideo**—Play and process movies with frame-by-frame control: `<CoreVideo/CoreVideo.h>`.
- **CoreMedia**—Handle time-based AV assets: `<CoreMedia/CoreMedia.h>`.
- **CoreImage**—Use pixel-accurate near-real-time image processing: `<CoreImage/CoreImage.h>`.
- **ImageIO**—Add and edit image metadata including GPS and EXIF support: `<ImageIO/ImageIO.h>`.
- **QuartzCore**—Add 2D graphics rendering support and access to view layers: `<QuartzCore/QuartzCore.h>`.

Requiring Cameras

If your application is built around a live camera feed, it should not install on any device that doesn't offer a built-in camera. You can mandate this by editing the Info.plist file, adding items to a `UIRequiredDeviceCapabilities` array. Add `still-camera` to assert that a camera, any camera, is available. You can narrow down device models and capabilities by specifying there must be an `auto-focus-camera`, a `front-facing-camera`, `camera-flash`, and `video-camera` as well. Required items are discussed further in Chapter 1, "Device-Specific Development."

Querying and Retrieving Cameras

iOS devices may offer zero, one, or two cameras. For multicamera devices, you can allow your user to select which camera to use, switching between the front and back. Listing 7-1 shows how to retrieve the number of cameras, to check whether the front and back cameras are available on a given device, and how to return `AVCaptureDevice` instances for each.

Listing 7-1 **Cameras**

```
+ (int) numberOfCameras
{
    return [AVCaptureDevice devicesWithMediaType:AVMediaTypeVideo].count;
}

+ (BOOL) backCameraAvailable
{
    NSArray *videoDevices =
        [AVCaptureDevice devicesWithMediaType:AVMediaTypeVideo];
    for (AVCaptureDevice *device in videoDevices)
        if (device.position == AVCaptureDevicePositionBack) return YES;
    return NO;
}
```

```objc
+ (BOOL) frontCameraAvailable
{
    NSArray *videoDevices =
        [AVCaptureDevice devicesWithMediaType:AVMediaTypeVideo];
    for (AVCaptureDevice *device in videoDevices)
        if (device.position == AVCaptureDevicePositionFront) return YES;
    return NO;
}

+ (AVCaptureDevice *)backCamera
{
    NSArray *videoDevices =
        [AVCaptureDevice devicesWithMediaType:AVMediaTypeVideo];
    for (AVCaptureDevice *device in videoDevices)
        if (device.position == AVCaptureDevicePositionBack)
            return device;

    // Return whatever is available if there's no back camera
    return [AVCaptureDevice defaultDeviceWithMediaType:AVMediaTypeVideo];
}

+ (AVCaptureDevice *)frontCamera
{
    NSArray *videoDevices =
        [AVCaptureDevice devicesWithMediaType:AVMediaTypeVideo];
    for (AVCaptureDevice *device in videoDevices)
        if (device.position == AVCaptureDevicePositionFront)
            return device;

    // Return whatever device is available if there's no back camera
    return [AVCaptureDevice defaultDeviceWithMediaType:AVMediaTypeVideo];
}
```

Establishing a Camera Session

After you select a device to work with and retrieve its `AVCaptureDevice` instance, you can establish a new camera session. Listing 7-2 shows the steps involved. A session includes creating a capture input, which grabs data from the selected camera, and a capture output, which sends buffer data to its delegate.

Listing 7-2 **Creating a Session**

```objc
// Create the capture input
AVCaptureDeviceInput *captureInput =
    [AVCaptureDeviceInput deviceInputWithDevice:device error:&error];
```

```
    if (!captureInput)
    {
        NSLog(@"Error establishing device input: %@", error);
        return;
    }

    // Create capture output.
    char *queueName = "com.sadun.tasks.grabFrames";
    dispatch_queue_t queue = dispatch_queue_create(queueName, NULL);
    AVCaptureVideoDataOutput *captureOutput =
        [[AVCaptureVideoDataOutput alloc] init];
    captureOutput.alwaysDiscardsLateVideoFrames = YES;
    [captureOutput setSampleBufferDelegate:self queue:queue];

    // Establish settings
    NSDictionary *settings = [NSDictionary
        dictionaryWithObject:
            [NSNumber numberWithUnsignedInt:kCVPixelFormatType_32BGRA]
        forKey:(NSString *)kCVPixelBufferPixelFormatTypeKey];
    [captureOutput setVideoSettings:settings];

    // Create a session
    _session = [[AVCaptureSession alloc] init];
    [_session addInput:captureInput];
    [_session addOutput:captureOutput];
```

After a camera session is created, you can start it and stop it by sending `startRunning` and `stopRunning` method calls, as shown here:

```
[session startRunning];
```

When running, a session sends regular buffer updates to its delegate. Here is where you can catch the raw data and convert it into a form that's better suited for image processing work. This method stores a `CIImage` instance as a retained property. Assuming you implement the capture routine in a helper class, you can retrieve this data from your application on demand. Listing 7-3 shows how you can capture output.

Listing 7-3 **Capturing Output**

```
- (void)captureOutput:(AVCaptureOutput *)captureOutput
    didOutputSampleBuffer:(CMSampleBufferRef)sampleBuffer
    fromConnection:(AVCaptureConnection *)connection
{
    @autoreleasepool
    {
        // Transfer into a Core Video image buffer
        CVImageBufferRef imageBuffer =
```

```
            CMSampleBufferGetImageBuffer(sampleBuffer);

    // Create a Core Image result
    CFDictionaryRef attachments =
        CMCopyDictionaryOfAttachments(kCFAllocatorDefault,
            sampleBuffer, kCMAttachmentMode_ShouldPropagate);
    self.ciImage = [[CIImage alloc]
        initWithCVPixelBuffer:imageBuffer
        options:(__bridge_transfer NSDictionary *)attachments];
    }
}
```

Core Image's `CIImage` class is similar to `UIImage` and can be converted back and forth as needed. You can initialize a `UIImage` instance with a Core Image instance as follows:

```
UIImage *newImage = [UIImage imageWithCIImage:self.ciImage];
return newImage;
```

> **Note**
>
> At the time this book was being written, the `imageWithCIImage:` method was not properly operational. The workaround methods in the sample code that accompanies this chapter were added because of that.

The capture-output delegate method uses the Core Image version for several reasons. First, it's the natural API recipient for Core Video pixel buffers. Converting to a `CIImage` takes only a single call. Second, it enables you to integrate your captured image with Core Image filtering and feature detection such as face detection. You'll likely want to add filters outside the capture routine because these can slow down processing; you don't want to add filtering or detection to every frame you capture.

Switching Cameras

After you establish a session, you can switch cameras without stopping or tearing down that session. Enter configuration mode, make your changes, and then commit the configuration updates, as shown in Listing 7-4. If your session is stopped, you may want to restart it so that it's clear to the user that the cameras have actually switched.

Listing 7-4   **Selecting from Available Cameras**

```
- (void) switchCameras
{
    if (![CameraImageHelper numberOfCameras] > 1) return;

    _isUsingFrontCamera = !_isUsingFrontCamera;
    AVCaptureDevice *newDevice = _isUsingFrontCamera ?
```

```objc
            [CameraImageHelper frontCamera] : [CameraImageHelper backCamera];

    [session beginConfiguration];

    // Remove existing inputs
    for (AVCaptureInput *input in [session inputs])
        [_session removeInput:input];

    // Change the input
    AVCaptureDeviceInput *captureInput =
        [AVCaptureDeviceInput deviceInputWithDevice: newDevice error:nil];
    [_session addInput:captureInput];

    [_session commitConfiguration];
}
```

Camera Previews

AVFoundation's `AVCaptureVideoPreviewLayer` offers a live image preview layer that you can embed into views. Preview layers are both simple to use and powerful in their effect. Listing 7-5 shows how to create a new preview layer and insert it into a view, matching the view's frame. The video gravity here defaults to resize-aspect, which preserves the video's aspect ratio while fitting it within a given layer's bounds.

You can retrieve the preview layer from a view by searching for a layer of the right class.

Listing 7-5 **Embedding and Retrieving Previews**

```objc
- (void) embedPreviewInView: (UIView *) aView
{
    if (!_session) return;

    AVCaptureVideoPreviewLayer *preview =
        [AVCaptureVideoPreviewLayer layerWithSession: session];
    preview.frame = aView.bounds;
    preview.videoGravity = AVLayerVideoGravityResizeAspect;
    [aView.layer addSublayer: preview];
}
- (AVCaptureVideoPreviewLayer *) previewInView: (UIView *) view
{
    for (CALayer *layer in view.layer.sublayers)
        if ([layer isKindOfClass:[AVCaptureVideoPreviewLayer class]])
            return (AVCaptureVideoPreviewLayer *)layer;

    return nil;
}
```

Laying Out a Camera Preview

One of the core challenges of working with live camera previews is keeping the video pointing up regardless of the device orientation. Although the camera connection does handle some orientation issues, it's easiest to work directly with the preview layer and transform it to the proper orientation. Listing 7-6 shows how this is done.

Listing 7-6 **Adding Camera Previews**

```
- (void) layoutPreviewInView: (UIView *) aView
{
    AVCaptureVideoPreviewLayer *layer =
        [self previewInView:aView];
    if (!layer) return;

    UIDeviceOrientation orientation =
        [UIDevice currentDevice].orientation;
    CATransform3D transform = CATransform3DIdentity;
    if (orientation == UIDeviceOrientationPortrait) ;
    else if (orientation == UIDeviceOrientationLandscapeLeft)
        transform = CATransform3DMakeRotation(-M_PI_2, 0.0f, 0.0f, 1.0f);
    else if (orientation == UIDeviceOrientationLandscapeRight)
        transform = CATransform3DMakeRotation(M_PI_2, 0.0f, 0.0f, 1.0f);
    else if (orientation == UIDeviceOrientationPortraitUpsideDown)
        transform = CATransform3DMakeRotation(M_PI, 0.0f, 0.0f, 1.0f);

    layer.transform = transform;
    layer.frame = aView.frame;
}
```

Camera Image Helper

Recipe 7-3 combines all the methods described in this section into a simple helper class. This class offers all the basic features needed to query for cameras, establish a session, create a preview, and retrieve an image. To use this class, create a new instance, start running a session, and maybe embed a preview into one of your main views. Make sure to lay out your preview again any time you autorotate the interface, so the preview will remain properly oriented.

Recipe 7-3 **Helper Class for Cameras**

```
@interface CameraImageHelper : NSObject
    <AVCaptureVideoDataOutputSampleBufferDelegate>

@property (nonatomic, strong)   AVCaptureSession *session;
@property (nonatomic, strong)   CIImage *ciImage;
```

```
@property (nonatomic, readonly) UIImage *currentImage;
@property (nonatomic, readonly) BOOL isUsingFrontCamera;

+ (int) numberOfCameras;
+ (BOOL) backCameraAvailable;
+ (BOOL) frontCameraAvailable;
+ (AVCaptureDevice *) backCamera;
+ (AVCaptureDevice *) frontCamera;

+ (id) helperWithCamera: (NSUInteger) whichCamera;

- (void) startRunningSession;
- (void) stopRunningSession;
- (void) switchCameras;

- (void) embedPreviewInView: (UIView *) aView;
- (AVCaptureVideoPreviewLayer *) previewInView: (UIView *) view;
- (void) layoutPreviewInView: (UIView *) aView;
@end
```

> **Get This Recipe's Code**
> To find this recipe's full sample project, point your browser to https://github.com/erica/iOS-6-Advanced-Cookbook and go to the folder for Chapter 7.

Recipe: EXIF

Exchangeable Image File Format (EXIF) was created by the Japan Electronic Industries Development Association (JEIDA). It offers a standardized set of image annotations that include camera settings (such as shutter speed and metering mode), date and time information, and an image preview. Other metadata standards include International Press Telecommunications Council (IPTC) and Adobe's Extensible Metadata Platform (XMP).

Orientation is the only meta-information currently exposed directly through the `UIImage` class. This information is accessible through its `imageOrientation` property; as you see in the next section, it bears a correspondence with EXIF orientation, but the value uses a different numbering system.

ImageIO

The ImageIO framework was first introduced to iOS in iOS 4. ImageIO specializes (as the name suggests) in reading and writing image data. Using "image sources" instead of standard file access enables you to access image metadata properties. These properties include basic values such as the image's width and height, as well as more nuanced information such as the camera's shutter speed and focal length that were used to capture the image. Readers of the last edition of this Cookbook asked me to add a recipe showing how to read and update EXIF and location (GPS) metadata. Recipe 7-4 addresses this by implementing a custom MetaImage class.

This class encapsulates a standard UIImage, augmenting it with a metadata properties dictionary. The properties derive from the image, exposing those values for reading and writing. To create that transparency, the properties must be initialized somewhere. So in this class, when you build instances by providing an image, the class generates the base categories for that data. Similarly, when created from a file, it reads whatever metadata has been stored with that image.

Querying Metadata

Listing 7-7 demonstrates how to query metadata for both UIImage and stored file sources. The first of these two functions reads in the image properties dictionary from the file at a given path. It creates an image source using a file URL and then copies the properties from that source. The properties return as a mutable dictionary.

The second function starts with an image instance rather than a file. Because image sources must work with data, the method converts the UIImage instance to JPEG data, which was chosen as the most common use case. This time, the image source is generated from the image data, but the rest of the work is otherwise identical. The function copies the source's properties and returns them as a mutable dictionary.

Data returned from the UIImage function is, expectedly, sparse. Generated pictures cannot offer details regarding white balance, focal planes, and metering. The metadata for a UIImage instance includes its depth, orientation, pixel dimensions, and color model.

Listing 7-7 **Retrieving Image Metadata**

```
// Read image properties at a file path
NSMutableDictionary *imagePropertiesDictionaryForFilePath(
    NSString *path)
{
    CFDictionaryRef options = (__bridge CFDictionaryRef)@{
    };

    CFURLRef url = (__bridge CFURLRef) [NSURL fileURLWithPath:path];
    CGImageSourceRef imageSource =
        CGImageSourceCreateWithURL(url, options);
    if (imageSource == NULL)
    {
        NSLog(@"Error: Could not establish \
            image source for file at path: %@", path);
        return nil;
    }
```

```objc
    CFDictionaryRef imagePropertiesDictionary =
        CGImageSourceCopyPropertiesAtIndex(imageSource, 0, NULL);
    CFRelease(imageSource);

    return [NSMutableDictionary dictionaryWithDictionary:
        (__bridge_transfer NSDictionary *)
            imagePropertiesDictionary];
}

// Read the properties for an image instance
NSMutableDictionary *imagePropertiesFromImage(UIImage *image)
{
    CFDictionaryRef options = (__bridge CFDictionaryRef)@{
    };
    NSData *data = UIImageJPEGRepresentation(image, 1.0f);
    CGImageSourceRef imageSource =
        CGImageSourceCreateWithData(
            (__bridge CFDataRef) data, options);
    if (imageSource == NULL)
    {
        NSLog(@"Error: Could not establish image source");
        return nil;
    }
    CFDictionaryRef imagePropertiesDictionary =
        CGImageSourceCopyPropertiesAtIndex(imageSource, 0, NULL);
    CFRelease(imageSource);

    return [NSMutableDictionary dictionaryWithDictionary:
        (__bridge_transfer NSDictionary *)
            imagePropertiesDictionary];
}
```

Wrapping UIImage

Although Listing 7-7 showed how to import and retrieve metadata, Recipe 7-4 details how the `MetaImage` wrapper implementation exports that data back to file. Its `writeToPath:` method builds a new file using an image destination, storing both the image and metadata.

This implementation first writes this data to a temporary file and then moves it into place. You'll probably want to tweak these details to match your normal practices. As with Listing 7-7, this wrapper method calls ImageIO's Core Foundation-style C function library, which is why there's so much ARC-style bridging in use.

In this implementation, two class properties (`gps` and `exif`) expose subdictionaries embedded into the primary metadata. Established by the wrapper class as mutable instances, they enable you to directly update dictionary values in preparation for writing metadata to an enhanced image file. Here's an example of how you might load an image, adjust its metadata, and store it to file.

```
MetaImage *mi = [MetaImage newImage:image];
mi.exif[@"UserComment"] = @"This is a test comment";
[mi writeToPath:destPath];
```

Both dictionary properties use standard keys. These keys are declared in the ImageIO framework and are based off industry standards. Refer to Apple's CGImageProperties reference for details. This document defines the image characteristics used by the framework, specifies the key constants you use to address them, and summarizes how they're typically used.

Recipe 7-4 **Exposing Image Metadata**

```
- (BOOL) writeToPath: (NSString *) path
{
    // Prepare to write to temporary path
    NSString *temporaryPath =
        [NSTemporaryDirectory() stringByAppendingPathComponent:
            [path lastPathComponent]];
    if ([[NSFileManager defaultManager]
        fileExistsAtPath:temporaryPath])
    {
        if (![[NSFileManager defaultManager]
            removeItemAtPath:temporaryPath error:nil])
        {
            NSLog(@"Could not establish temporary writing file");
            return NO;
        }
    }

    // Where to write to
    NSURL *temporaryURL = [NSURL fileURLWithPath:temporaryPath];
    CFURLRef url = (__bridge CFURLRef) temporaryURL;

    // What to write
    CGImageRef imageRef = self.image.CGImage;

    // Metadata
    NSDictionary *properties = [NSDictionary
        dictionaryWithDictionary:self.properties];
    CFDictionaryRef propertiesRef =
        (__bridge CFDictionaryRef) properties;

    // UTI -- See Chapter 2
    NSString *uti = preferredUTIForExtension(path.pathExtension);
```

```objc
    if (!uti) uti = @"public.image";
    CFStringRef utiRef = (__bridge CFStringRef) uti;

    // Create image destination
    CGImageDestinationRef imageDestination =
        CGImageDestinationCreateWithURL(url, utiRef, 1, NULL);

    // Save data
    CGImageDestinationAddImage(imageDestination, imageRef, propertiesRef);
    CGImageDestinationFinalize(imageDestination);

    // Clean up
    CFRelease(imageDestination);

    // Move file into place
    NSURL *destURL = [NSURL fileURLWithPath:path];

    BOOL success;
    NSError *error;

    // Remove previous file
    if ([[NSFileManager defaultManager] fileExistsAtPath:path])
    {
        success = [[NSFileManager defaultManager]
            removeItemAtPath:path error:&error];
        if (!success)
        {
            NSLog(@"Error: Could not overwrite file properly. \
                Original not removed.");
            return NO;
        }
    }

    success = [[NSFileManager defaultManager]
        moveItemAtURL: temporaryURL toURL:destURL error:&error];
    if (!success)
    {
        NSLog(@"Error: could not move new file into place: %@",
            error.localizedFailureReason);
        return NO;
    }

    return YES;
}
```

> **Get This Recipe's Code**
>
> To find this recipe's full sample project, point your browser to https://github.com/erica/iOS-6-Advanced-Cookbook and go to the folder for Chapter 7.

Image Orientations

Geometry represents the most difficult portion of working with AVFoundation camera feeds. Unlike using image pickers, which automatically handle these matters for you, you must deal with raw image buffers. The camera you use (front or back) and the orientation of the device influence how the image data is oriented, as shown in Table 7-1.

Table 7-1 **Mapping Device Orientation to Image Orientation**

Orientation	Home Button	Camera	Natural Output	Mirrored
Portrait	Bottom	Front	Left Mirrored	Right
LandscapeLeft	Right	Front	Down Mirrored	Down
PortraitUpsideDown	Top	Front	Right Mirrored	Left
LandscapeRight	Left	Front	Up Mirrored	Up
Portrait	Bottom	Back	Right	Left Mirrored
LandscapeLeft	Right	Back	Up	Up Mirrored
PortraitUpsideDown	Top	Back	Left	Right Mirrored
LandscapeRight	Left	Back	Down	Down Mirrored

The Natural Output column in Table 7-1 specifies the orientation produced by each device orientation and camera combination. For example, a device held in portrait, using the front camera, results in left-mirrored output. The most "natural" setup for an iPhone is created by holding the camera landscape left—that is, with the Home button to the right and using the back camera. This creates an up-oriented image.

The orientation equivalents in Table 7-2 tell iOS how to map EXIF-values to UIImage settings. You use this data to restore images to their proper presentation. Examine the first F in the table. This F is standing up and its leaves point to the right. An image with this orientation has raw data matching this orientation. Pixel data starts at the top left and moves left to right, row after row, eventually to the bottom right.

When users snap photos using other orientations or use a front camera instead of a back camera, the data is stored as it's captured from the camera's sensors. The top-left corner of the picture may not appear at pixel 0 of the image data. For example, say a user snaps a photo using the front camera, with the Home button at the right. Table 7-1 indicates the output will be down-mirrored. This is UI orientation 5, EXIF orientation 4. The raw picture data will not only be upside down, but it will also be mirrored.

Knowing how the underlying data representation (sideways, mirrored, upside down, and so on) maps to the intended representation (always with the image starting at the top left) is critical for image processing, especially for automatic detection routines. The last column in Table 7-2 shows the EXIF orientation equivalent for each `UIImageOrientation`. This conversion enables you to work with Core Image. CI uses EXIF not `UIImage` orientation.

Table 7-2 Image Orientations

Orientation	UIImageOrientation	EXIF Equivalent
XXXXXX XX XXXX XX XX	UIImageOrientationUp 0	Top Left 1
XX XX XXXX XX XXXXXX	UIImageOrientationDown 1	Bottom Right 3
XX XX XX XXXXXXXXXX	UIImageOrientationLeft 2	Right Top 6
XXXXXXXXXX XX XX XX	UIImageOrientationRight 3	Left Bottom 8
XXXXXX XX XXXX XX XX	UIImageOrientationUpMirrored 4	Top Right 2
XX XX XXXX XX XXXXXX	UIImageOrientationDownMirrored 5	Bottom Left 4
XXXXXXXXXX XX XX XX	UIImageOrientationLeftMirrored 6	Left Top 5

Orientation	UIImageOrientation	EXIF Equivalent
XX	UIImageOrientationRightMirrored	Right Bottom
XX XX	7	7
XXXXXXXXXX		

The following functions allow you to convert between the EXIF and `UIImage` orientation schemes:

```
NSUInteger exifOrientationFromUIOrientation(
    UIImageOrientation uiorientation)
{
    if (uiorientation > 7) return 1;
    int orientations[8] = {1, 3, 6, 8, 2, 4, 5, 7};
    return orientations[uiorientation];
}

UIImageOrientation imageOrientationFromEXIFOrientation(
    NSUInteger exiforientation)
{
    if ((exiforientation < 1) || (exiforientation > 8))
        return UIImageOrientationUp;
    int orientations[8] = {0, 4, 1, 5, 6, 2, 7, 3};
    return orientations[exiforientation];
}
```

Recipe: Core Image Filtering

Core Image (CI) filters enable you to process images quickly. Recipe 7-5 introduces CI filtering by applying a simple monochrome filter to images captured from the onboard camera. This filter is not applied to a live video feed but rather to individual images. A monochrome filter transforms an RGB image to a single channel and then applies a single color, which you specify as an input to the filter.

This recipe is powered by a `CADisplayLink`. Each time it fires, the `snap` method grabs the current `ciImage` from its helper (refer to Recipe 7-3). If the user has enabled filtering (using a simple bar button toggle), the code creates a new filter, sets the image as its input, adjusts the output color to red, and then retrieves and displays the output image.

You can adjust the display link's frame interval to trade off between smooth presentation and processing overhead. This is a critical factor, especially when using high-resolution cameras. Increase the frame interval to expand the number of frames that must pass before the display link calls `snap` again.

Recipe 7-5 Adding a Simple Core Image Filter

```objc
@implementation TestBedViewController

// Switch between cameras
- (void) switch: (id) sender
{
    [helper switchCameras];
}

// Enable/disable the filter
- (void) toggleFilter: (id) sender
{
    useFilter = !useFilter;
}

// Grab an image, optionally applying the filter,
// and display the results
- (void) snap
{
    // Orientation needed for CI image workaround
    UIImageOrientation orientation =
        currentImageOrientation(helper.isUsingFrontCamera, NO);

    if (useFilter) // monochrome - red
    {
        CIFilter *filter =
            [CIFilter filterWithName:@"CIColorMonochrome"];
        [filter setValue:helper.ciImage forKey:@"inputImage"];
        [filter setDefaults];
        [filter setValue:@1 forKey:@"inputIntensity"];
        [filter setValue:[CIColor colorWithRed:1.0f
            green:0.0f blue:0.0f] forKey:@"inputColor"];
        CIImage *outputImage = [filter valueForKey:kCIOutputImageKey];

        // Apply workaround to transform ciImage to UIImage
        if (outputImage)
            imageView.image = [UIImage imageWithCIImage:outputImage
                orientation:orientation];
        else NSLog(@"Missing image");
    }

    if (!useFilter) // no filter
    {
        // Convert using ciImage workaround
        imageView.image = [UIImage imageWithCIImage:helper.ciImage
            orientation:orientation];
    }
}
```

```
- (void)viewWillAppear:(BOOL)animated
{
    [super viewWillAppear:animated];

    // Establish the preview session
    helper = [CameraImageHelper helperWithCamera:kCameraBack];
    [helper startRunningSession];

    displayLink = [CADisplayLink displayLinkWithTarget:self
        selector:@selector(snap)];
    [displayLink addToRunLoop:[NSRunLoop currentRunLoop]
        forMode:NSRunLoopCommonModes];
}
@end
```

> **Get This Recipe's Code**
>
> To find this recipe's full sample project, point your browser to https://github.com/erica/iOS-6-Advanced-Cookbook and go to the folder for Chapter 7.

Recipe: Core Image Face Detection

Face detection was introduced in iOS 5 as one of iOS's splashier features. It's something that is both easy to implement and quite hard to get right, and that all comes down to basic geometry. Retrieving the feature set involves creating a new face-type detector and querying it for the detected features. This works by returning an array of zero or more `CIFaceFeature` objects, corresponding to the number of faces found in the scene.

```
- (NSArray *) featuresInImage
{
    NSDictionary *detectorOptions = [NSDictionary
        dictionaryWithObject:CIDetectorAccuracyLow
        forKey:CIDetectorAccuracy];

    CIDetector *detector = [CIDetector
        detectorOfType:CIDetectorTypeFace context:nil
        options:detectorOptions];

    NSUInteger orientation = detectorEXIF(helper.isUsingFrontCamera, NO);
    NSDictionary *imageOptions =
        [NSDictionary dictionaryWithObject:
            [NSNumber numberWithInt:orientation]
            forKey:CIDetectorImageOrientation];
    return [detector featuresInImage:ciImage options:imageOptions];
}
```

Chapter 7 Cameras

Several issues make this a particularly hard task. The detector depends on knowing the proper orientation to interpret the CI image passed to it. It cannot detect faces that are sideways or upside down. That's why this method includes a call to a `detectorEXIF` function. The problem is that, at least at the time this book was written, the required detector orientation doesn't always match the real-world image orientation, causing workarounds like this:

```
NSUInteger detectorEXIF(BOOL isUsingFrontCamera, BOOL shouldMirrorFlip)
{
    if (isUsingFrontCamera || deviceIsLandscape())
        return currentEXIFOrientation(isUsingFrontCamera,
            shouldMirrorFlip);

    // Only back camera portrait or upside down here.
    // Detection happens but the geometry is messed.
    int orientation = currentEXIFOrientation(!isUsingFrontCamera,
        shouldMirrorFlip);
    return orientation;
}
```

As you can tell, this is hacky, buggy, and a disaster waiting to happen.

The second problem lies in the geometry. Apple has not provided APIs that map coordinates into images while respecting the image orientation information. The coordinate (0,0) may refer to the top left, bottom left, top right, or bottom right. What's more, mirroring means the coordinate may be offset from the left side or the right, from the bottom or the top. The math is left up to you. Enter Listing 7-8. These methods help you convert point and rectangle results from their raw data coordinates into normalized image coordinates.

The `ExifOrientation` enumeration used here is not built in to iOS. It is defined to match the natural EXIF values shown in Listing 7-8. Notice that there are no cases for `kRightTop` and `kLeftBottom`. These are the two EXIF orientations not currently supported by the Core Image detector. Should Apple update its implementation, you need to add cases to support both.

Listing 7-8 Converting Geometry from EXIF to Image Coordinates

```
CGPoint pointInEXIF(ExifOrientation exif, CGPoint aPoint, CGRect rect)
{
    switch(exif)
    {
        case kTopLeft:
            return CGPointMake(aPoint.x,
                rect.size.height - aPoint.y);
        case kTopRight:
            return CGPointMake(rect.size.width - aPoint.x,
                rect.size.height - aPoint.y);
        case kBottomRight:
            return CGPointMake(rect.size.width - aPoint.x,
                aPoint.y);
```

```
        case kBottomLeft:
            return CGPointMake(aPoint.x, aPoint.y);

        case kLeftTop:
            return CGPointMake(aPoint.y, aPoint.x);
        case kRightBottom:
            return CGPointMake(rect.size.width - aPoint.y,
                rect.size.height - aPoint.x);

        default:
            return aPoint;
    }
}

CGSize sizeInEXIF(ExifOrientation exif, CGSize aSize)
{
    switch(exif)
    {
        case kTopLeft:
        case kTopRight:
        case kBottomRight:
        case kBottomLeft:
            return aSize;

        case kLeftTop:
        case kRightBottom:
            return CGSizeMake(aSize.height, aSize.width);
    }
}

CGRect rectInEXIF(ExifOrientation exif, CGRect inner, CGRect outer)
{
    CGRect rect;
    rect.origin = pointInEXIF(exif, inner.origin, outer);
    rect.size = sizeInEXIF(exif, inner.size);

    switch(exif)
    {
        case kTopLeft:
            rect = CGRectOffset(rect, 0.0f, -inner.size.height);
            break;
        case kTopRight:
            rect = CGRectOffset(rect, -inner.size.width,
                -inner.size.height);
            break;
        case kBottomRight:
            rect = CGRectOffset(rect, -inner.size.width, 0.0f);
            break;
```

```
            case kBottomLeft:
                break;
                
            case kLeftTop:
                break;
            case kRightBottom:
                rect = CGRectOffset(rect, -inner.size.width,
                    -inner.size.height);
                break;
            default:
                break;
    }
    
    return rect;
}
```

Recipe 7-6 demonstrates how to perform facial detection in your application. This method retrieves detected features, outlines the detected face bounds with a shaded rectangle, and highlights three specific features (left eye, right eye, and mouth) with circles. Figure 7-2 shows the detection in action.

Figure 7-2   Automatically detecting faces and their features.

Each member of the `CIFaceFeature` array returned by the detection can report several features of geometric interest. These include the rectangular bounds around the face and positions for the left and right eyes and the mouth. These subfeatures are not always created, so you should first test whether positions are available before using them. Recipe 7-6 demonstrates both test and use.

Recipe 7-6 also handles several workarounds due to Core Image's portrait EXIF eccentricities. Should Apple update the way Core Image detects faces, adjust the code accordingly for the two remaining orientations.

Recipe 7-6 **Detecting Faces**

```
- (void) snap
{
    ciImage = helper.ciImage;
    UIImage *baseImage = [UIImage imageWithCIImage:ciImage];
    CGRect imageRect = (CGRect){.size = baseImage.size};

    NSDictionary *detectorOptions = [NSDictionary
        dictionaryWithObject:CIDetectorAccuracyLow
        forKey:CIDetectorAccuracy];

    CIDetector *detector = [CIDetector
        detectorOfType:CIDetectorTypeFace
        context:nil options:detectorOptions];

    ExifOrientation detectOrientation =
        detectorEXIF(helper.isUsingFrontCamera, NO);

    NSDictionary *imageOptions = [NSDictionary
        dictionaryWithObject:[NSNumber numberWithInt:detectOrientation]
        forKey:CIDetectorImageOrientation];
    NSArray *features =
        [detector featuresInImage:ciImage options:imageOptions];

    UIGraphicsBeginImageContext(baseImage.size);
    [baseImage drawInRect:imageRect];

    for (CIFaceFeature *feature in features)
    {
        CGRect rect = rectInEXIF(detectOrientation,
            feature.bounds, imageRect);
        if (deviceIsPortrait() && helper.isUsingFrontCamera) // workaround
        {
            rect.origin = CGPointFlipHorizontal(rect.origin, imageRect);
            rect.origin =
                CGPointOffset(rect.origin, -rect.size.width, 0.0f);
        }
```

```objc
[[[UIColor blackColor] colorWithAlphaComponent:0.3f] set];
UIBezierPath *path = [UIBezierPath bezierPathWithRect:rect];
[path fill];

if (feature.hasLeftEyePosition)
{
    [[[UIColor redColor] colorWithAlphaComponent:0.5f] set];
    CGPoint position = feature.leftEyePosition;
    CGPoint pt = pointInEXIF(detectOrientation, position,
        imageRect);

    if (deviceIsPortrait() && helper.isUsingFrontCamera)
        pt = CGPointFlipHorizontal(pt, imageRect); // workaround

    UIBezierPath *path = [UIBezierPath bezierPathWithArcCenter:pt
        radius:30.0f startAngle:0.0f endAngle:2 * M_PI
        clockwise:YES];
    [path fill];
}

if (feature.hasRightEyePosition)
{
    [[[UIColor blueColor] colorWithAlphaComponent:0.5f] set];
    CGPoint position = feature.rightEyePosition;
    CGPoint pt = pointInEXIF(detectOrientation, position,
        imageRect);
    if (deviceIsPortrait() && helper.isUsingFrontCamera)
        pt = CGPointFlipHorizontal(pt, imageRect); // workaround

    UIBezierPath *path = [UIBezierPath bezierPathWithArcCenter:pt
        radius:30.0f startAngle:0.0f endAngle:2 * M_PI
        clockwise:YES];

    [path fill];
}

if (feature.hasMouthPosition)
{
    [[[UIColor greenColor] colorWithAlphaComponent:0.5f] set];
    CGPoint position = feature.mouthPosition;
    CGPoint pt = pointInEXIF(detectOrientation, position,
        imageRect);
    if (deviceIsPortrait() && helper.isUsingFrontCamera)
        pt = CGPointFlipHorizontal(pt, imageRect); // workaround

    UIBezierPath *path = [UIBezierPath bezierPathWithArcCenter:pt
        radius:30.0f startAngle:0.0f endAngle:2 * M_PI
```

```
                clockwise:YES];
     
        [path fill];
    }

}

    imageView.image = UIGraphicsGetImageFromCurrentImageContext();
    UIGraphicsEndImageContext();
}
```

> **Get This Recipe's Code**
>
> To find this recipe's full sample project, point your browser to https://github.com/erica/iOS-6-Advanced-Cookbook and go to the folder for Chapter 7.

Recipe: Sampling a Live Feed

You can use bitmap access to create real-time responses for the user. Recipe 7-7 samples the center of each image to retrieve a prevailing color. It works by extracting a 128-by-128-pixel swatch and performing some basic statistical image processing on it. The navigation bar's color continuously updates to display this most popular color from the center of the sampled feed.

This method converts each pixel from RGB to HSB, enabling the algorithm to retrieve a characteristic hue. It increases the histogram bucket for the hue and accumulates the saturation and brightness—that accumulation will eventually be divided by the number of samples in that bucket to create an average saturation and brightness for a given hue.

The most popular hue in the sample wins. The hue with the greatest hit count (that is, the mode) forms the basis of the new color, with the average saturation and brightness rounding out the creation. That color is assigned to tint the navigation bar.

Recipe 7-7 **Analyzing Bitmap Samples**

```
#define SAMPLE_LENGTH 128
- (void) pickColor
{
    // Retrieve the center 128x128 sample as bits
    UIImage *currentImage = helper.currentImage;
    CGRect sampleRect =
        CGRectMake(0.0f, 0.0f, SAMPLE_LENGTH, SAMPLE_LENGTH);
    sampleRect = CGRectCenteredInRect(sampleRect,
        (CGRect){.size = currentImage.size});
    UIImage *sampleImage =
        [currentImage subImageWithBounds:sampleRect];
```

```
NSData *bitData = sampleImage.bytes;
Byte *bits = (Byte *)bitData.bytes;

// Create the histogram and average sampling buckets
int bucket[360];
CGFloat sat[360], bri[360];

for (int i = 0; i < 360; i++)
{
    bucket[i] = 0; // histogram sample
    sat[i] = 0.0f; // average saturation
    bri[i] = 0.0f; // average brightness
}

// Iterate over each sample pixel, accumulating hsb info
for (int y = 0; y < SAMPLE_LENGTH; y++)
    for (int x = 0; x < SAMPLE_LENGTH; x++)
    {
        CGFloat r = ((CGFloat)bits[redOffset(x, y,
            SAMPLE_LENGTH)] / 255.0f);
        CGFloat g = ((CGFloat)bits[greenOffset(x, y,
            SAMPLE_LENGTH)] / 255.0f);
        CGFloat b = ((CGFloat)bits[blueOffset(x, y,
            SAMPLE_LENGTH)] / 255.0f);

        // Convert from RGB to HSV
        CGFloat h, s, v;
        rgbtohsb(r, g, b, &h, &s, &v);
        int hue = (hue > 359.0f) ? 0 : (int) h;

        // Collect metrics on a per-hue basis
        bucket[hue]++;
        sat[hue] += s;
        bri[hue] += v;
    }

// Retrieve the hue mode
int max = -1;
int maxVal = -1;
for (int i = 0; i < 360; i++)
{
    if (bucket[i] > maxVal)
    {
        max = i;
        maxVal = bucket[i];
    }
}
```

```
    // Create a color based on the mode hue, average sat & bri
    float h = max / 360.0f;
    float s = sat[max]/maxVal;
    float br = bri[max]/maxVal;

    CGFloat red, green, blue;
    hsbtorgb((CGFloat) max, s, br, &red, &green, &blue);

    UIColor *hueColor = [UIColor colorWithHue:h saturation:s
        brightness:br alpha:1.0f];

    // Display the selected hue
    self.navigationController.navigationBar.tintColor = hueColor;

    free(bits);
}
```

Get This Recipe's Code

To find this recipe's full sample project, point your browser to https://github.com/erica/iOS-6-Advanced-Cookbook and go to the folder for Chapter 7.

Converting to HSB

Recipe 7-8 depends on converting its colors to HSB from RGB using the function in Listing 7-9. This function is adapted from the author's Ph.D. advisor Jim Foley's seminal textbook on computer graphics, *Computer Graphics: Principles and Practice in C (2nd Edition)* (Addison-Wesley Professional, ISBN-13: 9780201848403).

Listing 7-9 **Converting Between RGB and HSB**

```
void rgbtohsb(CGFloat r, CGFloat g, CGFloat b, CGFloat *pH,
    CGFloat *pS, CGFloat *pV)
{
    CGFloat h,s,v;

    // From Foley and Van Dam
    CGFloat max = MAX(r, MAX(g, b));
    CGFloat min = MIN(r, MIN(g, b));

    // Brightness
    v = max;

    // Saturation
    s = (max != 0.0f) ? ((max - min) / max) : 0.0f;
```

```
    if (s == 0.0f) {
        // No saturation, so undefined hue
        h = 0.0f;
    } else {
        // Determine hue distances from...
        CGFloat rc = (max - r) / (max - min); // from red
        CGFloat gc = (max - g) / (max - min); // from green
        CGFloat bc = (max - b) / (max - min); // from blue

        if (r == max) h = bc - gc; // between yellow & magenta
        else if (g == max) h = 2 + rc - bc; // between cyan & yellow
        else h = 4 + gc - rc; // between magenta & cyan

        h *= 60.0f; // Convert to degrees
        if (h < 0.0f) h += 360.0f; // Make non-negative
    }

    if (pH) *pH = h;
    if (pS) *pS = s;
    if (pV) *pV = v;
}
```

Summary

This chapter introduced code-level ways to work with image capture hardware on iOS devices. You read how to survey onboard equipment and how to tweak the settings. You discovered how the AVFoundation framework offers direct access to feeds without using a standard image picker. You also explored how you can sample and process the data exposed by those feeds. Here are a few final thoughts to take away from this chapter:

- The OpenCV computer vision library has been widely ported to iOS. Consider making use of its features directly rather than reinventing the wheel. You might also want to explore Accelerate and Core Image.

- I probably spent more time working through the fussy details of the AVFoundation wrapper class than on any other examples in this Cookbook series. Just getting orientation conversions alone took hours. If you do find errors, please contact me so that I can fix the details on the sample code repository and in any future printings.

- Feeds from the newest iOS devices can be huge due to the high quality of the built-in cameras. The front camera generally offers a lower-resolution data feed than the back camera. Don't be afraid to subsample and use only those parts of the image that help your application.

8

Audio

The iOS device is a media master; its built-in iPod features expertly handle both audio and video. The iOS SDK exposes that functionality to developers. A rich suite of classes simplifies media handling via playback, search, and recording. This chapter introduces recipes that use those classes for audio, presenting media to your users and letting your users interact with that media. You see how to build audio players and audio recorders. You discover how to browse the iPod library and how to choose what items to play. The recipes you're about to encounter provide step-by-step demonstrations showing how to add these media-rich features to your own apps.

Recipe: Playing Audio with AVAudioPlayer

The `AVAudioPlayer` class plays back audio data. Part of the `AVFoundation` framework, it provides a simple-to-use class that offers numerous features, several of which are highlighted in Figure 8-1. With this class, you can load audio, play it, pause it, stop it, monitor average and peak levels, adjust the playback volume, and set and detect the current playback time. All these features are available with little associated development cost. As you are about to see, the `AVAudioPlayer` class provides a solid API.

Initializing an Audio Player

The audio playback features provided by `AVAudioPlayer` take little effort to implement in your code. Apple has provided an uncomplicated class that's streamlined for loading and playing files.

To begin, allocate a player and initialize it, either with data or with the contents of a local URL. This snippet uses a file URL to point to an audio file. It reports any error involved in creating and setting up the player. You can also initialize a player with data that's already stored in memory using `initWithData:error:`. That's handy for when you've already read data into memory (such as during an audio chat) rather than reading from a file stored on the device:

```
player = [[AVAudioPlayer alloc] initWithContentsOfURL:
    [NSURL fileURLWithPath:path] error:&error];

if (!player)
{
    NSLog(@"AVAudioPlayer could not be established: %@",
        error.localizedFailureReason);
    return;
}
```

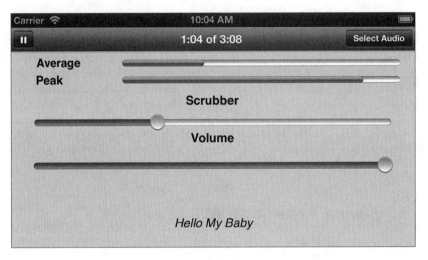

Figure 8-1   The features highlighted in this screenshot were built with a single class, AVAudioPlayer. This class provides time monitoring (in the title bar center), sound levels (average and peak), scrubbing and volume sliders, and play/pause control (at the right of the title bar).

When you've initialized the player, prepare it for playback. Calling prepareToPlay ensures that when you are ready to play the audio, that playback starts as quickly as possible. The call preloads the player's buffers and initializes the audio playback hardware:

```
[player prepareToPlay];
```

Pause playback at any time by calling pause. Pausing freezes the player's currentTime property. You can resume playback from that point by calling play again.

Halt playback entirely with stop. Stopping playback undoes the buffered setup you initially established with prepareToPlay. It does not, however, set the current time back to 0.0; you can pick up from where you left off by calling play again, just as you would with pause. You may experience starting delays as the player reloads its buffers. You normally stop playback before switching new media.

Monitoring Audio Levels

When you intend to monitor audio levels, start by setting the `meteringEnabled` property. Enabling metering lets you check levels as you play back or record audio:

```
player.meteringEnabled = YES;
```

The `AVAudioPlayer` class provides feedback for average and peak power, which you can retrieve on a per-channel basis. Query the player for the number of available channels (via the `numberOfChannels` property), and then request each power level by supplying a channel index. A mono signal uses channel 0, as does the left channel for a stereo recording.

In addition to enabling metering as a whole, you need to call `updateMeters` each time you want to test your levels; this AV player method updates the current meter levels. After you do so, use the `peakPowerForChannel:` and `averagePowerForChannel:` methods to read those levels. Recipe 8-6, which appears later in this chapter, shows the details of what's likely going on under the hood in the player when it requests those power levels. You can see that code requests the meter levels and then extracts either the peak or average power. The `AVAudioPlayer` class hides those details, simplifying access to these values.

The `AVAudioPlayer` measures power in Decibels, which is supplied in floating-point format. Decibels use a logarithmic scale to measure sound intensity. Power values range from 0 dB at the highest to some negative value representing less-than-maximum power. The lower the number (and they are all negative), the weaker the signal will be:

```
int channels = player.numberOfChannels;
[player updateMeters];
for (int i = 0; i < channels; i++)
{
    // Log the peak and average power
    NSLog(@"%d %0.2f %0.2f",
        [player peakPowerForChannel:i],
        [player averagePowerForChannel:i]);
}

// Show average and peak values on the two meters for primary channel
meter1.progress = pow(10, 0.05f * [player averagePowerForChannel:0]);
meter2.progress = pow(10, 0.05f * [player peakPowerForChannel:0]);
```

To query the audio player gain (that is, its volume), use the `volume` property. This property returns a floating-point number, here between 0.0 and 1.0, and applies specifically to the player volume rather than the system audio volume. You can set this property and read it. This snippet can be used with a target-action pair to update the volume when the user manipulates an onscreen volume slider:

```
- (void) setVolume: (id) sender
{
    // Set the audio player gain to the current slider value
    if (player) player.volume = volumeSlider.value;
}
```

Playback Progress and Scrubbing

Two properties, `currentTime` and `duration`, help monitor the playback progress of your audio. To find the current playback percentage, divide the current time by the total audio duration:

```
progress = player.currentTime / player.duration;
```

To scrub your audio, that is, let your user select the current playback position within the audio track, you may want to pause playback. Although the `AVAudioPlayer` class in current iOS releases can provide audio-based scrubbing hints, these hints can sound jerky. You may want to wait until the scrubbing finishes to begin playback at the new location. Test out live scrubbing using the sample code for this recipe, and you'll hear the relative roughness of this feature.

Make sure to implement at least two target-action pairs if you base your scrubber on a standard `UISlider`. For the first target-action item, mask `UIControlEventTouchDown` with `UIControlEventValueChanged`. These event types enable you to catch the start of a user scrub and whenever the value changes. If you choose to pause on scrub, respond to these events by pausing the audio player and provide some visual feedback for the newly selected time:

```
- (void) scrub: (id) sender
{
    // Pause the player -- optional
    [self.player pause];
    self.navigationItem.leftBarButtonItem =
        SYSBARBUTTON(UIBarButtonSystemItemPlay, @selector(play:));

    // Calculate the new current time
    player.currentTime = scrubber.value * player.duration;

    // Update the title with the current time
    self.title = [NSString stringWithFormat:@"%@ of %@",
        [self formatTime:self.player.currentTime],
        [self formatTime:self.player.duration]];
}
```

For the second target-action pair, this mask of three values—`UIControlEventTouchUpInside | UIControlEventTouchUpOutside | UIControlEventCancel`—enables you to catch release events and touch interruptions. Upon release, you want to start playing at the new time set by the user's scrubbing:

```
- (void) scrubbingDone: (id) sender
{
    // resume playback here
    [self play:nil];
}
```

Catching the End of Playback

Detect the end of playback by setting the player's delegate and catching the `audioPlayerDidFinishPlaying:successfully:` delegate callback. That method is a great place to clean up any details, such as reverting the Pause button back to a Play button. Apple provides several system bar button items specifically for media playback:

- `UIBarButtonSystemItemPlay`
- `UIBarButtonSystemItemPause`
- `UIBarButtonSystemItemRewind`
- `UIBarButtonSystemItemFastForward`

The Rewind and Fast Forward buttons provide the double-arrowed icons that are normally used to move playback to a previous or next item in a playback queue. You could also use them to revert to the start of a track or progress to its end. Unfortunately, the Stop system item is an X, used for stopping an ongoing load operation and not the standard filled square used on many consumer devices for stopping playback or a recording.

Recipe 8-1 puts all these pieces together to create the unified (albeit horribly designed) interface you saw in Figure 8-1. Here, the user can select audio, start playing it back, pause it, adjust its volume, scrub, and so forth.

Recipe 8-1 **Playing Back Audio with AVAudioPlayer**

```
// Set the meters to the current peak and average power
- (void) updateMeters
{
    // Retrieve current values
    [player updateMeters];

    // Show average and peak values on the two meters
    meter1.progress = pow(10, 0.05f * [player averagePowerForChannel:0]);
    meter2.progress = pow(10, 0.05f * [player peakPowerForChannel:0]);

    // And on the scrubber
    scrubber.value = (player.currentTime / player.duration);

    // Display the current playback progress in minutes and seconds
    self.title = [NSString stringWithFormat:@"%@ of %@",
        [self formatTime:player.currentTime],
        [self formatTime:player.duration]];
}

// Pause playback
- (void) pause: (id) sender
{
```

```objc
    if (player)
        [player pause];

    // Update the play/pause button
    self.navigationItem.leftBarButtonItem =
        SYSBARBUTTON(UIBarButtonSystemItemPlay, @selector(play:));

    // Disable interactive elements
    meter1.progress = 0.0f;
    meter2.progress = 0.0f;
    volumeSlider.enabled = NO;
    scrubber.enabled = NO;

    // Stop listening for meter updates
    [timer invalidate];
}

// Start or resume playback
- (void) play: (id) sender
{
    if (player)
        [player play];

    // Enable the interactive elements
    volumeSlider.value = player.volume;
    volumeSlider.enabled = YES;
    scrubber.enabled = YES;

    // Update the play/pause button
    self.navigationItem.leftBarButtonItem =
        SYSBARBUTTON(UIBarButtonSystemItemPause, @selector(pause:));

    // Start listening for meter updates
    timer = [NSTimer scheduledTimerWithTimeInterval:0.1f
        target:self selector:@selector(updateMeters)
        userInfo:nil repeats:YES];
}

// Update the volume
- (IBAction) setVolume: (id) sender
{
    if (player) player.volume = volumeSlider.value;
}

// Catch the end of the scrubbing
- (IBAction) scrubbingDone: (id) sender
{
```

```objc
    [self play:nil];
}

// Update playback point during scrubs
- (IBAction) scrub: (id) sender
{
    // Calculate the new current time
    player.currentTime = scrubber.value * player.duration;

    // Update the title, nav bar
    self.title = [NSString stringWithFormat:@"%@ of %@",
        [self formatTime:player.currentTime],
        [self formatTime:player.duration]];
}

// Prepare but do not play audio
- (BOOL) prepAudio
{
    NSError *error;
    if (![[NSFileManager defaultManager] fileExistsAtPath:path])
        return NO;

    // Establish player
    player = [[AVAudioPlayer alloc] initWithContentsOfURL:
        [NSURL fileURLWithPath:path] error:&error];
    if (!player)
    {
        NSLog(@"AVAudioPlayer could not be established: %@",
            error.localizedFailureReason);
        return NO;
    }

    [player prepareToPlay];
    player.meteringEnabled = YES;
    player.delegate = self;

    // Initialize GUI
    meter1.progress = 0.0f;
    meter2.progress = 0.0f;
    self.navigationItem.leftBarButtonItem =
        SYSBARBUTTON(UIBarButtonSystemItemPlay, @selector(play:));
    scrubber.enabled = NO;

    return YES;
}
```

```objc
// On finishing, return to quiescent state
- (void)audioPlayerDidFinishPlaying:(AVAudioPlayer *)player
    successfully:(BOOL)flag
{
    self.navigationItem.leftBarButtonItem = nil;
    scrubber.value = 0.0f;
    scrubber.enabled = NO;
    volumeSlider.enabled = NO;
    [self prepAudio];
}

// Select a media file
- (void) pick: (UIBarButtonItem *) sender
{
    self.navigationItem.rightBarButtonItem.enabled = NO;

    // Each of these media files is in the public domain via archive.org
    NSArray *choices = [@"Alexander's Ragtime Band*Hello My Baby*\
        Ragtime Echoes*Rhapsody In Blue*A Tisket A Tasket*In the Mood"
        componentsSeparatedByString:@"*"];
    NSArray *media = [@"ARB-AJ*HMB1936*ragtime*RhapsodyInBlue*Tisket*\
        InTheMood" componentsSeparatedByString:@"*"];

    UIActionSheet *actionSheet = [[UIActionSheet alloc]
        initWithTitle:@"Musical selections" delegate:nil
        cancelButtonTitle:IS_IPAD ? nil : @"Cancel"
        destructiveButtonTitle:nil otherButtonTitles:nil];
    for (NSString *choice in choices)
        [actionSheet addButtonWithTitle:choice];

    // See the Core Cookbook, Alerts Chapter for details on
    // the ModalSheetDelegate class
    ModalSheetDelegate *msd =
        [ModalSheetDelegate delegateWithSheet:actionSheet];
    actionSheet.delegate = msd;

    int answer = [msd showFromBarButtonItem:sender animated:YES];
    self.navigationItem.rightBarButtonItem.enabled = YES;

    if (IS_IPAD)
    {
        if (answer == -1) return; // cancel
        if (answer >= choices.count)
            return;
    }
    else
    {
```

```objc
        if (answer == 0) return; // cancel
        if (answer > choices.count)
            return;
        answer--;
    }

    // no action, if already playing
    if ([nowPlaying.text isEqualToString:[choices objectAtIndex:answer]])
        return;

    // stop any current item
    if (player)
        [player stop];

    // Load in the new audio and play it
    path = [[NSBundle mainBundle]
        pathForResource:[media objectAtIndex:answer] ofType:@"mp3"];
    nowPlaying.text = [choices objectAtIndex:answer];
    [self prepAudio];
    [self play:nil];
}
```

> **Get This Recipe's Code**
>
> To find this recipe's full sample project, point your browser to https://github.com/erica/iOS-6-Advanced-Cookbook and go to the folder for Chapter 8.

Recipe: Looping Audio

Loops help present ambient background audio without having to respond to delegate callbacks each time the loop finishes a single play through. You can use a loop to play an audio snippet several times or play it continuously. Recipe 8-2 demonstrates an audio loop that plays only during the presentation of a particular video controller, providing an aural backdrop for that controller.

You set the number of times an audio plays before the playback ends. A high number (such as 999999) essentially provides for an unlimited number of loops. For example, a 4-second loop would take more than 1,000 hours to play back fully with a loop number that high:

```objc
// Prepare the player and set the loops
[self.player prepareToPlay];
[self.player setNumberOfLoops:999999];
```

Recipe 8-2 uses looped audio for its primary view controller. Whenever its view is onscreen, the loop plays in the background. Hopefully, you choose a loop that's unobtrusive, sets the mood for your application, and smoothly transitions from the end of playback to the beginning.

This recipe uses a fading effect to introduce and hide the audio. It fades the loop into hearing when the view appears and fades it out when the view disappears. It accomplishes this with a simple approach. A loop iterates through volume levels, from 0.0 to 1.0 on appearing and 1.0 down to 0.0 on disappearing. A call to NSThread's built-in sleep functionality adds the time delays (one-tenth of a second between each volume change) without affecting the audio playback.

Recipe 8-2 **Creating Ambient Audio Through Looping**

```
- (BOOL) prepAudio
{
    // Check for the file.
    NSError *error;
    NSString *path = [[NSBundle mainBundle]
        pathForResource:@"loop" ofType:@"mp3"];
    if (![[NSFileManager defaultManager]
        fileExistsAtPath:path]) return NO;

    // Initialize the player
    player = [[AVAudioPlayer alloc] initWithContentsOfURL:
        [NSURL fileURLWithPath:path] error:&error];
    if (!player)
    {
        NSLog(@"Could not establish AV Player: %@",
            error.localizedFailureReason);
        return NO;
    }

    // Prepare the player and set the loops large
    [player prepareToPlay];
    [player setNumberOfLoops:999999];

    return YES;
}

- (void) fadeIn
{
    player.volume = MIN(player.volume + 0.05f, 1.0f);
    if (player.volume < 1.0f)
        [self performSelector:@selector(fadeIn)
            withObject:nil afterDelay:0.1f];
}
```

```objc
- (void) fadeOut
{
    player.volume = MAX(player.volume - 0.1f, 0.0f);
    if (player.volume > 0.05f)
        [self performSelector:@selector(fadeOut)
            withObject:nil afterDelay:0.1f];
    else
        [player pause];
}

- (void) viewDidAppear: (BOOL) animated
{
    // Start playing at no-volume
    player.volume = 0.0f;
    [player play];

    // fade in the audio over a second
    [self fadeIn];

    // Add the push button
    self.navigationItem.rightBarButtonItem.enabled = YES;
}

- (void) viewWillDisappear: (BOOL) animated
{
    [self fadeOut];
}

- (void) push
{
    // Create a simple new view controller
    UIViewController *vc = [[UIViewController alloc] init];
    vc.view.backgroundColor = [UIColor whiteColor];
    vc.title = @"No Sounds";

    // Disable the now-pressed right-button
    self.navigationItem.rightBarButtonItem.enabled = NO;

    // Push the new view controller
    [self.navigationController
        pushViewController:vc animated:YES];
}

- (void) loadView
{
    [super loadView];
    self.navigationItem.rightBarButtonItem =
```

```
            BARBUTTON(@"Push", @selector(push));
    self.title = @"Looped Sounds";
    [self prepAudio];
}
```

> **Get This Recipe's Code**
>
> To find this recipe's full sample project, point your browser to https://github.com/erica/iOS-6-Advanced-Cookbook and go to the folder for Chapter 8.

Recipe: Handling Audio Interruptions

When users receive phone calls during audio playback, that audio fades away. The standard answer/decline screen appears. As this happens, `AVAudioPlayer` delegates receive the `audioPlayerBeginInterruption:` callback, as shown in Recipe 8-3. The audio session deactivates, and the player pauses. You cannot restart playback until the interruption ends.

Should the user accept the call, the application delegate receives an `applicationWill-ResignActive:` callback. When the call ends, the application relaunches (with an `applicationDidBecomeActive:` callback). If the user declines the call or if the call ends without an answer, the delegate is instead sent `audioPlayerEndInterruption:`. You can resume playback from this method.

If it is vital that playback resumes after accepting a call, and the application needs to relaunch, you can save the current time, as shown in Recipe 8-3. The `viewDidAppear:` method in this recipe, which is called on most kinds of resumptions, checks for a stored interruption value in the user defaults. When it finds one, it uses this to set the current time for resuming playback.

This approach takes into account that the application relaunches rather than resumes after the call finishes. You do not receive the end interruption callback when the user accepts a call.

Recipe 8-3 **Storing the Interruption Time for Later Pickup**

```
- (BOOL) prepAudio
{
    NSError *error;
    NSString *path = [[NSBundle mainBundle]
        pathForResource:@"MeetMeInSt.Louis1904" ofType:@"mp3"];
    if (![[NSFileManager defaultManager]
        fileExistsAtPath:path]) return NO;

    // Initialize the player
    player = [[AVAudioPlayer alloc]
        initWithContentsOfURL:[NSURL fileURLWithPath:path]
        error:&error];
```

Recipe: Handling Audio Interruptions

```
    player.delegate = self;
    if (!player)
    {
        NSLog(@"Could not establish player: %@",
            error.localizedFailureReason);
        return NO;
    }

    [player prepareToPlay];

    return YES;
}

- (void)audioPlayerDidFinishPlaying:(AVAudioPlayer *)aPlayer
    successfully:(BOOL)flag
{
    // just keep playing
    [player play];
}

- (void)audioPlayerBeginInterruption:(AVAudioPlayer *)aPlayer
{
    // perform any interruption handling here
    fprintf(stderr, "Interruption Detected\n");
    [[NSUserDefaults standardUserDefaults]
        setFloat:[player currentTime] forKey:@"Interruption"];
}

- (void)audioPlayerEndInterruption:(AVAudioPlayer *)aPlayer
{
    // resume playback at the end of the interruption
    fprintf(stderr, "Interruption ended\n");
    [player play];

    // remove the interruption key. it won't be needed
    [[NSUserDefaults standardUserDefaults]
        removeObjectForKey:@"Interruption"];
}

- (void) viewDidAppear:(BOOL)animated
{
    [self prepAudio];
    // Check for previous interruption
    if ([[NSUserDefaults standardUserDefaults]
        objectForKey:@"Interruption"])
    {
        player.currentTime =
```

```
            [[NSUserDefaults standardUserDefaults]
                floatForKey:@"Interruption"];
        [[NSUserDefaults standardUserDefaults]
            removeObjectForKey:@"Interruption"];
    }

    // Start playback
    [player play];
}
```

Get This Recipe's Code

To find this recipe's full sample project, point your browser to https://github.com/erica/iOS-6-Advanced-Cookbook and go to the folder for Chapter 8.

Recipe: Recording Audio

The `AVAudioRecorder` class simplifies audio recording in your applications. It provides the same API friendliness as `AVAudioPlayer`, along with similar feedback properties. Together, these two classes leverage development for many standard application audio tasks.

Start your recordings by establishing an `AVAudioSession`. Use a play and record session if you intend to switch between recording and playback in the same application. Otherwise, use a simple record session (via `AVAudioSessionCategoryRecord`).

When you have a session, you can check its `inputAvailable` property. This property indicates that the current device has access to a microphone. The property replaces `inputIsAvailable`, which was deprecated in iOS 6. If you plan to deploy to pre-6 frameworks, make sure to use the older style property:

```
- (BOOL) startAudioSession
{
    // Prepare the audio session
    NSError *error;
    AVAudioSession *session = [AVAudioSession sharedInstance];

    if (![session
        setCategory:AVAudioSessionCategoryPlayAndRecord
        error:&error])
    {
        NSLog(@"Error setting session category: %@",
            error.localizedFailureReason);
        return NO;
    }
```

```objc
    if (![session setActive:YES error:&error])
    {
        NSLog(@"Error activating audio session: %@",
            error.localizedFailureReason);
        return NO;
    }

    // inputIsAvailable deprecated in iOS 6
    return session.inputAvailable;
}
```

Recipe 8-4 demonstrates the next step after creating the session. It sets up the recorder and provides methods for pausing, resuming, and stopping the recording.

To start recording, it creates a settings dictionary and populates it with keys and values that describe how the recording should be sampled. This example uses mono Linear PCM sampled 8,000 times a second, which is a fairly low sample rate. Here are a few examples of customizing formats:

- Set `AVNumberOfChannelsKey` to 1 for mono audio and 2 for stereo.
- Audio formats (`AVFormatIDKey`) that work well on iOS include `kAudioFormatLinearPCM` (large files) and `kAudioFormatAppleIMA4` (compact files).
- Standard `AVSampleRateKey` sampling rates include 8,000, 11,025, 22,050, and 44,100.
- For the linear PCM-only bit depth (`AVLinearPCMBitDepthKey`), use either 16 or 32 bits.

The code allocates a new `AVRecorder` instance and initializes it with both a file URL and the settings dictionary. When created, this code sets the recorder's delegate and enables metering. Metering for `AVAudioRecorder` instances works like metering for `AVAudioPlayer` instances. You must update the meter before requesting average and peak power levels:

```objc
- (void) updateMeters
{
    // Show the current power levels
    [self.recorder updateMeters];
    meter1.progress =
        pow(10, 0.05f * [recorder averagePowerForChannel:0]);
    meter2.progress =
        pow(10, 0.05f * [recorder peakPowerForChannel:0]);

    // Update the current recording time
    self.title = [NSString stringWithFormat:@"%
        [self formatTime:self.recorder.currentTime]];
}
```

This code also tracks the recording's `currentTime`. When you pause a recording, the current time stays still until you resume. Basically, the current time indicates the recording duration to date.

When you're ready to proceed with the recording, use `prepareToRecord` and start the recording with `record`. Issue pause to take a break in recording; resume again with another call to `record`. The recording picks up where it left off. To finish a recording, use `stop`. This produces a callback to `audioRecorderDidFinishRecording:successfully:`. That's where you can clean up your interface and finalize any recording details.

To retrieve recordings, set the `UIFileSharingEnabled` key to `YES` in the application's Info.plist. This allows your users to access all the recorded material through iTunes' device-specific Apps pane.

Recipe 8-4 **Audio Recording with AVAudioRecorder**

```
- (void)audioPlayerDidFinishPlaying:(AVAudioPlayer *)player
    successfully:(BOOL)flag
{
    // Prepare UI for recording
    self.title = nil;
    meter1.hidden = NO;
    meter2.hidden = NO;
    {
        // Return to play and record session
        NSError *error;
        if (![[AVAudioSession sharedInstance]
            setCategory:AVAudioSessionCategoryPlayAndRecord
            error:&error])
        {
            NSLog(@"Error: %@", error.localizedFailureReason);
            return;
        }
        self.navigationItem.rightBarButtonItem =
            BARBUTTON(@"Record", @selector(record));
    }
}

- (void)audioRecorderDidFinishRecording:(AVAudioRecorder *)aRecorder
    successfully:(BOOL)flag
{
    // Stop monitoring levels, time
    [timer invalidate];
    meter1.progress = 0.0f;
    meter1.hidden = YES;
    meter2.progress = 0.0f;
```

```objc
        meter2.hidden = YES;
        self.navigationItem.leftBarButtonItem = nil;
        self.navigationItem.rightBarButtonItem = nil;

        NSURL *url = recorder.url;
        NSError *error;

        // Start playback
        player = [[AVAudioPlayer alloc] initWithContentsOfURL:url
            error:&error];
        if (!player)
        {
            NSLog(@"Error establishing player for %@: %@",
                url, error.localizedFailureReason);
            return;
        }
        player.delegate = self;

        // Change audio session for playback
        if (![[AVAudioSession sharedInstance]
            setCategory:AVAudioSessionCategoryPlayback
            error:&error])
        {
            NSLog(@"Error updating audio session: %@",
                error.localizedFailureReason);
            return;
        }

        self.title = @"Playing back recording...";
        [player prepareToPlay];
        [player play];
}

- (void) stopRecording
{
    // This causes the didFinishRecording delegate method to fire
    [recorder stop];
}

- (void) continueRecording
{
    // resume from a paused recording
    [recorder record];
    self.navigationItem.rightBarButtonItem =
        BARBUTTON(@"Done", @selector(stopRecording));
```

```objc
    self.navigationItem.leftBarButtonItem =
        SYSBARBUTTON(UIBarButtonSystemItemPause, self,
        @selector(pauseRecording));
}

- (void) pauseRecording
{
    // pause an ongoing recording
    [recorder pause];
    self.navigationItem.leftBarButtonItem =
        BARBUTTON(@"Continue", @selector(continueRecording));
    self.navigationItem.rightBarButtonItem = nil;
}

- (BOOL) record
{
    NSError *error;

    // Recording settings
    NSMutableDictionary *settings =
        [NSMutableDictionary dictionary];
    settings[AVFormatIDKey] = @(kAudioFormatLinearPCM);
    settings[AVSampleRateKey] = @(8000.0f);
    settings[AVNumberOfChannelsKey] = @(1); // mono
    settings[AVLinearPCMBitDepthKey] = @(16);
    settings[AVLinearPCMIsBigEndianKey] = @NO;
    settings[AVLinearPCMIsFloatKey] = @NO;

    // File URL
    NSURL *url = [NSURL fileURLWithPath:FILEPATH];

    // Create recorder
    recorder = [[AVAudioRecorder alloc] initWithURL:url
        settings:settings error:&error];
    if (!recorder)
    {
        NSLog(@"Error establishing recorder: %@",
            error.localizedFailureReason);
        return NO;
    }

    // Initialize delegate, metering, etc.
    recorder.delegate = self;
    recorder.meteringEnabled = YES;
    meter1.progress = 0.0f;
    meter2.progress = 0.0f;
    self.title = @"0:00";
```

Recipe: Recording Audio

```
    if (![recorder prepareToRecord])
    {
        NSLog(@"Error: Prepare to record failed");
        return NO;
    }

    if (![recorder record])
    {
        NSLog(@"Error: Record failed");
        return NO;
    }

    // Set a timer to monitor levels, current time
    timer = [NSTimer scheduledTimerWithTimeInterval:0.1f
        target:self selector:@selector(updateMeters)
        userInfo:nil repeats:YES];

    // Update the navigation bar
    self.navigationItem.rightBarButtonItem =
        BARBUTTON(@"Done", @selector(stopRecording));
    self.navigationItem.leftBarButtonItem =
        SYSBARBUTTON(UIBarButtonSystemItemPause, self,
        @selector(pauseRecording));

    return YES;
}

- (BOOL) startAudioSession
{
    // Prepare the audio session
    NSError *error;
    AVAudioSession *session = [AVAudioSession sharedInstance];

    if (![session setCategory:AVAudioSessionCategoryPlayAndRecord
        error:&error])
    {
        NSLog(@"Error setting session category: %@",
            error.localizedFailureReason);
        return NO;
    }

    if (![session setActive:YES error:&error])
    {
        NSLog(@"Error activating audio session: %@",
            error.localizedFailureReason);
        return NO;
    }
```

```
        return session.inputIsAvailable;
}

- (void) viewDidLoad
{
    if ([self startAudioSession])
        self.navigationItem.rightBarButtonItem =
            BARBUTTON(@"Record", @selector(record));
    else
        self.title = @"No Audio Input Available";
}
```

> **Get This Recipe's Code**
>
> To find this recipe's full sample project, point your browser to https://github.com/erica/iOS-6-Advanced-Cookbook and go to the folder for Chapter 8.

Recipe: Recording Audio with Audio Queues

In addition to the `AVAudioPlayer` class, Audio Queues can handle recording and playing tasks in your applications. Audio Queues were needed for recording before the `AVAudioRecorder` class debuted. Using queues directly helps demonstrate what's going on under the hood of the `AVAudioRecorder` class. As a bonus, queues offer full access to the raw underlying audio in `HandleInputBuffer`. You can do all kinds of interesting things such as signal processing and analysis.

Recipe 8-5 records audio at the Audio Queue level, providing a taste of the C-style functions and callbacks used. This code is heavily based on Apple sample code and specifically showcases functionality that is hidden behind the `AVAudioRecorder` wrapper.

The settings used in Recipe 8-5's `setupAudioFormat:` method have been tested and work reliably on the iOS family of devices. It's easy, however, to mess up these parameters when trying to customize your audio quality. If you don't have the parameters set up just right, the queue may fail with little feedback. A quick Google search provides copious settings examples.

Recipe 8-5 Recording with Audio Queues: The Recorder.m Implementation

```
// Write out current packets as the input buffer is filled
static void HandleInputBuffer (void *aqData,
    AudioQueueRef inAQ, AudioQueueBufferRef inBuffer,
    const AudioTimeStamp *inStartTime,
    UInt32 inNumPackets,
    const AudioStreamPacketDescription *inPacketDesc)
{
    RecordState *pAqData = (RecordState *) aqData;
```

```objc
    if (inNumPackets == 0 &&
        pAqData->dataFormat.mBytesPerPacket != 0)
        inNumPackets = inBuffer->mAudioDataByteSize /
            pAqData->dataFormat.mBytesPerPacket;

    if (AudioFileWritePackets(pAqData->audioFile, NO,
        inBuffer->mAudioDataByteSize, inPacketDesc,
        pAqData->currentPacket, &inNumPackets,
        inBuffer->mAudioData) == noErr)
    {
        pAqData->currentPacket += inNumPackets;
        if (pAqData->recording == 0) return;
        AudioQueueEnqueueBuffer (pAqData->queue, inBuffer,
            0, NULL);
    }
}

@implementation Recorder

// Set up the recording format as low quality mono AIFF
- (void)setupAudioFormat:(AudioStreamBasicDescription*)format
{
    format->mSampleRate = 8000.0;
    format->mFormatID = kAudioFormatLinearPCM;
    format->mFormatFlags = kLinearPCMFormatFlagIsBigEndian |
        kLinearPCMFormatFlagIsSignedInteger |
        kLinearPCMFormatFlagIsPacked;

    format->mChannelsPerFrame = 1; // mono
    format->mBitsPerChannel = 16;
    format->mFramesPerPacket = 1;
    format->mBytesPerPacket = 2;
    format->mBytesPerFrame = 2;
    format->mReserved = 0;
}

// Begin recording
- (BOOL) startRecording: (NSString *) filePath
{
    // Many of these calls mirror the process for AVAudioRecorder

    // Set up the audio format and the url to record to
    [self setupAudioFormat:&recordState.dataFormat];
    CFURLRef fileURL = CFURLCreateFromFileSystemRepresentation(
        NULL, (const UInt8 *) [filePath UTF8String],
        [filePath length], NO);
    recordState.currentPacket = 0;
```

```
// Initialize the queue with the format choices
OSStatus status;
status = AudioQueueNewInput(&recordState.dataFormat,
    HandleInputBuffer, &recordState,
    CFRunLoopGetCurrent(),kCFRunLoopCommonModes, 0,
    &recordState.queue);
if (status) {
    fprintf(stderr, "Could not establish new queue\n");
    return NO;
}

// Create the output file
status = AudioFileCreateWithURL(fileURL,
    kAudioFileAIFFType, &recordState.dataFormat,
    kAudioFileFlags_EraseFile, &recordState.audioFile);
if (status)
{
    fprintf(stderr, "Could not create file to record audio\n");
    return NO;
}

// Set up the buffers
DeriveBufferSize(recordState.queue, recordState.dataFormat,
    0.5, &recordState.bufferByteSize);
for(int i = 0; i < NUM_BUFFERS; i++)
{
    status = AudioQueueAllocateBuffer(recordState.queue,
        recordState.bufferByteSize, &recordState.buffers[i]);
    if (status) {
        fprintf(stderr, "Error allocating buffer %d\n", i);
        return NO;
    }
    status = AudioQueueEnqueueBuffer(recordState.queue,
        recordState.buffers[i], 0, NULL);
    if (status) {
        fprintf(stderr, "Error enqueuing buffer %d\n", i);
        return NO;
    }
}

// Enable metering
UInt32 enableMetering = YES;
status = AudioQueueSetProperty(recordState.queue,
    kAudioQueueProperty_EnableLevelMetering,
    &enableMetering,sizeof(enableMetering));
```

```objc
        if (status)
        {
            fprintf(stderr, "Could not enable metering\n");
            return NO;
        }

        // Start the recording
        status = AudioQueueStart(recordState.queue, NULL);
        if (status)
        {
            fprintf(stderr, "Could not start Audio Queue\n");
            return NO;
        }

        recordState.currentPacket = 0;
        recordState.recording = YES;
        return YES;
    }

    // Return the average power level
    - (CGFloat) averagePower
    {
        AudioQueueLevelMeterState state[1];
        UInt32   statesize = sizeof(state);
        OSStatus status;
        status = AudioQueueGetProperty(recordState.queue,
            kAudioQueueProperty_CurrentLevelMeter, &state, &statesize);
        if (status)
        {
            fprintf(stderr, "Error retrieving meter data\n");
            return 0.0f;
        }
        return state[0].mAveragePower;
    }

    // Return the peak power level
    - (CGFloat) peakPower
    {
        AudioQueueLevelMeterState state[1];
        UInt32   statesize = sizeof(state);
        OSStatus status;
        status = AudioQueueGetProperty(recordState.queue,
            kAudioQueueProperty_CurrentLevelMeter, &state, &statesize);
        if (status)
        {
            fprintf(stderr, "Error retrieving meter data\n");
```

```objc
        return 0.0f;
    }
    return state[0].mPeakPower;
}


// There's generally about a one-second delay before the
// buffers fully empty
- (void) reallyStopRecording
{
    AudioQueueFlush(recordState.queue);
    AudioQueueStop(recordState.queue, NO);
    recordState.recording = NO;
    
    for(int i = 0; i < NUM_BUFFERS; i++)
    {
        AudioQueueFreeBuffer(recordState.queue,
            recordState.buffers[i]);
    }
    
    AudioQueueDispose(recordState.queue, YES);
    AudioFileClose(recordState.audioFile);
}

// Stop the recording after waiting just a second
- (void) stopRecording
{
    [self performSelector:@selector(reallyStopRecording)
        withObject:NULL afterDelay:1.0f];
}

// Pause after allowing buffers to catch up
- (void) reallyPauseRecording
{
    if (!recordState.queue) {
        fprintf(stderr, "Nothing to pause\n"); return;}
    OSStatus status = AudioQueuePause(recordState.queue);
    if (status)
    {
        fprintf(stderr, "Error pausing audio queue\n");
        return;
    }
}

// Pause the recording after waiting a half second
- (void) pause
{
```

```objc
    [self performSelector:@selector(reallyPauseRecording)
        withObject:NULL afterDelay:0.5f];
}

// Resume recording from a paused queue
- (BOOL) resume
{
    if (!recordState.queue)
    {
        fprintf(stderr, "Nothing to resume\n");
        return NO;
    }

    OSStatus status = AudioQueueStart(recordState.queue, NULL);
    if (status)
    {
        fprintf(stderr, "Error restarting audio queue\n");
        return NO;
    }

    return YES;
}

// Return the current recording duration
- (NSTimeInterval) currentTime
{
    AudioTimeStamp outTimeStamp;
    OSStatus status = AudioQueueGetCurrentTime (
        recordState.queue, NULL, &outTimeStamp, NULL);
    if (status)
    {
        fprintf(stderr, "Error: Could not retrieve current time\n");
        return 0.0f;
    }

    // 8000 samples per second
    return outTimeStamp.mSampleTime / 8000.0f;
}

// Return whether the recording is active
- (BOOL) isRecording
{
    return recordState.recording;
}
@end
```

> **Get This Recipe's Code**
>
> To find this recipe's full sample project, point your browser to https://github.com/erica/iOS-6-Advanced-Cookbook and go to the folder for Chapter 8.

Recipe: Picking Audio with the MPMediaPickerController

The `MPMediaPickerController` class provides an audio equivalent for the image-picking facilities of the `UIImagePickerController` class. It allows users to choose an item or items from their music library including music, podcasts, and audio books. The iPod-style interface allows users to browse via playlists, artists, songs, albums, and more.

To use this class, allocate a new picker and initialize it with the kinds of media to be used. For audio, you can choose from `MPMediaTypeMusic`, `MPMediaTypePodcast`, `MPMediaTypeAudioBook`, and `MPMediaTypeAnyAudio`. Video types are also available for movies, TV shows, video podcasts, music videos, and iTunesU. These are flags and can be or'ed together to form a mask:

```
MPMediaPickerController *mpc = [[MPMediaPickerController alloc]
    initWithMediaTypes:MPMediaTypeAnyAudio];
mpc.delegate = self;
mpc.prompt = @"Please select an item";
mpc.allowsPickingMultipleItems = NO;
[self presentViewController:mpc animated:YES];
```

Set a delegate that conforms to `MPMediaPickerControllerDelegate` and optionally set a prompt. The prompt is text that appears at the top of the media picker, as shown in Figure 8-2. When you choose to allow multiple item selection, the Cancel button on the standard picker is replaced by the label Done. Normally, the dialog ends when a user taps a track. With multiple selection, users can keep picking items until they press the Done button. Selected items are updated to use grayed text.

The `mediaPicker:didPickMediaItems:` delegate callback handles the completion of a user selection. The `MPMediaItemCollection` instance that is passed as a parameter can be enumerated by accessing its `items`. Each item is a member of the `MPMediaItem` class and can be queried for its properties, as shown in Recipe 8-6. Recipe 8-6 uses a media picker to select multiple music tracks. It logs the items the user selected by artist and title.

Consult Apple's `MPMediaItem` class documentation for the available properties for media items, the type they return, and whether they can be used to construct a media property predicate. Building queries and using predicates is discussed in the next section.

Keep in mind that media item properties are not Objective-C properties. You must retrieve them using the `valueForProperty:` method, although it's easy to create a wrapper class that does provide these items as real properties via Objective-C method calls.

Recipe: Picking Audio with the MPMediaPickerController 287

Figure 8-2 In this multiple selection media picker, already selected items appear in gray. Users tap Done when finished. An optional prompt field (here, Please Select an Item) appears above the normal picker elements.

Recipe 8-6   **Selecting Music Items from the iPod Library**

```
- (void)mediaPicker: (MPMediaPickerController *)mediaPicker
    didPickMediaItems:(MPMediaItemCollection *)mediaItemCollection
{
    for (MPMediaItem *item in [mediaItemCollection items])
        NSLog(@"[%@] %@",
            [item valueForProperty:MPMediaItemPropertyArtist],
            [item valueForProperty:MPMediaItemPropertyTitle]);
    [self performDismiss];
}

- (void)mediaPickerDidCancel:(MPMediaPickerController *)mediaPicker
{
    if (IS_IPHONE)
        [self dismissViewControllerAnimated:YES completion:nil];
}
```

```
// Popover was dismissed
- (void)popoverControllerDidDismissPopover:
    (UIPopoverController *)aPopoverController
{
    popoverController = nil;
}

// Start the picker session
- (void) action: (UIBarButtonItem *) bbi
{
    MPMediaPickerController *mpc =
        [[MPMediaPickerController alloc]
            initWithMediaTypes:MPMediaTypeMusic];
    mpc.delegate = self;
    mpc.prompt = @"Please select an item";
    mpc.allowsPickingMultipleItems = YES;
    [self presentController:mpc];
}
```

> **Get This Recipe's Code**
>
> To find this recipe's full sample project, point your browser to https://github.com/erica/iOS-6-Advanced-Cookbook and go to the folder for Chapter 8.

Creating a Media Query

Media Queries enable you to filter your iPod library contents, limiting the scope of your search. Table 8-1 lists the nine class methods that `MPMediaQuery` provides for predefined searches. Each query type controls the grouping of the data returned. Each collection is organized as tracks by album, by artist, by audio book, and so on.

This approach reflects the way that iTunes works on the desktop. In iTunes, you select a column to organize your results, but you search by entering text into the application's Search field.

Table 8-1 **Query Types**

Class Method	Global?	Filter	Group Type
`albumsQuery`	No	`MPMediaTypeMusic`	`MPMediaGroupingAlbum`
`artistsQuery`	No	`MPMediaTypeMusic`	`MPMediaGroupingArtist`
`audiobooksQuery`	No	`MPMediaTypeAudioBook`	`MPMediaGroupingTitle`

Class Method	Global?	Filter	Group Type
compilations⮕Query	No	MPMediaTypeAny | MPMediaItemPropertyIs⮕Compilation	MPMediaGroupingAlbum
composersQuery	Yes	MPMediaTypeAny	MPMediaGrouping Composer
genresQuery	Yes	MPMediaTypeAny	MPMediaGroupingGenre
playlistsQuery	Yes	MPMediaTypeAny	MPMediaGrouping Playlist
podcastsQuery	No	MPMediaTypePodcast	MPMediaGrouping PodcastTitle
songsQuery	No	MPMediaTypeMusic	MPMediaGroupingTitle

Building a Query

Count the number of albums in your library using an album query. This snippet creates that query and then retrieves an array, each item of which represents a single album. These album items are collections of individual media items. A collection may contain a single track or many:

```
MPMediaQuery *query = [MPMediaQuery albumsQuery];
NSArray *collections = query.collections;
NSLog(@"You have %d albums in your library\n", collections.count);
```

Many iOS users have extensive media collections often containing hundreds or thousands of albums, let alone individual tracks. A simple query like this one may take several seconds to run and return a data structure that represents the entire library.

A search using a different query type returns collections organized by that type. You can use a similar approach to recover the number of artists, songs, composers, and so on.

Using Predicates

A media property predicate efficiently filters the items returned by a query. For example, you might want to find only those songs whose title matches the phrase "road." The following snippet creates a new songs query and adds a filter predicate to search for that phrase. The predicate is constructed with a value (the search phrase), a property (searching the song title), and a comparison type (in this case, "contains"). Use `MPMedia-PredicateComparisonEqualTo` for exact matches and `MPMediaPredicate-ComparisonContains` for substring matching:

```
MPMediaQuery *query = [MPMediaQuery songsQuery];

// Construct a title comparison predicate
MPMediaPropertyPredicate *mpp = [MPMediaPropertyPredicate
    predicateWithValue:@"road"
    forProperty:MPMediaItemPropertyTitle
```

```
        comparisonType:MPMediaPredicateComparisonContains];
[query addFilterPredicate:mpp];

// Recover the collections
NSArray *collections = query.collections;
NSLog(@"You have %d matching tracks in your library\n",
    collections.count);

// Iterate through each item, logging the song and artist
for (MPMediaItemCollection *collection in collections)
{
    for (MPMediaItem *item in [collection items])
    {
        NSString *song = [item valueForProperty:
            MPMediaItemPropertyTitle];
        NSString *artist = [item valueForProperty:
            MPMediaItemPropertyArtist];
        NSLog(@"%@, %@", song, artist);
    }
}
```

> **Note**
>
> If you'd rather use regular predicates with your media collections than media property predicates, I created an `MPMediaItem` properties category (http://github.com/erica/MPMediaItem-Properties). This category enables you to apply standard `NSPredicate` queries against collections, such as those returned by a multiple-item selection picker.

Recipe: Using the MPMusicPlayerController

Cocoa Touch includes a simple-to-use music player class that works seamlessly with media collections. Despite what its name implies, the `MPMusicPlayerController` class is not a view controller. It provides no onscreen elements for playing back music. Instead, it offers an abstract controller that handles playing and pausing music.

It publishes optional notifications when its playback state changes. The class offers two shared instances: the `iPodMusicPlayer` and an `applicationMusicPlayer`. Always use the former. It provides reliable state change feedback, which you want to catch programmatically.

Initialize the player controller by calling `setQueueWithItemCollection:` with an `MPMediaItemCollection`:

```
[[MPMusicPlayerController iPodMusicPlayer]
    setQueueWithItemCollection: songs];
```

Alternatively, you can load a queue with a media query. For example, you might set a `playlistsQuery` matching a specific playlist phrase or an artist query to search for songs by a given artist. Use `setQueueWithQuery:` to generate a queue from an `MPMediaQuery` instance.

If you want to shuffle playback, assign a value to the controller's `shuffleMode` property. Choose from `MPMusicShuffleModeDefault`, which respects the user's current setting, `MPMusicShuffleModeOff` (no shuffle), `MPMusicShuffleModeSongs` (song-by-song shuffle), and `MPMusicShuffleModeAlbums` (album-by-album shuffle). A similar set of options exists for the music's `repeatMode`.

After you set the item collection, you can play, pause, skip to the next item in the queue, go back to a previous item, and so forth. To rewind without moving back to a previous item, issue `skipToBeginning`. You can also seek within the currently playing item, moving the playback point forward or backward.

Recipe 8-7 offers a simple media player that shows the currently playing song (along with its artwork, if available). When run, the user selects a group of items using an `MPMediaPicker Controller`. This item collection is returned and assigned to the player, which begins playing back the group.

A pair of observers uses the default notification center to watch for two key changes: when the current item changes and when the playback state changes. To catch these changes, you must manually request notifications. This allows you to update the interface with new "now playing" information when the playback item changes:

```
[[MPMusicPlayerController iPodMusicPlayer]
    beginGeneratingPlaybackNotifications];
```

You may undo this request by issuing `endGeneratingPlaybackNotifications`, or you can simply allow the program to tear down all observers when the application naturally suspends or terminates. Because this recipe uses the iPod music player, playback continues after leaving the application unless you specifically stop it. Playback is not affected by the application suspension or teardown:

```
- (void) applicationWillResignActive: (UIApplication *) application
{
    // Stop player when the application quits
    [[MPMusicPlayerController iPodMusicPlayer] stop];
}
```

In addition to demonstrating playback control, Recipe 8-7 shows how to display album art during playback. It uses the same kind of `MPItem` property retrieval used in previous recipes. In this case, it queries for `MPMediaItemPropertyArtwork` and, if artwork is found, it uses the `MPMediaItemArtwork` class to convert that artwork to an image of a given size.

Recipe 8-7 Simple Media Playback with the iPod Music Player

```
#define PLAYER [MPMusicPlayerController iPodMusicPlayer]

#pragma mark PLAYBACK
- (void) pause
{
    // Pause playback
    [PLAYER pause];
    toolbar.items = [self playItems];
}

- (void) play
{
    // Restart play
    [PLAYER play];
    toolbar.items = [self pauseItems];
}

- (void) fastforward
{
    // Skip to the next item
    [PLAYER skipToNextItem];
}

- (void) rewind
{
    // Skip to the previous item
    [PLAYER skipToPreviousItem];
}


#pragma mark STATE CHANGES
- (void) playbackItemChanged: (NSNotification *) notification
{
    // Update title and artwork
    self.title = [PLAYER.nowPlayingItem
        valueForProperty:MPMediaItemPropertyTitle];
    MPMediaItemArtwork *artwork = [PLAYER.nowPlayingItem
        valueForProperty: MPMediaItemPropertyArtwork];
    imageView.image = [artwork imageWithSize:[imageView frame].size];
}

- (void) playbackStateChanged: (NSNotification *) notification
{
    // On stop, clear title, toolbar, artwork
    if (PLAYER.playbackState == MPMusicPlaybackStateStopped)
```

Recipe: Using the MPMusicPlayerController

```
    {
        self.title = nil;
        toolbar.items = nil;
        imageView.image = nil;
    }
}

#pragma mark MEDIA PICKING
- (void)mediaPicker: (MPMediaPickerController *)mediaPicker
    didPickMediaItems:(MPMediaItemCollection *)mediaItemCollection
{
    // Set the songs to the collection selected by the user
    songs = mediaItemCollection;

    // Update the playback queue
    [PLAYER setQueueWithItemCollection:songs];

    // Display the play items in the toolbar
    [toolbar setItems:[self playItems]];

    // Clean up the picker
    [self performDismiss];
;
}


- (void)mediaPickerDidCancel:(MPMediaPickerController *)mediaPicker
{
    // User has canceled
    [self performDismiss];
}

- (void) pick: (UIBarButtonItem *) bbi
{
    // Select the songs for the playback queue
    MPMediaPickerController *mpc = [[MPMediaPickerController alloc]
        initWithMediaTypes:MPMediaTypeMusic];
    mpc.delegate = self;
    mpc.prompt = @"Please select items to play";
    mpc.allowsPickingMultipleItems = YES;
    [self presentViewController:mpc];
}

#pragma mark INIT VIEW
- (void) viewDidLoad
{
    self.navigationItem.rightBarButtonItem = BARBUTTON(@"Pick",
        @selector(pick));
```

```
    // Stop any ongoing music
    [PLAYER stop];

    // Add observers for state and item changes
    [[NSNotificationCenter defaultCenter] addObserver:self
        selector:@selector(playbackStateChanged)
        name:MPMusicPlayerControllerPlaybackStateDidChangeNotification
        object:PLAYER];
    [[NSNotificationCenter defaultCenter] addObserver:self
        selector:@selector(playbackItemChanged)
        name:MPMusicPlayerControllerNowPlayingItemDidChangeNotification
        object:PLAYER];
    [PLAYER beginGeneratingPlaybackNotifications];
}
@end
```

Get This Recipe's Code

To find this recipe's full sample project, point your browser to https://github.com/erica/iOS-6-Advanced-Cookbook and go to the folder for Chapter 8.

Summary

This chapter introduced many ways to handle audio media, including playback and recording. You saw recipes that worked with high-level Objective-C classes and those that worked with lower-level C functions. You read about media pickers, controllers, and more. Here are some final thoughts to take from this chapter:

- Apple remains in the process of building its AV media playback classes. They are becoming more and more stunningly powerful over time. Due to time and space limitations, this chapter didn't address many technologies that power audio on iOS: AVFoundation, OpenAL, Core Audio, Audio Units, or Core MIDI. These topics are huge and each worthy of its own book. For more information on these, see Chris Adamson and Kevin Avila's book, *Learning Core Audio: A Hands-On Guide to Audio Programming for Mac and iOS* (Addison-Wesley Professional).

- This chapter did not discuss `ALAssetsLibrary`, although examples of using this class are covered elsewhere in this book and in the Core Cookbook. You might want to look into the `MPMediaItemPropertyAssetURL` and `AVAssetExportSession` classes for working with audio library assets.

- Audio Queue provides powerful low-level audio routines, but they're not for the faint of heart or for anyone who just wants a quick solution. If you need the kind of fine-grained audio control that Audio Queues bring, Apple supplies extensive documentation on achieving your goals.

- When creating local assets using recording features, make sure to enable `UIFileSharingEnabled` in your Info.plist to allow users to access and manage the files they create through iTunes.
- The `MPMusicPlayerController` class introduces a simple way to interact with music from your onboard iTunes library. Be sure to master both `AVAudioPlayer` use for local data files and `MPMusicPlayerController`, which interacts with the user's iTunes media collection.

9

Connecting to the Address Book

In addition to standard user interface controls and media components that you see on any computer, the iOS SDK provides a number of tightly focused developer solutions specific to iOS delivery. Among the most useful of these is the AddressBook framework, enabling you to programmatically access and manage the contacts database. This chapter introduces the Address Book and demonstrates how to use its frameworks in your applications. You read about accessing information on a contact-by-contact basis, how to modify and update contact information, and how to use predicates to find just the contact you're interested in. This chapter also covers the GUI classes that provide interactive solutions for picking, viewing, and modifying contacts. By reading this chapter, you can discover the Address Book from the bottom up.

The AddressBook Frameworks

The iOS SDK provides not one, but two AddressBook frameworks: AddressBook.framework and AddressBookUI.framework. As their names suggest, they occupy distinct niches in the iOS SDK. AddressBook provides low-level Core Foundation (C-based) objects and routines for accessing contact information from the iPhone's onboard databases. AddressBookUI offers high-level Objective-C–based `UIViewController` browser objects to present to users. Both frameworks are small. They provide just a few classes and data types.

AddressBook UI

The AddressBookUI framework provides several precooked view controllers that interact with the onboard contacts database. These interfaces include a general people picker, a contact viewer, and a contact editor. You set a delegate and then push these controllers onto your navigation stack or display them modally, as shown in the recipes in this chapter.

Like other special-purpose controllers discussed throughout this book, the AddressBookUI controllers are not general or flexible. Apple intends you to use them as provided, with little or no customization from the developer. What's more, they require a certain degree of low-level programming prowess. As you see in this chapter, these classes interact with the underlying Address Book in circuitous ways.

AddressBook and Its Databases

The AddressBook framework offers centralized access to iOS's Contacts database. This database stores information about individual contacts, including names, addresses, phone numbers, IM details, and more. Built using a Core Foundation-style API, the AddressBook framework enables you to query, modify, and add contacts from your applications.

The iOS contact data resides in the user's home Library folder. On the Macintosh-based iPhone simulator, you can freely access these files in ~/Library/Application Support/iPhone Simulator/*iOS Version*/Library/AddressBook. The files you find there—specifically, AddressBook.sqlitedb and AddressBookImages.sqlitedb—use standard SQLite to store contact information and, in the latter file, optional contact images.

On iOS, the same files live in /var/mobile/Library/AddressBook—that is, out of the application sandbox and therefore out of the reach of your direct inspection. Use the two AddressBook frameworks to query or modify the user's contact information rather than attempting to access these files through code.

> **Note**
> By default, starting in OS X 10.7, a user's Library folder is hidden in Finder. Access it via the Terminal or by holding the Option key while using Finder's Go menu. Alternatively, enter chflags nohidden ~/Library/ in Terminal to make the folder visible in Finder.

Records

In the Core Foundation C-based AddressBook framework, the `ABRecordRef` type provides the core contact structure. This record stores all information for each contact, including name, e-mail, phone numbers, an optional image, and so forth. Every record corresponds to a complete Address Book contact. Create a new record by calling `ABPersonCreate()`:

```
ABRecordRef person = ABPersonCreate();
```

Each record is stored in a single central Address Book. You can access this data by creating a new Address Book instance. How you do this has changed in iOS 6. Prior to iOS 6, you'd call `ABAddressBookCreate()`. Now you call `ABAddressbookCreateWithOptions()`, which takes two arguments. The first is reserved, so pass `NULL`; the second is a `CFErrorRef` pointer, for example, `ABAddressBookCreateWithOptions(NULL, &errorRef)`.

Should the create request fail, you can easily cast the `CFErrorRef` to an `NSError` to determine the failure reason.

Despite the name, the `ABAddressbookCreateWithOptions()` function does not create a new Address Book; it builds a local object that loads its data from the system Address Book. This function offers the primary entry point for you to access Address Book data directly. What's more, until you create an Address Book or a record instance, many of the constants you need to use in your applications will be undefined.

Apple has many warnings along these lines in its documentation: "The value of these constants is undefined until one of the following functions has been called: `ABAddressBookCreate()`, `ABPersonCreate()`, `ABGroupCreate()`."

The Custom ABStandin Class

Rather than create an Address Book instance each time data is accessed, the samples in this chapter use a custom `ABStandin` class. This class provides several basic Address Book access utilities and stores a static shared `ABAddressBookRef` instance. It is implemented as follows:

```
// Shared reference
static ABAddressBookRef shared = NULL;

@implementation ABStandin

// C-function callback updates address book when changed
void addressBookUpdated(ABAddressBookRef reference,
    CFDictionaryRef dictionary, void *context)
{
    ABAddressBookRevert(reference);
}

// Return the current address book
+ (ABAddressBookRef) addressBook
{
    if (shared) return shared;

    // Create the new address book
    CFErrorRef errorRef;
    shared = ABAddressBookCreateWithOptions(NULL, &errorRef);
    if (!shared)
    {
        NSError *error = (__bridge_transfer NSError *)errorRef;
        NSLog(@"Error creating new address book object: %@",
            error.localizedFailureReason);
        return nil;
    }
```

```objc
    // Register for automatic updates when information changes
    ABAddressBookRegisterExternalChangeCallback(
        shared, addressBookUpdated, NULL);
    return shared;
}

// Load the current address book with updates
+ (ABAddressBookRef) currentAddressBook
{
    if (!shared)
        return [self addressBook];

    ABAddressBookRevert(shared);
    return shared;
}

// Thanks Frederic Bronner. Save the address book out
+ (BOOL) save: (NSError **) error
{
    CFErrorRef errorRef;
    if (shared)
    {
        BOOL success = ABAddressBookSave(shared, &errorRef);
        if (!success)
        {
            if (error)
                *error = (__bridge_transfer NSError *)errorRef;
            return NO;
        }
        return YES;
    }
    return NO;
}

// Test authorization status
+ (BOOL) authorized
{
    ABAuthorizationStatus status =
        ABAddressBookGetAuthorizationStatus();
    return (status == kABAuthorizationStatusAuthorized);
}

// Place access request
+ (void) requestAccess
{
    // Post a notification that the app is authorized
    if ([self authorized])
```

```objc
    {
        NSNotification *note = [NSNotification notificationWithName:
            kAuthorizationUpdateNotification object:@YES];
        [[NSNotificationCenter defaultCenter] postNotification:note];
        return;
    }

    // Build a completion handler for the next step
    ABAddressBookRequestAccessCompletionHandler handler =
    ^(bool granted, CFErrorRef errorRef){
        // Respond to basic error condition
        if (errorRef)
        {
            NSError *error = (__bridge NSError *) errorRef;
            NSLog(@"Error requesting Address Book access: %@",
                error.localizedFailureReason);
            return;
        }

        // Post notification on main thread for success/fail
        dispatch_async(dispatch_get_main_queue(), ^{
            NSNotification *note =
                [NSNotification notificationWithName:
                        kAuthorizationUpdateNotification
                    object:@(granted)];
            [[NSNotificationCenter defaultCenter]
                postNotification:note];
        });
    };

    ABAddressBookRequestAccessWithCompletion(shared, handler);
}

+ (void) showDeniedAccessAlert
{
    UIAlertView *alertView = [[UIAlertView alloc]
        initWithTitle:@"Cannot Access Contacts"
        message:@"Please enable in Settings > Privacy > Contacts."
        delegate:nil cancelButtonTitle:nil
        otherButtonTitles:@"Okay", nil];
    dispatch_async(dispatch_get_main_queue(), ^{
        [alertView show];
    });
}
@end
```

In earlier editions of this book, Address Book objects were created on-the-fly and autoreleased. The code in this edition assumes that a single static instance is available at all times, which is automatically updated whenever the Address Book data changes. A method called `save:` updates the persistent Address Book data store with any updates you have made to the shared in-memory Address Book instance.

The big change to this edition is represented by the new iOS 6 restrictions that limit access to users' contacts. Users must explicitly allow access to their Address Book, and when they do (or do not) are not generally asked again. If a user has denied access, (you can test with `ABAddressBookGetAuthorizationStatus()`), you must offer directions to them on how to grant that access. In this case, the `showDeniedAccessAlert` method sends them to the Settings app.

Always provide some sort of instructions that enable users to work around previously denied permissions and never assume they know how to do this on their own. Users deny access for all sorts of reasons: mistapped buttons, a general distrust, or because they weren't paying attention to what they were being asked. Apple does not provide an "ask again" API. Instead, have your application react to the current status on launch.

That is the only time you need to check. Updating permissions in Settings automatically kills your application if it is running in the background. It will relaunch rather than resume the next time the user accesses it. This detail ensures that there will never be a situation, at least in iOS 6, in which you must update your GUI while the app is running to respond to a change in permissions. Apple may change this behavior in the future.

This class uses a notification system that updates the application regarding current authorization status. This enables you to use the current status to enable or disable affected GUI elements. You won't want to offer an Add Contact button, for example, when your app isn't authorized to respond to those requests:

```
[[NSNotificationCenter defaultCenter]
    addObserverForName:kAuthorizationUpdateNotification
    object:nil queue:[NSOperationQueue mainQueue]
    usingBlock:^(NSNotification *note)
    {
        NSNumber *granted = note.object;
        [self enableGUI:granted.boolValue];
    }];
```

Querying the Address Book

You can ask the Address Book for the number of objects currently stored in its database using a single function call. Although the name refers to Persons, this method counts all entries, including businesses, and returns a simple count.

```
+ (int) contactsCount
{
    ABAddressBookRef addressBook = [ABStandin addressBook];
    return ABAddressBookGetPersonCount(addressBook);
}
```

Wrapping the AddressBook Framework

Because it's C-based, the AddressBook framework can be fussy and difficult to work with if you're more comfortable using Objective-C. To deal with this, this chapter introduces a simple wrapper called `ABContact`. This wrapper class hides the CF nature of the class and replaces it with an Objective-C interface. Objective-C wrappers simplify integration between the C-based Address Book calls and normal Cocoa Touch development and memory management. You can find the full source for this, and other associated wrapper classes, in the sample code repository for this chapter.

The `ABContact` class hides an auto-synthesized internal `ABRecordRef`, the CF type that corresponds to each contact record. The remaining portion of the wrapper involves nothing more than generating properties and methods that enable you to reach into the `ABRecordRef` to set and access its subrecords:

```
@interface ABContact : NSObject
@property (nonatomic, readonly) ABRecordRef record;
@end
```

A second helper class `ABContactsHelper` wraps the AddressBook framework at the global level. It enables you to modify and query the database as a whole. Take its `contacts` method, for example. As with the record count, the Address Book can also return an array of contacts. You recover individual records by calling the `ABAddressBookCopyArrayOfAllPeople()` function, which returns a CF array of `ABRecordRef` objects.

The following `ABContactsHelper` method wraps this function to retrieve a record array. It then adds each contact record to a new `ABContact` wrapper and returns a standard NSArray populated with Objective-C contact instances:

```
+ (NSArray *) contacts
{
    ABAddressBookRef addressBook = [ABStandin addressBook];
    NSArray *thePeople =
        (__bridge_transfer NSArray *)
            ABAddressBookCopyArrayOfAllPeople(addressBook);
    NSMutableArray *array = [NSMutableArray array];
    for (id person in thePeople)
        [array addObject:[ABContact
            contactWithRecord:(__bridge ABRecordRef)person]];
    return array;
}
```

Working with Record Functions

The majority of `ABRecordRef` functions use the `ABPerson` prefix. This prefix corresponds to the `ABPerson` class available on the Macintosh but not on iOS. So, although the function calls are `ABPerson`-centric, all the data affected by these calls are actually `ABRecordRef` instances.

The reason for this becomes clearer when you notice that the same `ABRecordRef` structure is used in the AddressBook framework to represent both people (individual contacts, whether people or businesses) and groups (collections of contacts, such as work colleagues and personal friends). The SDK provides `ABGroup` functions and `ABPerson` ones, enabling you to add members and access membership lists. You read more about groups later in this section.

You already encountered the `ABPersonCreate()` function earlier in this section. As its prefix indicates, it's used to create a contact record. The SDK offers almost two dozen other `ABPerson`-named functions and a handful each of `ABGroup` and `ABRecord` functions. Together, these functions enable you to set and retrieve values within a contact, compare records, test for conforming data, and more.

Retrieving and Setting Strings

Each `ABRecord` stores about a dozen simple string values that represent, among other items, a person's name, title, job, and organization. You retrieve *single-valued strings* by copying field values from the record. Single-valued means that a person can own just one of each of these items. A person has only one first name, one birthday, and one nickname—at least as far as the AddressBook framework is concerned. Contrast this with phone numbers, e-mail addresses, and social-network identities. These latter are called *multivalued* and refer to items that naturally occur in multiples.

Recovering Strings

As you'd expect, working with single-valued elements is simpler than multivalued ones. You have just one item to set and to retrieve. All single-valued elements can be accessed using the same `ABRecordCopyValue()` function. The `ABContact` wrapper specializes this access based on the data type returned, such as for strings or dates. The following method copies a record value, casts it to a string, and returns that content:

```
- (NSString *) getRecordString:(ABPropertyID) anID
{
    NSString *result =
        (__bridge_transfer NSString *)
            ABRecordCopyValue(_record, anID);
    return result;
}
```

To determine which value to copy, it requires a property constant (`ABPropertyID`) that identifies the requested field in the record. The sample calls that follow use identifiers for the first and last name fields:

```
// Sample uses
- (NSString *) firstname
    {return [self getRecordString:kABPersonFirstNameProperty];}
- (NSString *) lastname
    {return [self getRecordString:kABPersonLastNameProperty];}
```

String Fields

The string-based fields you can recover in this fashion follow; their names are self-explanatory. These identifiers are declared as constant integers in the `ABPerson.h` header file and have remained stable now for generations of iOS. Each item identifies a single string-based field in an `ABRecordRef` record:

- kABPersonFirstNameProperty
- kABPersonLastNameProperty
- kABPersonMiddleNameProperty
- kABPersonPrefixProperty
- kABPersonSuffixProperty
- kABPersonNicknameProperty
- kABPersonFirstNamePhoneticProperty
- kABPersonLastNamePhoneticProperty
- kABPersonMiddleNamePhoneticProperty
- kABPersonOrganizationProperty
- kABPersonJobTitleProperty
- kABPersonDepartmentProperty
- kABPersonNoteProperty

Setting String Properties

Setting string-based properties proves to be just as simple as retrieving them. Cast the string you want to set to a `CFStringRef`. You need to use Core Foundation objects with Address Book calls. Then, use `ABRecordSetValue()` to store the data into the record.

These calls do not update the Address Book. They change only the data within the record. If you want to store a user's contact information, you must write that information back to the Address Book. Use the `save:` method, shown earlier in this section:

```
- (BOOL) setString: (NSString *) aString
    forProperty:(ABPropertyID) anID
{
    CFErrorRef errorRef = NULL;
```

Chapter 9 Connecting to the Address Book

```
    BOOL success = ABRecordSetValue(_record, anID,
        (__bridge CFStringRef) aString, &errorRef);
    if (!success)
    {
        NSError *error = (__bridge_transfer NSError *) errorRef;
        NSLog(@"Error: %@", error.localizedFailureReason);
    }
    return success;
}
// Examples of use
- (void) setFirstname: (NSString *) aString
    {[self setString: aString forProperty:
        kABPersonFirstNameProperty];}

- (void) setLastname: (NSString *) aString
    {[self setString: aString
        forProperty: kABPersonLastNameProperty];}
```

Working with Date Properties

In addition to the string properties you just saw, the Address Book stores three key dates: an optional birthday, the date the record was created, and the date the record was last modified. These items use the following property constants:

- kABPersonBirthdayProperty
- kABPersonCreationDateProperty
- kABPersonModificationDateProperty

Access these items exactly as you would with strings, but cast to and from NSDate instances instead of NSString instances. Don't try to modify the latter two properties; allow the Address Book to handle them for you:

```
// Return a date-time field from a record
- (NSDate *) getRecordDate:(ABPropertyID) anID
{
    return (__bridge_transfer NSDate *)
        ABRecordCopyValue(_record, anID);
}

// Get the contact's birthday
- (NSDate *) birthday
    {return [self getRecordDate:kABPersonBirthdayProperty];}
```

```
// Set a date-time field in a record
- (BOOL) setDate: (NSDate *) aDate forProperty:(ABPropertyID) anID
{
    CFErrorRef errorRef = NULL;
    BOOL success = ABRecordSetValue(_record, anID,
        (__bridge CFDateRef) aDate, &errorRef);
    if (!success)
    {
        NSError *error = (__bridge_transfer NSError *) errorRef;
        NSLog(@"Error: %@", error.localizedFailureReason);
    }
    return success;
}

// Set the contact's birthday
- (void) setBirthday: (NSDate *) aDate
    {[self setDate: aDate forProperty: kABPersonBirthdayProperty];}
```

Multivalue Record Properties

Each person may have multiple e-mail addresses, phone numbers, and important dates (beyond the birthday singleton) associated with his or her contact. `ABPerson` uses a multivalue structure to store lists of these items.

Each multivalue item is actually an array of pairs, which can be a bit confusing when you first start using this feature:

`@[pair1, pair2, pair3,...]`

Further, `ABContact` represents these pairs as a two-item dictionary. The *value* refers to the actual item that's added to the Address Book. The *label* specifies the role of that item:

`@[{label:label1, value:value1}, {label:label2, value:value2}, ...]`

For example, a person may have a work phone, a home phone, and a mobile phone. Each of these roles (work, home, and mobile) are labels. The actual phone numbers are the values associated with the roles:

```
@[{label:WorkPhone, value:303-555-1212},
    {label:HomePhone, value:720-555-1212}, ...]
```

Although you'd think it would be far easier for Apple to use a single dictionary using this information as key/value pairs, this is not how AddressBook is implemented.

Here's how you can create the `ABContact` dictionaries that populate each of these arrays. You pass a value and label, building up a dictionary from those elements. These elements are used when calling `ABMultiValueAddValueAndLabel()`.

```
+ (NSDictionary *) dictionaryWithValue: (id) value
    andLabel: (CFStringRef) label
{
    NSMutableDictionary *dict = [NSMutableDictionary dictionary];
    if (value) dict[@"value"] = value;
    if (label) dict[@"label"] = (__bridge NSString *)label;
    return dict;
}
```

Multivalue Labels

The AddressBook framework uses three generic labels, which can be used for any data class. These are kABWorkLabel, kABHomeLabel, and kABOtherLabel. In addition, you can find type-specific labels for URLs, Social Media, Instant Messaging, Phone Numbers, Dates, and Relationships. Each of these items is documented in the AddressBook documentation and may change over time. The Social Media elements (Twitter, Facebook, LinkedIn, Sina Weibo, and so on) represent recent additions to the iOS SDK.

Most multivalue items now enable you to add custom labels in addition to the system supplied ones. For example, you can add a contact's "Receptionist" (custom) in addition to Friend (system). The method calls in the following snippet are implemented in the ABContact wrapper:

```
// System-supplied
[person addRelationItem:contact1.compositeName
    withLabel:kABPersonFriendLabel];
// Custom
 [person addRelationItem:contact2.compositeName
    withLabel:(__bridge CFStringRef)@"Receptionist"];
```

> **Note**
>
> "Related names" is now fully supported in iOS 6. The kABPersonRelatedNamesProperty constant helps store names and their relationships (for example, Mary BallWashington might be stored in George Washington's contact using the kABPersonMotherLabel). See ABPerson.h for a full list of predefined relation constants.

Retrieving Multivalue Arrays

Recover each array from the record via its property identifier, returning a CFArrayRef, which you can use directly or bridge to an NSArray. The multivalue property identifiers you work with are as follows:

- kABPersonEmailProperty
- kABPersonPhoneProperty

- `kABPersonURLProperty`
- `kABPersonDateProperty`
- `kABPersonAddressProperty`
- `kABPersonSocialProfileProperty`
- `kABPersonInstantMessageProperty`

Each multivalue type plays an important role in storing data back to the record. The type is used to allocate memory and determine the size for each field within the record.

The first three of these items (e-mail, phone, and URL) store `multistrings`—that is, arrays of strings. Their associated type is the `kABMultiStringPropertyType`.

The next item, the date property, stores an array of dates using `kABMultiDateTimePropertyType`.

The last three items, namely the address, social profile, and instant-message properties, consist of arrays of dictionaries and use `kABMultiDictionaryPropertyType`. Each of these dictionaries has its own complexities and custom keys. (They are discussed later in this section.)

Retrieving Multivalue Property Values

To retrieve an array of values for any of these properties, copy the property out of the record (using `ABRecordCopyValue()`) and then break it down into its component array. The Address Book provides a function that copies the array from the property into a standard `CFArrayRef`:

```
- (NSArray *) arrayForProperty: (ABPropertyID) anID
{
    CFTypeRef theProperty =
        ABRecordCopyValue(_record, anID);
    if (!theProperty) return nil;
    NSArray *items = (__bridge_transfer NSArray *)
        ABMultiValueCopyArrayOfAllValues(theProperty);
    CFRelease(theProperty);
    return items;
}
```

Although you might think you've retrieved all the information with those two calls, you have not. Value retrieval alone is not sufficient for working with multivalued items. As mentioned earlier, each dictionary stored in a multivalue array uses a label and a value. Figure 9-1 shows part of an Address Book contact page. Grouped items use labels to differentiate the role for each e-mail, phone number, and so forth. This contact has three phone numbers and three e-mail address, each of which displays a label indicating the value's role.

Figure 9-1 Multivalue items consist of both a label (for example, main, Google, and mobile for these phone numbers, or home, work, and Google for these e-mail addresses) and a value.

Retrieving Multivalue Labels

You must transfer the labels from the property and the values to retrieve all the information stored for each multivalue property. The following method copies each label by index and then adds it to a labels array. Together, the labels and the values compose a complete multivalue collection:

```
- (NSArray *) labelsForProperty: (ABPropertyID) anID
{
    CFTypeRef theProperty = ABRecordCopyValue(_record, anID);
    if (!theProperty) return nil;
    NSMutableArray *labels = [NSMutableArray array];
    for (int i = 0;
         i < ABMultiValueGetCount(theProperty); i++)
    {
        NSString *label = (__bridge_transfer NSString *)
            ABMultiValueCopyLabelAtIndex(theProperty, i);
        [labels addObject:label];
    }
    CFRelease(theProperty);
    return labels;
}
```

Combine the retrieved labels and values together as follows to create the final multivalued array of dictionaries:

```
- (NSArray *) dictionaryArrayForProperty: (ABPropertyID) aProperty
{
    NSArray *valueArray = [self arrayForProperty:aProperty];
    NSArray *labelArray = [self labelsForProperty:aProperty];

    int num = MIN(valueArray.count, labelArray.count);
    NSMutableArray *items = [NSMutableArray array];
    for (int i = 0; i < num; i++)
    {
        NSMutableDictionary *md = [NSMutableDictionary dictionary];
        md[@"value"] = valueArray[i];
        md[@"label"] = labelArray[i];
        [items addObject:md];
    }
    return items;
}
```

Storing Multivalue Data

Saving into multivalue objects works the same way as the retrieval functions you just read about, but in reverse. To store multivalued items into a record, you transform your Cocoa Touch objects into a form the record can work with, namely the two-item value-label pair.

Custom Dictionaries

Although the value and label items are passed to the AddressBook functions as separate parameters, it's convenient to store them together in a single dictionary. This makes it easier to retrieve them in tandem when they are later needed. The following method tests a dictionary for conformance:

```
+ (BOOL) isMultivalueDictionary: (NSDictionary *) dictionary
{
    if (dictionary.allKeys.count != 2)
        return NO;
    if (!dictionary[@"value"])
        return NO;
    if (!dictionary[@"label"])
        return NO;
    return YES;
}
```

The following method expects an array of these custom dictionaries, whose objects correspond to the value and label retrieved from the original multivalued property. This code iterates through that array of dictionaries and adds each value and label to the mutable multivalue object, an instance of `ABMutableMultiValueRef`, which is created by calling `ABMultiValueCreateMutable()` with a property type:

```
- (ABMutableMultiValueRef) copyMultiValueFromArray:
    (NSArray *) anArray withType: (ABPropertyType) aType
{
    ABMutableMultiValueRef multi = ABMultiValueCreateMutable(aType);
    for (NSDictionary *dict in anArray)
    {
        if (![ABContact isMultivalueDictionary:dict])
            continue;
        ABMultiValueAddValueAndLabel(
            multi, (__bridge CFTypeRef) dict[@"value"],
            (__bridge CFTypeRef) dict[@"label"], NULL);
    }
    return multi;
}
```

The `ABPropertyType` supplied as the method parameter specifies what kind of values is stored. For example, you can populate an e-mail property with strings and labels using `kABMultiStringPropertyType` or an Address Book using `kABMultiDictionaryPropertyType`.

Working with Multivalue Items

After you create a multivalue item using a method like the one just shown, you can then add it to a contact by setting a record value. The following method assigns a multivalue item to a property ID using the standard `ABRecordSetValue()` call. There is essentially no difference between this call and the calls that set a single date or string property. All the work is done in creating the multivalue item in the first place, not in the actual assignment:

```
- (BOOL) setMultiValue: (ABMutableMultiValueRef) multi
    forProperty: (ABPropertyID) anID
{
    CFErrorRef errorRef = NULL;
    BOOL success = ABRecordSetValue(_record, anID, multi, &errorRef);
    if (!success)
    {
        NSError *error = (__bridge_transfer NSError *) errorRef;
        NSLog(@"Error: %@", error.localizedFailureReason);
    }
    return success;
}
```

You might call this general assignment method from a specialized setter, such as the one shown here. This method assigns an array of e-mail dictionaries by first building up a multivalued item and then setting the value:

```
- (void) setEmailDictionaries: (NSArray *) dictionaries
{
    // kABWorkLabel, kABHomeLabel, kABOtherLabel
    ABMutableMultiValueRef multi =
        [self copyMultiValueFromArray:dictionaries
            withType:kABMultiStringPropertyType];
    [self setMultiValue:multi
        forProperty:kABPersonEmailProperty];
    CFRelease(multi);
}
```

Addresses, Social Profile, and Instant-Message Properties

Three properties (addresses, social profiles, and instant-message IDs) use multivalued dictionaries rather than strings or dates. This adds an extra step to the creation of a multivalue array. You must populate a set of dictionaries and then add them to an array along with their labels before building the multivalued item and adding it to a contact record. Figure 9-2 illustrates this additional layer. As the figure shows, e-mail multivalue items consist of an array of label-value pairs, where each value is a single string. In contrast, addresses use a separate dictionary, which is bundled into the value item.

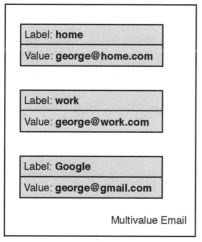
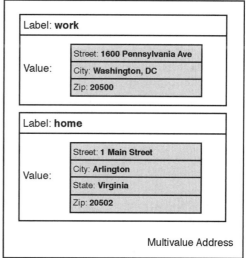

Figure 9-2   Unlike multivalue e-mail, which stores a single string for each value, multivalue addresses contain an entire address dictionary.

Here's an example that demonstrates the steps involved in creating a two-address multivalue item. This code builds the dictionaries and then adds them, along with their labels, to a base array. The array created by this code corresponds to the multivalue address object shown on the right side of Figure 9-2:

```
// Create the array that will store all the address
// value-label dictionaries
NSMutableArray *addresses = [NSMutableArray array];

// Create White House Address and add it to the array
NSDictionary *wh_addy = [ABContact
    addressWithStreet:@"1600 Pennsylvania Avenue"
    withCity:@"Washington, DC" withState:nil
    withZip:@"20500" withCountry:nil withCode:nil];
[addresses addObject:[ABContact dictionaryWithValue:wh_addy
    andLabel:kABWorkLabel]];

// Create a home address and add it to the array
NSDictionary *home_addy = [ABContact
    addressWithStreet:@"1 Main Street" withCity:@"Arlington"
    withState:@"Virginia" withZip:@"20502"
    withCountry:nil withCode:nil];
[addresses addObject:[ABContact dictionaryWithValue:home_addy
    andLabel:kABHomeLabel]];
```

This code relies on convenience methods to create both the address dictionaries and the value/label dictionaries used for the multivalue array. The following methods produce the label/value dictionaries for these three types. Notice how the keys for each dictionary are predefined, using Address Book key constants:

```
+ (NSDictionary *) addressWithStreet: (NSString *) street
    withCity: (NSString *) city
    withState:(NSString *) state
    withZip: (NSString *) zip
    withCountry: (NSString *) country
    withCode: (NSString *) code
{
    NSMutableDictionary *md = [NSMutableDictionary dictionary];
    if (street) md[
        (__bridge NSString *) kABPersonAddressStreetKey] = street;
    if (city) md[
        (__bridge NSString *) kABPersonAddressCityKey] = city;
    if (state) md[
        (__bridge NSString *) kABPersonAddressStateKey] = state;
    if (zip) md[
        (__bridge NSString *) kABPersonAddressZIPKey] = zip;
    if (country) md[
        (__bridge NSString *) kABPersonAddressCountryKey] = country;
```

```
        if (code) md[
            (__bridge NSString *) kABPersonAddressCountryCodeKey] = code;
        return md;
}
+ (NSDictionary *) socialWithURL: (NSString *) url
    withService: (NSString *) serviceName
    withUsername: (NSString *) username
    withIdentifier: (NSString *) key
{
    NSMutableDictionary *md = [NSMutableDictionary dictionary];
    if (url) md[
        (__bridge NSString *) kABPersonSocialProfileURLKey] = url;
    if (serviceName) md[
        (__bridge NSString *) kABPersonSocialProfileServiceKey] = serviceName;
    if (username) md[
        (__bridge NSString *) kABPersonSocialProfileUsernameKey] = username;
    if (key) md[
        (__bridge NSString *) kABPersonSocialProfileUserIdentifierKey] = key;
    return md;
}

+ (NSDictionary *) imWithService: (CFStringRef) service
    andUser: (NSString *) userName
{
    NSMutableDictionary *im = [NSMutableDictionary dictionary];
    if (service) im[(__bridge NSString *)
        kABPersonInstantMessageServiceKey] = (__bridge NSString *) service;
    if (userName) im[(__bridge NSString *)
        kABPersonInstantMessageUsernameKey] = userName;
    return im;
}
```

Address Book Images

Each record in the Address Book may be associated with an optional image. You can copy image data to and from each record. The `ABPersonHasImageData()` function indicates whether data is available for a given record. Use this to test whether you can retrieve image data.

Image data is stored as Core Foundation data, which can be bridged to `NSData`. As the `UIImage` class fully supports converting images into data and creating images from data, you just need to cast that data as needed. You might use the `UIImagePNGRepresentation()` function, for example, to transform a `UIImage` instance into an `NSData` representation. Address Book supports JPEG, JPG, BMP, PNG, and GIF data. To reverse, use `imageWithData:` to create a new image from `NSData`.

Apple details target size limitations for Contacts in various support documents. The iCloud-specific document http://support.apple.com/kb/HT4489 limits contacts to 25,000 entries, each of which may contain a 224 KB contact photo. (The entire contact is limited to 256 KB total.) The entire contacts book should not exceed 100 MB of photo data and 24 MB of card information:

```
// Retrieve image
- (UIImage *) image
{
    if (!ABPersonHasImageData(_record)) return nil;
    CFDataRef imageData = ABPersonCopyImageData(_record);
    if (!imageData) return nil;

    NSData *data = (__bridge_transfer NSData *)imageData;
    UIImage *image = [UIImage imageWithData:data];
    return image;
}

// Set image
- (void) setImage: (UIImage *) image
{
    CFErrorRef errorRef = NULL;
    BOOL success;

    if (image == nil) // remove
    {
        if (!ABPersonHasImageData(_record)) return;
        success = ABPersonRemoveImageData(_record, &errorRef);
        if (!success)
        {
            NSError *error = (__bridge_transfer NSError *) errorRef;
            NSLog(@"Error: %@", error.localizedFailureReason);
        }
        return;
    }

    NSData *data = UIImagePNGRepresentation(image);
    success = ABPersonSetImageData(
        record, (__bridge CFDataRef) data, &errorRef);
    if (!success)
    {
        NSError *error = (__bridge_transfer NSError *) errorRef;
        NSLog(@"Error: %@", error.localizedFailureReason);
    }
    return;
}
```

Creating, Adding, and Deleting Records

The `ABPersonCreate()` function returns a new `ABRecordRef` instance. This record exists outside the Address Book and represents a freestanding data structure. To date, all the examples in this chapter have modified individual records, but none so far has actually saved a record to the Address Book. Keep that in mind as you look at this convenience method that returns a newly initialized contact:

```
+ (id) contact
{
    ABRecordRef person = ABPersonCreate();
    id contact = [ABContact contactWithRecord:person];
    CFRelease(person);
    return contact;
}
```

Writing to the Address Book

To write new information to the Address Book takes two steps. First, add the record and then save the Address Book. New iOS developers often forget the second step, leading to an Address Book that appears to resist changes. This method adds a new contact to the Address Book but does not save the changes. Remember to call the custom AddressBook class's `save:` method, shown earlier, after adding the contact to make the new contact stick:

```
+ (BOOL) addContact: (ABContact *) aContact
    withError: (NSError **) error
{
    ABAddressBookRef addressBook = [ABStandin addressBook];
    BOOL success;
    CFErrorRef errorRef = NULL;

    success = ABAddressBookAddRecord(
        addressBook, aContact.record, &errorRef);
    if (!success)
    {
        if (error)
            *error = (__bridge_transfer NSError *)errorRef;
        return NO;
    }

    return YES;
}
```

Removing Records

Removing a record from the Address Book requires a save step, just like adding a record. When removed, the record still exists as an object, but it no longer is stored in the Address Book database. Here's how you can remove a record from the Address Book. Again, call [ABStandin save:&error] after:

```
- (BOOL) removeSelfFromAddressBook: (NSError **) error
{
    CFErrorRef errorRef = NULL;
    BOOL success;

    ABAddressBookRef addressBook = [ABStandin addressBook];

    success = ABAddressBookRemoveRecord(
        addressBook, self.record, &errorRef);
    if (!success)
    {
        if (error)
            *error = (__bridge_transfer NSError *)errorRef;
        return NO;
    }

    return success;
}
```

Searching for Contacts

The default AddressBook framework enables you to perform a prefix search across records. This function returns an array of records whose composite names (typically first name appended by last name, but localizable to countries where that pattern is reversed) match the supplied string:

```
NSArray *array = (__bridge_transfer NSArray *)
    ABAddressBookCopyPeopleWithName(addressBook, CFSTR("Eri"));
```

It's easier to use predicates and properties to perform searches. Use the custom ABContact class properties with NSPredicate instances to quickly filter for wanted matches. The following code matches a string against a contact's first name, middle name, last name, and nickname. The predicate uses property names to define how it matches or rejects contacts. It does this using a case and diacritical insensitive match ([cd]) that compares against all points within each string (contains), not just the start (begins with):

```
+ (NSArray *) contactsMatchingName: (NSString *) fname
{
    NSPredicate *pred;
    NSArray *contacts = [ABContactsHelper contacts];
    pred = [NSPredicate predicateWithFormat:
        @"firstname contains[cd] %@ OR lastname contains[cd] %@ OR\
        nickname contains[cd] %@ OR middlename contains[cd] %@",
        fname, fname, fname, fname];
    return [contacts filteredArrayUsingPredicate:pred];
}
```

> **Note**
>
> Apple's *Predicate Programming Guide* offers a comprehensive introduction to predicate basics. It demonstrates how to create predicates and use them in your application.

Sorting Contacts

At times, you may want to sort your contacts in some way. The following methods mimic string sorting, but do so using the `ABContact` class's composite name property that combines the first, middle, and last names. You can easily adapt this approach to sort using other keys:

```
- (BOOL) isEqualToString: (ABContact *) aContact
{
    return [self.compositeName isEqualToString:aContact.compositeName];
}

- (NSComparisonResult) caseInsensitiveCompare: (ABContact *) aContact
{
    return [self.compositeName
        caseInsensitiveCompare:aContact.compositeName];
}
```

Working with Groups

Groups collect contacts into related sets, such as friends, colleagues, drinking buddies, and other natural groupings. Each group is nothing more than another `ABRecord`, but with a few special properties. Groups don't store names, addresses, and phone numbers. Instead, they store a reference to other contact records.

Counting Groups

You can count the number of groups in the current Address Book by retrieving the group records, as shown in this method. There's no direct way to query the number of groups as there is with the number of person contacts. The following methods are part of an `ABGroup` wrapper class that provides an Objective-C wrapper for Address Book groups like `ABContact` wraps Address Book contacts:

```
+ (int) numberOfGroups
{
    ABAddressBookRef addressBook = [ABStandin addressBook];
    NSArray *groups = (__bridge_transfer NSArray *)
        ABAddressBookCopyArrayOfAllGroups(addressBook);
    int ncount = groups.count;
    return ncount;
}
```

Creating Groups

Create groups using the `ABGroupCreate()` function. This function returns an AB-RecordRef in the same way that `ABPersonCreate()` does. The difference lies in the record type. For groups, this property is set to `kABGroupType` instead of `kABPersonType`:

```
+ (id) group
{
    ABRecordRef grouprec = ABGroupCreate();
    id group = [ABGroup groupWithRecord:grouprec];
    CFRelease(grouprec);
    return group;
}
```

Adding and Removing Members

Add and remove members of a group by calling `ABGroupAddMember()` and `ABGroupRemoveMember()`. These calls affect records only and, as always, are not applied globally until you save the Address Book:

```
- (BOOL) addMember: (ABContact *) contact
    withError: (NSError **) error
{
    CFErrorRef errorRef = NULL;
    BOOL success;

    success = ABGroupAddMember(self.record,
        contact.record, &errorRef);
    if (!success)
    {
        if (error)
            *error = (__bridge_transfer NSError *)errorRef;
        return NO;
    }

    return YES;
}

- (BOOL) removeMember: (ABContact *) contact
    withError: (NSError **) error
{
    CFErrorRef errorRef = NULL;
    BOOL success;

    success = ABGroupRemoveMember(self.record,
        contact.record, &errorRef);
    if (!success)
    {
```

```
        if (error)
            *error = (__bridge_transfer NSError *)errorRef;
        return NO;
    }

    return YES;
}
```

Listing Members

Each member of a group is a person, or using this chapter's `ABContact` terminology, a contact. This method scans through a group's members and returns an array of `ABContact` instances, each initialized with the `ABRecordRef` for a group member:

```
- (NSArray *) members
{
    NSArray *contacts = (__bridge_transfer NSArray *)
        ABGroupCopyArrayOfAllMembers(self.record);
    NSMutableArray *array =
        [NSMutableArray arrayWithCapacity:contacts.count];
    for (id contact in contacts)
        [array addObject:[ABContact contactWithRecord:
            (__bridge ABRecordRef)contact]];
    return array;
}
```

Group Names

Every group has a name. It is the primary group property that you can set and retrieve. It uses the `kABGroupNameProperty` identifier, and it otherwise works just like the contacts properties:

```
- (NSString *) getRecordString:(ABPropertyID) anID
{
    return (__bridge_transfer NSString *)
        ABRecordCopyValue(_record, anID);
}

- (NSString *) name
{
    return [self getRecordString:kABGroupNameProperty];
}

- (void) setName: (NSString *) aString
{
    CFErrorRef errorRef = NULL;
    BOOL success;
```

```
    success = ABRecordSetValue(_record, kABGroupNameProperty,
        (__bridge CFStringRef) aString, &errorRef);
    if (!success)
    {
        NSError *error = (__bridge_transfer NSError *) errorRef;
        NSLog(@"Error: %@", error.localizedFailureReason);
    }
}
```

ABContact, ABGroup, and ABContactsHelper

The sample code that accompanies this section includes the source for three wrapper classes that have been demonstrated through these sections. The snippets shown throughout this discussion highlight the techniques used in these classes. Because of the length and overall redundancy of the classes, a single recipe listing has been omitted from this section. You can find the complete classes in the sample code folder for this chapter on the github repository.

The custom `ABContact` class is somewhat based on the Mac-only `ABPerson` class. It provides a more Cocoa Touch interface with built-in Objective-C 2.0 properties than the C-style property queries that Apple's `ABPerson` uses. All the contact-specific methods you've seen to date in this section derive from this class.

A second class called `ABGroup` wraps all the group functionality for `ABRecordRef` instances. It offers Objective-C access to group creation and management. Use this class to build new classes and add and remove members.

The final class, `ABContactsHelper`, provides Address Book-specific methods. Use this class to search through the Address Book, retrieve arrays of records, and so forth. Although a few basic searches are included across names and phone numbers, you can easily expand this class, which is hosted at http://github.com/erica, for more complex queries.

Recipe: Searching the Address Book

Predicate-based searches are both fast and effective. Recipe 9-1 is built on `ABContact`'s predicate queries. It presents a search table that displays a scrolling table of contacts. This table responds to user search bar queries with live listing updates.

When a user taps a row, this recipe displays an `ABPersonViewController` instance. This class offers a contact overview, displaying the details for a given record, similar to the view shown in Figure 9-1. To use this view controller, you allocate and initialize it and set its `displayedPerson` property. As you'd expect, this property stores an `ABRecordRef`, which can easily be retrieved from an `ABContact` instance.

Person view controllers offer a limited delegate. By setting the `personViewDelegate` property, you can subscribe to the `personViewController:shouldPerformDefaultActionForPerson:` method. This method triggers when users select certain items in the view, including phone numbers, e-mail addresses, URLs, and addresses. Return YES to perform the default action (dialing, e-mailing, and so on) or NO to skip.

Recipe 9-1 uses this callback to display the value for the selected item in the debug console. It's a convenient way to determine what the user touched and where his or her point of interest lies.

Although you can interact with other display elements, like the contact note and ringtone, these items do not produce callbacks.

To extend this recipe to allow editing, set the person view controller's allowsEditing property to YES. This provides the Edit button that appears at the top right of the display. When tapped, the Edit button triggers the same editing features in the person view that you normally see in the Contacts application.

Recipe 9-1 **Selecting and Displaying Contacts with Search**

```
// Return the number of table sections
- (NSInteger)numberOfSectionsInTableView:(UITableView *)aTableView
{
    // One section for this example
    return 1;
}

// Return the number of rows per section
- (NSInteger)tableView:(UITableView *)aTableView
    numberOfRowsInSection:(NSInteger)section
{
    if (aTableView == self.tableView)
        return matches.count;

    // On cancel, matches are restored in the search bar delegate
    matches = [ABContactsHelper
        contactsMatchingName:searchBar.text];
    matches = [matches sortedArrayUsingSelector:
        @selector(caseInsensitiveCompare:)];
    return matches.count;


    matches = [matches sortedArrayUsingSelector:
        @selector(caseInsensitiveCompare:)];
    return matches.count;
}

// Produce a cell for the given index path
- (UITableViewCell *)tableView:(UITableView *)aTableView
    cellForRowAtIndexPath:(NSIndexPath *)indexPath
```

```objc
{
    // Dequeue or create a cell
    UITableViewCellStyle style = UITableViewCellStyleDefault;
    UITableViewCell *cell =
        [aTableView dequeueReusableCellWithIdentifier:@"BaseCell"];
    if (!cell)
        cell = [[UITableViewCell alloc]
            initWithStyle:style reuseIdentifier:@"BaseCell"];

    ABContact *contact = [matches objectAtIndex:indexPath.row];
    cell.textLabel.text = contact.compositeName;
    return cell;
}

- (BOOL)personViewController:
      (ABPersonViewController *) personViewController
    shouldPerformDefaultActionForPerson:(ABRecordRef)person
    property:(ABPropertyID)property
    identifier:(ABMultiValueIdentifier)identifierForValue
{
    // Reveal the item that was selected
    if ([ABContact propertyIsMultiValue:property])
    {
        NSArray *array =
            [ABContact arrayForProperty:property inRecord:person];
        NSLog(@"%@", [array objectAtIndex:identifierForValue]);
    }
    else
    {
        id object =
            [ABContact objectForProperty:property inRecord:person];
        NSLog(@"%@", [object description]);
    }

    return NO;
}

// Respond to user taps by displaying the person viewcontroller
- (void) tableView:(UITableView *)tableView
    didSelectRowAtIndexPath:(NSIndexPath *)indexPath
{
    ABContact *contact = [matches objectAtIndex:indexPath.row];
    ABPersonViewController *pvc =
        [[ABPersonViewController alloc] init];
    pvc.displayedPerson = contact.record;
    pvc.personViewDelegate = self;
    // pvc.allowsEditing = YES; // optional editing
```

```
    [self.navigationController
        pushViewController:pvc animated:YES];
}
```

> **Get This Recipe's Code**
> To find this recipe's full sample project, point your browser to https://github.com/erica/iOS-6-Advanced-Cookbook and go to the folder for Chapter 9.

Recipe: Accessing Contact Image Data

Recipe 9-2 expands on Recipe 9-1 by adding contact image thumbnails to each table cell, where available. It does this by creating a new thumbnail from the contact's data. When image data is available, that image is rendered onto the thumbnail. When it is not, the thumbnail is left blank. Upon being drawn, the image is assigned to the cell's `imageView`.

Figure 9-3 shows the interface for this recipe. In this screen shot, a search is in progress (matching against the letter "e"). The records that match the search each display their image thumbnail.

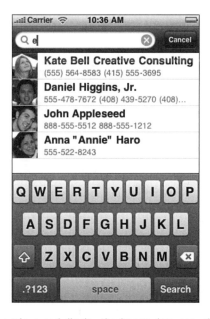

Figure 9-3 You can easily retrieve and display the image data associated with Address Book contacts.

Notice how simple it is to create and use thumbnails in this recipe. It takes just a few lines of code to build a new image context, draw into it (for contacts with images), and save it out to a `UIImage` instance.

Recipe 9-2 **Displaying Address Book Images in Table Cells**

```
- (UITableViewCell *)tableView:(UITableView *)aTableView
    cellForRowAtIndexPath:(NSIndexPath *)indexPath
{
    // Dequeue or create a cell
    UITableViewCellStyle style =  UITableViewCellStyleSubtitle;
    UITableViewCell *cell =
        [aTableView dequeueReusableCellWithIdentifier:@"BaseCell"];
    if (!cell)
        cell = [[UITableViewCell alloc]
            initWithStyle:style reuseIdentifier:@"BaseCell"];

    ABContact *contact = [matches objectAtIndex:indexPath.row];
    cell.textLabel.text = contact.compositeName;
    cell.detailTextLabel.text = contact.phonenumbers;

    CGSize small = CGSizeMake(48.0f, 48.0f);
    UIGraphicsBeginImageContext(small);
    UIImage *image = contact.image;
    if (image)
        [image drawInRect:(CGRect){.size = small}];
    cell.imageView.image = UIGraphicsGetImageFromCurrentImageContext();
    UIGraphicsEndImageContext();

    return cell;
}
```

Get This Recipe's Code

To find this recipe's full sample project, point your browser to https://github.com/erica/iOS-6-Advanced-Cookbook and go to the folder for Chapter 9.

Recipe: Picking People

The AddressBookUI framework offers a people picker controller, which enables you to browse your entire contacts list. It works similarly to the individual contact screen but does not require an a priori contact selection. The `ABPeoplePickerNavigationController` class presents the interactive browser, as shown in Figure 9-4.

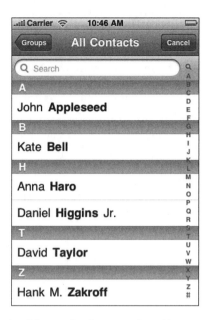

Figure 9-4 The iPhone people picker navigation control enables users to search through the contacts database and select a person or organization.

Allocate and display the controller before presenting it modally. Make sure to set the peoplePickerDelegate property, which enables you to catch user interactions with the view:

```
- (void) action: (UIBarButtonItem *) bbi
{
    ABPeoplePickerNavigationController *ppnc =
        [[ABPeoplePickerNavigationController alloc] init];
    ppnc.peoplePickerDelegate = self;
    [self presentViewController:ppnc animated:YES completion:nil];
}
```

When you declare the ABPeoplePickerNavigationControllerDelegate protocol, your class must implement the following three methods. These methods respond to users when they tap a contact, or any of a contact's properties, or when the user taps Cancel:

- **peoplePickerNavigationController:shouldContinueAfterSelectingPerson:**—
When users tap a contact, you have two choices. You can accept the person as the final selection and dismiss the modal view, or you can navigate to the individual display. To pick just the person, this method returns NO. To continue to the individual screen, return YES. The second argument contains the selected person, in case you want to stop after selecting any ABPerson record.

- **peoplePickerNavigationController:shouldContinueAfterSelectingPerson:pro perty:identifier:**—This method does not get called until the user has progressed to an individual contact display screen. Then, it's up to you whether to return control to your program (return NO) or to continue (return YES). You can determine which property has been tapped and to recover its value using the code from Recipe 9-1. Although this method should be optional, it is not at the time of writing this book.

- **peoplePickerNavigationControllerDidCancel:**—When a user taps Cancel, you still want a chance to dismiss the modal view. This method catches the cancel event, enabling you to use it to perform the dismissal.

Recipe 9-3 presents the simplest possible people picking example. It presents the picker and waits for a user to select a contact. When the user does so, it dismisses the picker and changes the view controller title (in the navigation bar) to show the composite name of the selected person. Returning NO from the primary callback means the property callback will never be called. You must still include it in your code because all three methods are required.

Recipe 9-3 **Picking People**

```
- (BOOL)peoplePickerNavigationController:
        (ABPeoplePickerNavigationController *)peoplePicker
    shouldContinueAfterSelectingPerson:(ABRecordRef)person
{
    self.title = [[ABContact contactWithRecord:person] compositeName];
    [self dismissViewControllerAnimated:YES completion:nil];
    return NO;
}


- (BOOL)peoplePickerNavigationController:
        (ABPeoplePickerNavigationController *)peoplePicker
    shouldContinueAfterSelectingPerson:(ABRecordRef)person
    property:(ABPropertyID)property
    identifier:(ABMultiValueIdentifier)identifier
{
    // required method that is never called in the people-only-picking
    [self dismissViewControllerAnimated:YES completion:nil];
    return NO;
}

- (void)peoplePickerNavigationControllerDidCancel:
    (ABPeoplePickerNavigationController *)peoplePicker
{
    [self dismissViewControllerAnimated:YES completion:nil];
}
```

```
- (void) action: (UIBarButtonItem *) bbi
{
    ABPeoplePickerNavigationController *ppnc =
        [[ABPeoplePickerNavigationController alloc] init];
    ppnc.peoplePickerDelegate = self;
    [self presentViewController:ppnc animated:YES completion:nil];
}
```

Get This Recipe's Code

To find this recipe's full sample project, point your browser to https://github.com/erica/iOS-6-Advanced-Cookbook and go to the folder for Chapter 9.

Recipe: Limiting Contact Picker Properties

When you need users to pick a certain kind of property, such as an e-mail address, you won't want to present users with a person's street address or fax number. Limit the picker's displayed properties to show just those items you want the users to select from. Figure 9-5's picker has been limited to e-mail selection.

Figure 9-5 The people picker's displayed properties enable you to choose which properties to present to users (in this case, e-mail only).

To make this happen, choose the displayed properties by submitting an array of property types to the controller. Set the picker's `displayedProperties` property. Recipe 9-4 offers two picking options: one for e-mail, the other for phone numbers. Although these examples use a single property for the properties array, you can choose to display any number of properties.

Recipe 9-4 **Choosing Display Properties**

```
- (BOOL)peoplePickerNavigationController:
        (ABPeoplePickerNavigationController *)peoplePicker
   shouldContinueAfterSelectingPerson:(ABRecordRef)person
{
   // Continue onto the detail screen
   return YES;
}

- (BOOL)peoplePickerNavigationController:
        (ABPeoplePickerNavigationController *)peoplePicker
   shouldContinueAfterSelectingPerson:(ABRecordRef)person
   property:(ABPropertyID)property
   identifier:(ABMultiValueIdentifier)identifier
{
   // Guaranteed to only be working with e-mail or phone here
   [self dismissViewControllerAnimated:YES completion:nil];
   NSArray *array =
      [ABContact arrayForProperty:property inRecord:person];
   self.title = (NSString *)[array objectAtIndex:identifier];
   return NO;
}

- (void)peoplePickerNavigationControllerDidCancel:
     (ABPeoplePickerNavigationController *)peoplePicker
{
   // Respond to cancel
   [self dismissViewControllerAnimated:YES completion:nil];
}

- (void) email: (UIBarButtonItem *) bbi
{
   ABPeoplePickerNavigationController *ppnc =
       [[ABPeoplePickerNavigationController alloc] init];
   ppnc.peoplePickerDelegate = self;
   [ppnc setDisplayedProperties:
        [NSArray arrayWithObject:@(kABPersonEmailProperty)]];
   [self presentViewController:ppnc animated:YES completion:nil];
}
```

```
- (void) phone: (UIBarButtonItem *) bbi
{
    ABPeoplePickerNavigationController *ppnc =
        [[ABPeoplePickerNavigationController alloc] init];
    ppnc.peoplePickerDelegate = self;
    [ppnc setDisplayedProperties:
        [NSArray arrayWithObject:@(kABPersonPhoneProperty)]];
    [self presentViewController:ppnc animated:YES completion:nil];
}
```

> **Get This Recipe's Code**
>
> To find this recipe's full sample project, point your browser to https://github.com/erica/iOS-6-Advanced-Cookbook and go to the folder for Chapter 9.

Recipe: Adding and Removing Contacts

Allow your users to create new contacts with the `ABNewPersonViewController` class. This view controller offers an editing screen that simplifies the interactive creation of a new Address Book entry. After allocating and initializing the view controller, start by creating a new contact and assigning it to the `displayedPerson` property. If you want, you can prefill the contact's record with properties first.

Next, assign the `newPersonViewDelegate`, making sure to declare the `ABNewPersonViewControllerDelegate` protocol in the delegate class. Delegates receive one new callback: `newPersonViewController:didCompleteWithNewPerson:`. This callback is sent for both selection and cancel events. Check the `person` parameter to determine which case applies, as shown in Recipe 9-5. If the person is `nil`, the user tapped Cancel, and the new contact should not be added to the Address Book. If the user taps Done after editing the new contact, add the contact data to the Address Book and save it.

To remove a contact from the Address Book, the user is first presented with standard people picker navigation controller and solicited for a selection. When the user selects a name, the callback method asks them to confirm deletion before continuing. When confirmed, the Address Book must be saved. Removing a record from the Address Book requires a save step, just like adding a record.

Recipe 9-5 **Using the New Person View Controller**

```
#pragma mark NEW PERSON DELEGATE METHODS
- (void)newPersonViewController:
        (ABNewPersonViewController *)newPersonViewController
    didCompleteWithNewPerson:(ABRecordRef)person
{
```

```objc
    if (person)
    {
        ABContact *contact = [ABContact contactWithRecord:person];
        self.title = [NSString stringWithFormat:
            @"Added %@", contact.compositeName];

        NSError *error;
        BOOL success =
            [ABContactsHelper addContact:contact withError:&error];
        if (!success)
        {
            NSLog(@"Could not add contact. %@",
                error.localizedFailureReason);
            self.title = @"Error.";
        }

        [ABStandin save:nil];
    }
    else
        self.title = @"Cancelled";

    [self.navigationController popViewControllerAnimated:YES];
}

#pragma mark PEOPLE PICKER DELEGATE METHODS
- (BOOL)peoplePickerNavigationController:
        (ABPeoplePickerNavigationController *)peoplePicker
    shouldContinueAfterSelectingPerson:(ABRecordRef)person
{
    [self dismissViewControllerAnimated:YES completion:nil];
    ABContact *contact = [ABContact contactWithRecord:person];

    NSString *query = [NSString stringWithFormat:
        @"Really delete %@?",  contact.compositeName];
    if ([self ask:query])
    {
        self.title = [NSString stringWithFormat:@"Deleted %@",
            contact.compositeName];
        [contact removeSelfFromAddressBook:nil];
        [ABStandin save:nil];
    }

    return NO;
}

- (BOOL)peoplePickerNavigationController:
        (ABPeoplePickerNavigationController *)peoplePicker
```

```objc
    shouldContinueAfterSelectingPerson:(ABRecordRef)person
    property:(ABPropertyID)property
    identifier:(ABMultiValueIdentifier)identifier
{
    [self dismissViewControllerAnimated:YES completion:nil];
    return NO;
}

- (void)peoplePickerNavigationControllerDidCancel:
    (ABPeoplePickerNavigationController *)peoplePicker
{
    [self dismissViewControllerAnimated:YES completion:nil];
}

#pragma mark Respond to User Requests
- (void) add
{
    // create a new view controller
    ABNewPersonViewController *npvc =
        [[ABNewPersonViewController alloc] init];

    // Create a new contact
    ABContact *contact = [ABContact contact];
    npvc.displayedPerson = contact.record;

    // Set delegate
    npvc.newPersonViewDelegate = self;

    [self.navigationController
        pushViewController:npvc animated:YES];
}

- (void) remove
{
    ABPeoplePickerNavigationController *ppnc =
        [[ABPeoplePickerNavigationController alloc] init];
    ppnc.peoplePickerDelegate = self;
    [self presentlViewController:ppnc animated:YES completion:nil];
}
```

Get This Recipe's Code

To find this recipe's full sample project, point your browser to https://github.com/erica/iOS-6-Advanced-Cookbook and go to the folder for Chapter 9.

Modifying and Viewing Individual Contacts

The `ABPersonViewController`'s `allowsEditing` property enables your user to edit contacts. You create a controller, assign an Address Book record as its `displayedPerson`, set its editable property, and display it. Here's a typical example of this feature in a callback method. Upon resolving to a particular contact, the application creates a new person presentation and pushes it onto the navigation stack:

```
- (void)unknownPersonViewController:
        (ABUnknownPersonViewController *)unknownPersonView
    didResolveToPerson:(ABRecordRef)person
{
    // Handle cancel events
    if (!person) return;

    ABPersonViewController *abpvc = [[ABPersonViewController alloc] init];
    abpvc.displayedPerson = person;
    abpvc.allowsEditing = YES;
    abpvc.personViewDelegate = self;

    [self.navigationController pushViewController:abpvc animated:YES];
}
```

User Edits

When the `allowsEditing` property has been set to `YES`, an Edit button appears at the top right of the displayed person presentation. Users may tap this button and modify contact details. Edits are automatically propagated to the Address Book. You do not need to explicitly add calls to save the data; this is handled for you by the Cocoa Touch AddressBookUI implementation. Unlike the Contacts application, the Delete Contact option will not appear in red at the bottom of the controller.

If you want to allow the user to view the contact information without enabling edits, switch the `allowsEditing` property value to `NO`. Otherwise, leave the presentation code as is. No other modifications are necessary.

Delegate Methods

The person view controller has only one delegate callback method. You specify if tapping items, such as e-mail addresses or URLs, should perform actions based on those taps (that is, opening Mail or linking to Safari). Return `YES` or `NO` as wanted. When set to `YES`, be aware that control may pass out of the application in response to these actions:

```
#pragma mark PERSON DELEGATE
- (BOOL)personViewController:
        (ABPersonViewController *)personViewController
    shouldPerformDefaultActionForPerson:(ABRecordRef)person
```

```
    property:(ABPropertyID)property
    identifier:(ABMultiValueIdentifier)identifierForValue
{
    return NO;
}
```

Recipe: The "Unknown" Person Controller

What happens when you have some information such as an e-mail address or a phone number, but you don't have a contact to associate with it yet? The `ABUnknownPersonViewController` enables you to add that information to a new or existing contact. This class exists to associate known properties with unknown contacts. It works like this.

You allocate and initialize the view controller and then create and prefill a record with whatever properties you like. For example, Recipe 9-6 builds a random "monster" image. You can add more items if you want, but typically, this controller is used to add a single bit of information. The unknown person controller specializes in that kind of add-one-item utility. Figure 9-6 shows this recipe's controller in action.

Figure 9-6 You can add images to contacts with the unknown person controller as well as e-mail addresses, phone numbers, and other text-based data.

Users can add the image to an existing contact or create a new contact. Tapping Back returns without adding the image.

The Create New Contact and Add to Existing Contact buttons are controlled by the controller's `allowsAddingToAddressBook` property. When a user taps on the new contact button, a form appears that is prefilled with the properties from the passed record. If you disable this adding property, these buttons do not appear.

The `allowsAction` property enables users to tap on form elements in this interface. This enables them to connect to an e-mail address, to phone numbers, to web pages, and so forth. By tapping any element, you permit users to branch out of your app to make a call or move to Mobile Safari, and so on. This provides a handy way to add interactive contact information and URLs into an already-defined view controller.

The `alternateName` and `message` properties on the controller provide text that fills the name and organization fields. Although you can populate these fields with data via the record, the options do not transfer to contacts. Therefore, they provide a nice way to prompt the user without side effects.

Set the `unknownPersonViewDelegate` property and declare the `ABUnknownPersonViewControllerDelegate` protocol to receive the `unknownPersonViewController:didResolveToPerson:` method. Called when the user taps Done, this method enables you to recover the record that the data was saved to. It also provides a place in which you can programmatically pop the view controller, knowing that the user has finished interaction with the dialog.

> **Note**
> The Monster ID project art used with this recipe consists of a collection of body part art that can be compiled together to form random pictures. It was developed by Andreas Gohr and was inspired by a web post by Don Park and the combinatoric critters. Built up by adding predrawn arms, legs, a body, and so forth, the resulting composite image produces a full creature.

Recipe 9-6 **Working with the Unknown Controller**

```
#pragma mark UNKNOWN PERSON DELEGATE
- (void)unknownPersonViewController:
        (ABUnknownPersonViewController *)unknownPersonView
    didResolveToPerson:(ABRecordRef)person
{
    // Handle cancel events
    if (!person) return;

    ABPersonViewController *abpvc =
        [[ABPersonViewController alloc] init];
    abpvc.displayedPerson = person;
    abpvc.allowsEditing = YES;
    abpvc.personViewDelegate = self;
```

```objc
    [self.navigationController
        pushViewController:abpvc animated:YES];
}

#pragma mark PERSON DELEGATE
- (BOOL)personViewController:
        (ABPersonViewController *)personViewController
    shouldPerformDefaultActionForPerson:(ABRecordRef)person
    property:(ABPropertyID)property
    identifier:(ABMultiValueIdentifier)identifierForValue
{
    return NO;
}

- (BOOL)unknownPersonViewController:
        (ABUnknownPersonViewController *)personViewController
    shouldPerformDefaultActionForPerson:(ABRecordRef)person
    property:(ABPropertyID)property
    identifier:(ABMultiValueIdentifier)identifier
{
    return YES;
}

#pragma mark Action
- (void) assignAvatar
{
    ABUnknownPersonViewController *upvc =
        [[ABUnknownPersonViewController alloc] init];
    upvc.unknownPersonViewDelegate = self;

    ABContact *contact = [ABContact contact];
    contact.image = imageView.image;

    upvc.allowsActions = NO;
    upvc.allowsAddingToAddressBook = YES;
    upvc.message = @"Who looks like this?";
    upvc.displayedPerson = contact.record;

    [self.navigationController pushViewController:upvc animated:YES];
}
```

Get This Recipe's Code

To find this recipe's full sample project, point your browser to https://github.com/erica/iOS-6-Advanced-Cookbook and go to the folder for Chapter 9.

Summary

This chapter introduced Address Book core functionality for both the AddressBook and AddressBookUI frameworks. You've read about how these work in this heavily technical chapter. After all that work, here are a few final thoughts about the frameworks you just encountered:

- Although useful, the low-level Address Book functions can prove frustrating to work with directly. The various helper classes that accompany this chapter may help make your life easier. All the wrappers have been updated for iOS 6 and its user-permission challenges.

- Accessing and modifying an Address Book image works like any other field. Supply image data instead of strings, dates, or multivalue arrays. Don't hesitate to use image data in your contacts applications.

- The view controllers provided by the AddressBookUI framework work seamlessly with the underlying AddressBook routines. There's no need to roll your own GUIs for most common Address Book interaction tasks.

- Do you work with Vcards? The AddressBook framework provides two Vcard-specific functions that enable you to create Vcard data from an array of records (`ABPersonCreateVCardRepresentationWithPeople()`) and transform Vcard data into ABRecordRefs (`ABPersonCreatePeopleInSourceWithVCardRepresentation ()`).

- The unknown person controller provides a great way to store specific information (such as a company's e-mail address, an important Web site, and so on) into a contact while allowing user discretion for where (or whether) to place that information.

10
Location

Location is meaningful. Where you compute is fast becoming just as important as how you compute and what you compute. iOS is constantly on the go, traveling with its users throughout the course of the day. Core Location infuses iOS with on-demand geopositioning. MapKit adds interactive in-application mapping enabling users to view and manipulate annotated maps. With Core Location and MapKit, you can develop applications that help users meet up with friends, search for local resources, or provide location-based streams of personal information. This chapter introduces these location-aware frameworks and shows you how you can integrate them into your iOS applications.

Authorizing Core Location

Apple requires users to both enable and authorize Core Location. These are two separate settings, leading to two separate checks that your application must perform. Users enable global location services in Settings > Privacy > Location (see Figure 10-1).

Testing for Location Services

Test whether location services have been enabled by checking the `locationServicesEnabled` class method. This returns a Boolean value indicating whether the user has globally opted out of location services:

```
// Check for location services
if (![CLLocationManager locationServicesEnabled])
{
    NSLog(@"User has disabled location services");
    // Display instructions on enabling location services
    return;
}
```

Users authorize location services on an app-by-app basis, as shown in Figure 10-2. This prompt appears the first time an application accesses location services. Users authorize or deny access by tapping Don't Allow or OK.

Figure 10-1   Users can enable and disable Location Services across their device in Settings > Privacy > Location. Wi-Fi is not needed on GPS-enabled devices.

Figure 10-2   Users authorize location services on an app-by-app basis. This prompt is displayed just once, and the user's choice is stored to Settings > General > Privacy > *Application Name*.

Should users deny access, an application has no recourse other than to instruct them on how to manually restore it. Specifically, tell your users to amend their authorization choices in Settings > Privacy > Location Services.

You can guard against this possibility in two ways. First, you can test authorization status at launch using the `authorizationStatus` class method. When "not determined," the user has not yet been prompted to decide. You may want to display some sort of precheck instructions, explaining why they should tap OK:

```
if ([CLLocationManager authorizationStatus] ==
    kCLAuthorizationStatusNotDetermined)
{
    // Perform any instructions, e.g. tell user that the
    // app is about to request permission
}
```

Second, you can update your Info.plist to add a short one-sentence description of how location services will be used. Assign your text to the `NSLocationUsageDescription` key. You see this key used in Figure 10-2. The custom description (App Demonstrates Location Capabilities to Users) enables the prompt to better explain *why* the user should grant access. Similar keys exist for other privileged services, including access to the photo library, calendar, contacts, and reminders.

Resetting Location and Privacy

When testing, you may want to try various situations—that is, the user has granted access, the user has denied access, the user grants access and then revokes it, and so on. You need to use two separate Settings panes to produce these scenarios.

To switch the current authorization from on to off or vice versa, visit Settings > Privacy > Location (see Figure 10-3). Scroll down the screen. There you can find a list of applications that have requested location access. A toggle appears next to each application name.

To restore privacy settings to their original pristine condition, that is, before the authorization prompt has yet displayed, go to Settings > General > Reset. Tap Reset Location & Privacy and then Reset Warnings (see Figure 10-4). This option restores privacy to its original pristine conditions—that is, the user will be prompted again for permission when the app is next launched. Its use is especially convenient for developers testing apps.

342 Chapter 10 Location

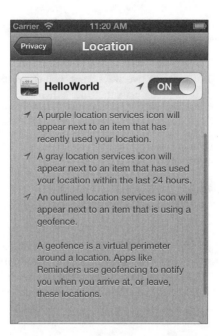

Figure 10-3 Toggle privacy settings on a per-application basis in Settings > Privacy > Location. Scroll down to find the list of applications.

Figure 10-4 Resetting Location & Privacy enables applications to reprompt users for permission to access location information.

Checking User Permissions

The same authorization status check that enables you to know whether the user has yet to be prompted also enables you to test whether an app has been granted or denied permission. If granted, you can begin Core Location updates:

```
// Check user permissions
if ([CLLocationManager authorizationStatus] ==
        kCLAuthorizationStatusDenied)
{
    NSLog(@"User has denied location services");
    return;
}

if ([CLLocationManager authorizationStatus] ==
    kCLAuthorizationStatusAuthorized)
{
    // The app need not query the user. It is already
    // authorized.
}

manager = [[CLLocationManager alloc] init];
[manager startUpdatingLocation];
```

> **Note**
> Increase GPS accuracy by ensuring the unit's date, time, and time zone are set correctly in Settings > General > Date & Time. Choose Set Automatically.

Testing Core Location Features

The `CLLocationManager` class offers several class methods that test whether Core Location features are enabled. These methods include the following:

- `authorizationStatus` specifies whether users have authorized location services for the application. If the user has agreed to allow the app to use location, this value returns YES.

- `locationServicesEnabled` indicates whether location services are enabled on the device at all. Users can enable and disable services entirely from Settings > Location Services or can enable/disable services on an application-by-application basis.

- `significantLocationChangeMonitoringAvailable` lets your app know whether the device can provide low-power/low-accuracy updates that occur when the primary cell tower association changes on the device. Significant location change monitoring offers a great way to roughly track vehicular travel and spot nearby attractions using little juice.

- `headingAvailable` establishes whether the location manager can provide heading events using an onboard compass.

- `regionMonitoringAvailable` specifies whether region monitoring is available on the current device; `regionMonitoringEnabled` indicates whether the user has enabled or overridden the feature. Region monitoring lets applications define geographical regions to track movement into and out from set areas. Applications can use this to associate actions with real-world location. For example, if you enter a region near your grocery store, applications can notify you to pick up an extra quart of milk. This is called *geofencing*.

Always test whether Core Location features are available before attempting to use them. Some of these limitations are device-specific (for example, you cannot use GPS on first-generation iPhones), whereas others are set by the user. If your user denies access to some of these, your application will be limited to whatever features (if any) remain authorized. Your app behavior in such situations will be tested during the App Store approval process.

> **Note**
>
> Use the `CLLocationCoordinate2DIsValid()` function to test whether a coordinate provides a valid latitude/longitude pair. The latitude needs to fall between –90 and 90 degrees and its longitude between –180 and 180 degrees.

Recipe: Core Location in a Nutshell

Core Location is easy to use, as demonstrated by the following steps. They walk you through a process of setting up your program to request location data that's representative of normal use. These steps and Recipe 10-1 provide just one example of using Core Location's services, showing how you might pinpoint a user's location:

1. Add the Core Location framework to your project, and optionally edit your Info.plist to add a location usage description key.

2. Check whether the user has enabled Core Location by testing the `CLLocationManager` class's `locationServicesEnabled` class value. Users have the option to switch off Core Location from General > Location Services in the Settings application.

3. Allocate a location manager. Set the manager's delegate to your primary view controller or application delegate. Optionally, set its wanted distance filter and accuracy. The distance filter specifies a minimum distance in meters. The device must move at least this distance before it can register a new update. If you set the distance for 5 meters, for example, you will not receive new events until the device has moved that far. If you plan to test by walking, you probably want to reduce that number.

    The accuracy property specifies the degree of precision that you're requesting. To be clear, the location manager does not guarantee any actual accuracy. Setting the requested accuracy asks the manager to (attempt to) retrieve at least that level. When you do not need precision, the manager will deliver its results using whatever technology is available.

When you do need precision, the `desiredAccuracy` property informs the manager of that need. You'll find a high level of accuracy especially important for walking and running applications. A lower accuracy level may work for driving in a car or for locating users within large geographical boundaries like cities, states, and countries.

4. Start locating. Tell the location manager to start updating the location. Delegate callbacks let you know when a new location has been found. This can take many seconds or up to a minute to occur.

5. Handle the location event delegate callbacks. You'll deal with two types of callbacks: successes that return `CLLocation` data (`locationManager:didUpdateLocations:`, which replaces `locationManager:didUpdateTo-Location:fromLocation:` in iOS 6) and failures that do not (`locationManager:didFailWithError:`). Add these delegate methods to your code to catch location updates. In Recipe 10-1, the successful location logs an information overview (`description`) that includes the current latitude and longitude results.

Depending on your requested accuracy, you may receive three or four location callbacks based on the various location methods used and the requested accuracy, so take this nonlinearity into account.

6. Start moving and wait. Callbacks arrive asynchronously, as location data becomes available. The location information returned to your application includes positioning information along with accuracy measures that you can use to evaluate precision.

When possible, test your Core Location applications on the device and not in the simulator. Although the simulator can now support a variety of location scenarios, on-device testing will provide you with the best results. Deploying Recipe 10-1 to the device allows you to test results as you walk or drive around with your iOS unit. If testing while driving, it's best if you're the passenger and not the driver.

Recipe 10-1 **Using Core Location to Retrieve Latitude and Longitude**

```
- (void)locationManager:(CLLocationManager *)manager
    didFailWithError:(NSError *)error
{
    if ([CLLocationManager authorizationStatus] ==
        kCLAuthorizationStatusDenied)
    {
        [self doLog:@"User has denied location services"];
        return;
    }
    
    [self doLog:@"Location manager error: %@",
        error.localizedFailureReason];
    return;
}
```

```
- (void)locationManager:(CLLocationManager *)manager
    didUpdateLocations:(NSArray *) locations
{
    [self doLog:@"%@\n",
        [[locations lastObject] description]];
}

- (void) startCL
{
    // Test for location services
    if (![CLLocationManager locationServicesEnabled])
    {
        [self doLog:@"User has disabled location services"];
        return;
    }

    // Test for authorization
    if ([CLLocationManager authorizationStatus] ==
        kCLAuthorizationStatusDenied)
    {
        [self doLog:@"User has denied location services"];
        return;
    }

    manager = [[CLLocationManager alloc] init];
    manager.delegate = self;
    manager.desiredAccuracy = kCLLocationAccuracyBest;
    manager.distanceFilter = 5.0f; // in meters
    [manager startUpdatingLocation];
}
```

> **Get This Recipe's Code**
>
> To find this recipe's full sample project, point your browser to https://github.com/erica/iOS-6-Advanced-Cookbook and go to the folder for Chapter 10.

Location Properties

Each `CLLocation` instance returned by the updated location callback contains a number of properties that describe the device as it travels. Location objects can combine their various properties into a single text result, as used in Recipe 10-1, via the `description` instance method. Alternatively, you can pull out each value on a property-by-property basis. Location properties include the following:

- **altitude**—This property returns the currently detected altitude. It returns a floating-point number in meters above sea level. Speaking as a resident of the Mile High City, I can assure you the accuracy of this value is minimal at best. Use these results with caution.

- **coordinate**—Recover the device's detected geoposition through the `coordinate` property. A coordinate is a structure with two fields, `latitude` and `longitude`, both of which store a floating-point value. Positive values for latitude lie north of the equator; negative ones south of the equator. Positive longitudes lie east of the meridian; negative longitudes west of the meridian.

- **course**—Use the `course` value to determine the general direction in which the device is heading. This value, which is 0 degrees for North, 90 degrees for East, 180 degrees for South, and 270 degrees for West, roughly approximates the direction of travel. For better accuracy, use headings (`CLHeading` instances) rather than courses. Headings provide access to magnetic and true North readings via the magnetometer.

- **horizontalAccuracy**—This property indicates the accuracy (that is, the uncertainty or measurement error) of the current coordinates in meters. Think of the coordinates that are returned as the center of a circle and the horizontal accuracy as its radius. The true device location falls somewhere in that circle. The smaller the circle, the more accurate the location. The larger the circle, the less accurate it is. Negative accuracy values indicate a measurement failure.

- **verticalAccuracy**—This property offers an altitude equivalent for horizontal accuracy. It returns the accuracy related to the true value of the altitude, which may (in theory) vary between the altitude minus that amount to the altitude plus that amount. In practice, altitude readings are extremely inaccurate, and the vertical accuracy typically bears little relationship to reality.

- **speed**—This value returns the speed of the device in meters per second.

- **timestamp**—This property identifies the time at which the location measurement took place. It returns an `NSDate` instance set to the time when the location was determined by Core Location.

Tracking Speed

The built-in `speed` property returned by each `CLLocation` instance enables you to track the device's velocity over time. It reports this value in meters-per-second. You can easily compute KPH by multiplying that value by 3.6. For miles per hour, multiply by 2.23693629:

```
- (void)locationManager:(CLLocationManager *)manager
    didUpdateLocations: (NSArray *) locations
{
    CLLocation *newLocation = [locations lastObject];

    // If a speed is detected, log that data in miles per hour
    if (newLocation.speed > 0.0f)
```

```
    {
        NSString *speedFeedback = [NSString stringWithFormat:
            @"Speed is %0.1f MPH, %0.1f KPH",
            2.23693629 * newLocation.speed,
            3.6 * newLocation.speed];
        NSLog(@"%@", speedFeedback);
    }
}
```

Recipe: Geofencing

A `CLRegion` object defines a geographical area centered on a given point, with some radius. Regions enable you to register for events whenever a user crosses its boundaries. Create new regions by calling `initCircularRegionWithCenter:radius:identifier:`. You can then pass that region to a location manager instance via `startMonitoringForRegion:`. Call this method for each region you want to monitor. Each identifier must be unique; otherwise, the new region will overwrite any old one.

Upon crossing a region boundary, the location manager delegate receives either the `locationManager:didEnterRegion:` or `locationManager:didExitRegion:` callback. If your app is not running, the system will relaunch it when registered boundaries are crossed, and the `UIApplicationLaunchOptionsLocationKey` will appear in the launch options dictionary. The first successful location lock that falls within the boundary triggers the callback.

This means that boundary-crossing signals are not real time and not 100% reliable. You may encounter situations in which a user crosses into the region and out of it again without registering a callback because no location was found during that time. In other words, use this feature with due caution.

Recipe 10-2 demonstrates several basic tasks you might need to do. The `listMonitored Regions` method lists all the regions currently monitored by your application. Each app can register up to 20 regions at a time. These regions persist beyond each run, so the results may reflect regions added previously.

The second method `clearMonitoredRegions` halts monitoring. It stops ongoing monitoring for each region associated with the application. It iterates through the items in the location manager's `monitoredRegions` property, requesting a stop for each region it finds. By clearing these regions, you ensure that the app will not relaunch when the device enters any of the previously monitored areas.

The third method is `markAndMonitor`. This method sets a fence at the current location using a radius of 50 meters. This range is easiest to test using an automobile, although you'd be better testing this feature as a passenger rather than a driver. The method generates its region name based on the number of existing regions. You must ensure that region names won't conflict with existing registrations.

The final two methods implement callbacks for entering and leaving regions. Here, they're minimally implemented—simply logging updates. An app that cares about region monitoring naturally offers more robust and interesting implementations.

Recipe 10-2   **Using Core Location to Geofence**

```
// List all regions registered to the current location manager
- (void) listMonitoredRegions
{
    for (CLRegion *eachRegion in [manager monitoredRegions])
        [self doLog:@"Region: %@", eachRegion];
}

// Remove all region monitors
- (void) clearMonitoredRegions
{
    for (CLRegion *eachRegion in [manager monitoredRegions])
    {
        [self doLog:@"Stopping monitor for %@", eachRegion];
        [manager stopMonitoringForRegion:eachRegion];
    }
}

// Create a new region monitor at the current location
- (void) markAndMonitor
{
    if (!mostRecentLocation)
    {
        [self doLog:@"No location. Sorry"];
        return;
    }
    
    [self doLog:@"Setting Geofence"];
    NSString *geofenceName = [NSString stringWithFormat:
        @"Region #%d", manager.monitoredRegions.count + 1];
    CLRegion *region = [[CLRegion alloc]
        initCircularRegionWithCenter:
            mostRecentLocation.coordinate
        radius:50.0f identifier:geofenceName];
    [manager startMonitoringForRegion:region];
}

// Callback for entering region
- (void) locationManager:(CLLocationManager *)manager
    didEnterRegion:(CLRegion *)aRegion
```

```
{
    [self doLog:@"Entered region %@", aRegion.identifier];
}

// Callback for departing region
- (void)locationManager:(CLLocationManager *)manager
    didExitRegion:(CLRegion *)aRegion
{
    [self doLog:@"Leaving region %@", aRegion.identifier];
}
```

> **Get This Recipe's Code**
> To find this recipe's full sample project, point your browser to https://github.com/erica/iOS-6-Advanced-Cookbook and go to the folder for Chapter 10.

Recipe: Keeping Track of "North" by Using Heading Values

The onboard location manager can return a computed `course` value that indicates the current direction of travel—that is, North, South, Southeast, and so on. These values take the form of a floating-point number between 0 and 360, with 0 degrees indicating North, 90 degrees being East, and so forth. This computed value is derived from tracking a user's location over time. Newer units have a better way to determine a user's course. Recent devices provide an onboard magnetometer, which can return both magnetic North and true North values.

Not every iOS device supports headings; although at this point, most modern units do. A magnetometer was first released on the iPhone 3GS. Test each device for this ability before subscribing to heading callbacks. If the location manager can generate heading events, the `headingAvailable` property returns `YES`. Use this result to control your `startUpdatingHeading` requests:

```
if (CLLocationManager.headingAvailable)
    [manager startUpdatingHeading];
```

Cocoa Touch enables you to filter heading callbacks just as you do with distance ones. Set the location manager's `headingFilter` property to a minimal angular change, specified as a floating-point number. For example, if you don't want to receive feedback until the device has rotated at least 5 degrees, set the property to 5.0. All heading values use degrees, between 0.0 and 360.0. To convert a heading value to radians, divide by 180.0 and multiply it by Pi.

Heading callbacks return a `CLHeading` object. You can query the heading for two properties: `magneticHeading` and `trueHeading`. The former returns the relative location of magnetic North, the latter true North. True North always points to the geographic North Pole. Magnetic

North corresponds to the pole of the Earth's geomagnetic field, which changes over time. The iPhone uses a computed offset (called a *declination*) to determine the difference between these two.

On an enabled device, magnetic heading updates are available even if the user has switched off location updates in the Settings application. What's more, users are not prompted to give permission to use heading data. Magnetic heading information cannot compromise user privacy so it remains freely available to your applications.

You can use the `trueHeading` property only with location detection. The iPhone requires a device's location to compute the declination needed to determine true North. Declinations vary by geoposition. The declination for Los Angeles is different from Perth's, which is different from Moscow's, and London's, and so forth. Some locations cannot use magnetometer readings. Certain anomalous regions, such as Michipicoten Island in Lake Superior and Grants, New Mexico, offer iron deposits and lava flows that interfere with normal magnetic compass use. Metallic and magnetic sources, such as your computer, car, or refrigerator, may also affect the magnetometer. Several "metal detector" applications in App Store leverage this quirk.

The `headingAccuracy` property provides an error value. This number indicates a plus or minus range that the actual heading falls within. A smaller error bar indicates a more accurate reading. A negative value represents an error in reading the heading.

You can retrieve raw magnetic values along the X, Y, and Z axes using the x, y, and z `CLHeading` properties. These values are measured in microteslas and normalized into a range that Apple states is –128 to 128. (The actual range is more likely to be –128 to 127 based on standard bit math.) Each axis value represents an offset from the magnetic field lines tracked by the device's built-in magnetometer.

Recipe 10-3 uses `CLHeading` data to rotate a small image view with an arrow pointer. The rotation ensures that the arrow always points North. Figure 10-5 shows the interface in action.

Recipe 10-3 **Detecting the Direction of North**

```
// Allow calibration
- (BOOL)locationManagerShouldDisplayHeadingCalibration:
    (CLLocationManager *)manager
{
    return YES;
}

// Respond to heading updates
- (void)locationManager:(CLLocationManager *)manager
    didUpdateHeading:(CLHeading *)newHeading
{
    CGFloat heading = -M_PI * newHeading.magneticHeading / 180.0f;
    imageView.transform = CGAffineTransformMakeRotation(heading);
}
```

```
- (void) startCL
{
    if (![CLLocationManager locationServicesEnabled])
    {
        [self doLog:@"User has disabled location services"];
        return;
    }

    if ([CLLocationManager authorizationStatus] == kCLAuthorizationStatusDenied)
    {
        [self doLog:@"User has denied location services"];
        return;
    }

    manager = [[CLLocationManager alloc] init];
    manager.delegate = self;
    if ([CLLocationManager headingAvailable])
        [manager startUpdatingHeading];
    else
        imageView.alpha = 0.0f;
}
```

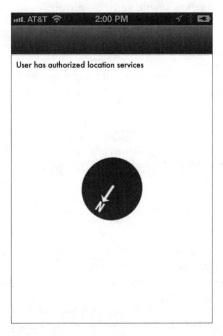

Figure 10-5   The iPhone's built-in magnetometer and the code from Recipe 10-4 ensure that this arrow always points North.

> **Get This Recipe's Code**
> To find this recipe's full sample project, point your browser to https://github.com/erica/iOS-6-Advanced-Cookbook and go to the folder for Chapter 10.

Recipe: Forward and Reverse Geocoding

The phrase *reverse geocoding* means transforming latitude and longitude information into human-recognizable address information; *forward geocoding* takes a user-readable address and converts it to latitude and longitude values. MapKit offers both forward and reverse geocoder routines as part of its `CLGeocoder` class.

The `CLPlacemark` class stores all this information. Its `location` property contains the latitude and longitude `coordinate` values. Its real-world properties (`country`, `postalCode`, `locality`, and so on) store the human-readable elements. This one class can be fully populated in either direction—from a real-world address or a coordinate.

Recipe 10-4 demonstrates how to perform both forward and reverse geocoding, using `CLGeocoder`'s block-based encoding approach. The user's device must have network access to perform geocoding.

For forward geocoding, supply the address string. For reverse, create a `CLLocation` instance and pass that. The completion block is passed an array of matching placemarks. If that array is nil, an error explains why.

Apple enumerates several rules for using this class. First, perform only one geocoding request at a time. Second, always reuse existing geocoded results when possible, rather than recalculate already known details. For real-world updates, do not send more than one request per minute, and do not send any request until the user has moved at least some significant distance. Finally, perform geocoding only when there is a user around to see the results (that is, when the application is not suspended or running in the background).

The `CLPlacemark` object also offers individual properties with the same information outside the dictionary structure. These properties include the following:

- `subThoroughfare` stores the street number (for example, the "1600" for 1600 Pennsylvania Avenue).
- `thoroughfare` contains the street name (for example, Pennsylvania Avenue).
- `sublocality`, when available, refers to the local neighborhood name or a landmark (for example, White House).
- `subAdministrativeArea` is typically the local county, parish, or other administrative area.
- `locality` stores the city (for example, Washington, DC).
- `administrativeArea` corresponds to the state, such as Maryland or Virginia.

- `postalCode` is the postal or ZIP code (for example, 20500).
- `country` is self-explanatory, storing the country name, such as the United States.
- `ISOcountryCode` provides an abbreviated country name, such as "US."

The `addressDictionary` stores an AddressBook framework-style version of the location. Additional properties include `inlandWater` and `ocean`, which describe any local lakes or oceans associated with the placemark and, charmingly, `areasOf-Interest` local to the placemark, such as national parks or attractions.

Recipe 10-4   **Recovering Address Information from Coordinates and Descriptions**

```
- (void) reverseGeocode: (id) sender
{
    // Starting location
    CLLocation *location = [[CLLocation alloc]
        initWithLatitude:37.33168400
        longitude:-122.03075800];
    CLGeocoder *geocoder = [[CLGeocoder alloc] init];
    [geocoder reverseGeocodeLocation:location
        completionHandler:^(NSArray *placemarks, NSError *error)
      {
          if (!placemarks)
          {
              [self doLog:@"Error retrieving placemarks: %@",
                  error.localizedFailureReason];
              return;
          }

          [self doLog:@"Placemarks from Location: %f, %f",
              location.coordinate.latitude,
              location.coordinate.longitude];

          for (CLPlacemark *placemark in placemarks)
          {
              [self doLog:@"%@", placemark.description];
          }
      }];

}
- (void) geocode: (id) sender
{
    // Retrieve coordinates from an address string
    CLGeocoder *geocoder = [[CLGeocoder alloc] init];
    NSString *address = @"1 Infinite Loop, Cupertino, CA 95014";
```

```
    [geocoder geocodeAddressString:address completionHandler:
        ^(NSArray *placemarks, NSError *error)
    {
        if (!placemarks)
        {
            [self doLog:@"Error retrieving placemarks: %@",
                error.localizedFailureReason];
            return;
        }

        [self doLog:@"Placemarks from Description (%@):",
            address];

        for (CLPlacemark *placemark in placemarks)
        {
            [self doLog:@"%@", placemark.description];
        }
    }];
}
```

Get This Recipe's Code

To find this recipe's full sample project, point your browser to https://github.com/erica/iOS-6-Advanced-Cookbook and go to the folder for Chapter 10.

Recipe: Viewing a Location

The `MKMapView` class presents users with interactive maps built on the coordinates and scale you provide. The following code snippet sets a map's region to a Core Location coordinate, showing 0.1 degrees of latitude and longitude around 1 Infinite Loop. In the United States, a region with that range corresponds to the size of a relatively small city or large town, approximately 5 by 7 miles. Figure 10-6 (left) shows that 0.1 degree-by-0.1 degree range on a map view:

```
// 1 Infinite Loop
CLLocation *location = [[CLLocation alloc]
    initWithLatitude:37.33168400 longitude:-122.03075800];
mapView.region = MKCoordinateRegionMake(
    location.coordinate, MKCoordinateSpanMake(0.1f, 0.1f));
```

Region size changes occur due to the curvature of the earth. At the equator, 1 degree of longitude corresponds to approximately 69 miles (~111 kilometers). This shrinks to 0 at the poles. Latitude is not affected by position. One degree of latitude is always approximately 69 miles (~111 km).

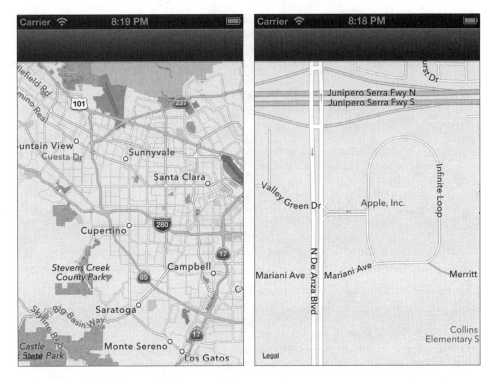

Figure 10-6 A coordinate region of one-tenth of a degree latitude by one-tenth of a degree longitude covers an area the size of a smallish city or large town, approximately 5 to 7 miles on a side (left). Shrinking that region down to 0.005 degrees on a side produces a street-level display (right).

To view map data on a neighborhood level, cut the coordinate span down to 0.01 by 0.01. For a street-by-street level, you can use a smaller span, say one-half of that, 0.005 degrees latitude by 0.005 degrees longitude. Figure 10-6 (right) shows an Infinite Loop at this range.

You can avoid dealing with latitude and longitude degrees and create regions by specifying distance in meters. This snippet sets the view region to a 500-by-500 meter square around the central coordinate. That roughly approximates the 0.005 by 0.005 degree lat/lon span, showing a street-by-street presentation:

```
mapView.region = MKCoordinateRegionMakeWithDistance(
    location.coordinate, 500.0f, 500.0f);
```

Finding the Best Location Match

Recipe 10-5 performs an on-demand location search using a timed approach. When the user taps the Find Me button, the code starts a 10-second timer. During this search, it attempts to find the best possible location. It uses the horizontal accuracy returned by each location hit to choose and retain the most accurate geoposition. When the time ends, the view controller zooms in its map view, revealing the detected location.

Recipe 10-5 displays the current user location both during and after the search. It does this by setting the `showsUserLocation` property to YES. When enabled, this property produces a pulsing pushpin that initially appears at the center of the map view at the device location. That location is detected with Core Location and updates according to user movement. You can test this in Simulator using the Debug > Location options.

Whenever this property is enabled, the map view tasks Core Location with finding the device's current location. So long as this property remains set to YES, the map continues to track and periodically updates the user location. A pulsing circle that surrounds the pushpin indicates the most recent search accuracy. Keep in mind that the map makes no effort to place the user location into the visible portion of the map view; you need to take care of that yourself.

After the location is set, the Recipe 10-5 permits the user to start interacting with the map. Enabling the `zoomEnabled` property means users can pinch, drag, and otherwise interact with and explore the displayed map. This recipe waits until the full search completes before allowing this interaction, ensuring that the user location remains centered until control returns to the user.

Upon finishing the search, the recipe stops requesting location callbacks by calling `stopUpdatingLocation`. At the same time, it permits the map view to continue tracking the user, leaving the `showsUserLocation` property set to YES.

After unsubscribing to updates, the view controller instance sets its location manager delegate to `nil`. This assignment prevents any outstanding callbacks from reaching the controller after the timer finishes. Otherwise, the user and the outstanding callbacks might compete for control of the screen.

Recipe 10-5 Presenting User Location Within a Map

```
- (void)locationManager:(CLLocationManager *)manager
    didFailWithError:(NSError *)error
{
    NSLog(@"Location manager error: %@",
        error.localizedDescription);
}

- (void)locationManager:(CLLocationManager *)manager
    didUpdateLocations: (NSArray *) locations
{
    CLLocation *newLocation = [locations lastObject];

    // Keep track of the best location found
    if (!bestLocation)
        bestLocation = newLocation;
    else if (newLocation.horizontalAccuracy <
```

358   Chapter 10   Location

```objc
            bestLocation.horizontalAccuracy)
            bestLocation = newLocation;

    mapView.region = MKCoordinateRegionMake(
        bestLocation.coordinate,
        MKCoordinateSpanMake(0.1f, 0.1f));
    mapView.showsUserLocation = YES;
    mapView.zoomEnabled = NO;
}

// Search for n seconds to get the best location during that time
- (void) tick: (NSTimer *) timer
{
    if (++timespent == MAX_TIME)
    {
        // Invalidate the timer
        [timer invalidate];

        // Stop the location task
        [manager stopUpdatingLocation];
        manager.delegate = nil;

        // Restore the find me button
        self.navigationItem.rightBarButtonItem =
            BARBUTTON(@"Find Me", @selector(findme));

        if (!bestLocation)
        {
            // no location found
            self.title = @"";
            return;
        }

        // Note the accuracy in the title bar
        self.title = [NSString stringWithFormat:@"%0.1f meters",
            bestLocation.horizontalAccuracy];

        // Update the map and allow user interaction
        [mapView setRegion:
            MKCoordinateRegionMakeWithDistance(
                bestLocation.coordinate, 500.0f, 500.0f)
            animated:YES];
```

```objc
            mapView.showsUserLocation = YES;
            mapView.zoomEnabled = YES;
        }
        else
            self.title =
                [NSString stringWithFormat:@"%d secs remaining",
                    MAX_TIME - timespent];
}

// Perform user-request for location
- (void) findme
{
    // disable right button
    self.navigationItem.rightBarButtonItem = nil;
    
    // Search for the best location
    timespent = 0;
    bestLocation = nil;
    manager.delegate = self;
    [manager startUpdatingLocation];
    [NSTimer scheduledTimerWithTimeInterval:1.0f
        target:self selector:@selector(tick:)
        userInfo:nil repeats:YES];
}

- (void) loadView
{
    self.view = [[UIView alloc] init];
    
    // Add a map view
    mapView = [[MKMapView alloc] initWithFrame:self.view.bounds];
    [self.view addSubview:mapView];
    
    if (!CLLocationManager.locationServicesEnabled)
    {
        NSLog(@"User has opted out of location services");
        return;
    }
    else
    {
        // User generally allows location calls
        manager = [[CLLocationManager alloc] init];
        manager.desiredAccuracy = kCLLocationAccuracyBest;
```

```
        self.navigationItem.rightBarButtonItem =
            BARBUTTON(@"Find Me", @selector(findme));
    }
}
```

> **Get This Recipe's Code**
>
> To find this recipe's full sample project, point your browser to https://github.com/erica/iOS-6-Advanced-Cookbook and go to the folder for Chapter 10.

Recipe: User Location Annotations

Recipe 10-5 provided a way to visually track a location event as it focused over time. Recipe 10-6 kicks this idea up a notch to track a device as it moves over time. Instead of sampling locations over time and picking the best result, it employs an easier approach while achieving similar results. Recipe 10-6 hands responsibility for user location to the map view and its `userLocation` property.

To enable this, you establish the `showsUserLocation` property and set the map view's tracking mode to follow the user:

```
mapView = [[MKMapView alloc] init];
mapView.showsUserLocation = YES;
mapView.userTrackingMode = MKUserTrackingModeFollow;
```

Recipe 10–6 checks the user location once every few seconds. It updates the map view to reflect that location in several ways. First, it keeps the map centered on the user's current position. Second, it adds a custom annotation to the user pin to display the current coordinates. Finally, it attempts to find a human-readable place mark to associate with it. If it finds one, it displays that placemark in a text view at the bottom of the screen, as shown in Figure 10-7.

Annotations are pop-up views that attach to locations on the map. They offer a title and a subtitle, which you can set as wanted. Figure 10-7 shows the annotation built by Recipe 10-6. The `MKUserLocation` class provides direct access to the user location pin and its associated annotation. The annotation offers two readable and writable properties called `title` and `subtitle`. Set these properties as wanted. Recipe 10-6 sets the title to Location Coordinates and the subtitle to a string containing the latitude and longitude.

This class greatly simplifies annotation editing, but in this case you are limited to working with the map view's user location property. The more general case for annotations proves more complicated. It is detailed in Recipe 10-7, which follows after this section.

Recipe: User Location Annotations

Figure 10-7 Maps can provide their own user location information without tasking Core Location directly.

Recipe 10-6 Tracking the Device Through the MapView

```
// Search for n seconds to get the best location during that time
- (void) tick: (NSTimer *) timer
{
    self.title = @"Searching...";
    if (mapView.userLocation)
    {
        // Check for valid coordinate
        CLLocationCoordinate2D coord =
            mapView.userLocation.location.coordinate;
        if (!coord.latitude && !coord.longitude) return;

        // Update titles
        self.title = @"Found!";
        [mapView setRegion:MKCoordinateRegionMake(coord,
            MKCoordinateSpanMake(0.005f, 0.005f)) animated:NO];
        mapView.userLocation.title = @"Location Coordinates";
        mapView.userLocation.subtitle = [NSString stringWithFormat:
            @"%f, %f", coord.latitude, coord.longitude];
```

```objc
        // Attempt to retrieve placemarks
        CLGeocoder *geocoder = [[CLGeocoder alloc] init];
        [geocoder reverseGeocodeLocation:
                mapView.userLocation.location
            completionHandler:^(
                NSArray *placemarks, NSError *error)
        {
            if (!placemarks)
            {
                NSLog(@"Error retrieving placemarks: %@",
                    error.localizedFailureReason);
                return;
            }

            NSMutableString *marks = [NSMutableString string];
            for (CLPlacemark *placemark in placemarks)
            {
                [marks appendFormat:@"\n%@", placemark.description];
                textView.alpha = 0.75f;
                textView.text = marks;
            }
        }];
    }
}

- (void) loadView
{
    self.view = [[UIView alloc] init];

    // Add map
    mapView = [[MKMapView alloc] init];
    mapView.showsUserLocation = YES;
    mapView.userTrackingMode = MKUserTrackingModeFollow;
    [self.view addSubview:mapView];

    if (!CLLocationManager.locationServicesEnabled)
    {
        NSLog(@"User has opted out of location services");
        return;
    }
    else
    {
        [NSTimer scheduledTimerWithTimeInterval:5.0f target:self
            selector:@selector(tick:) userInfo:nil repeats:YES];
    }
}
```

> **Get This Recipe's Code**
>
> To find this recipe's full sample project, point your browser to https://github.com/erica/iOS-6-Advanced-Cookbook and go to the folder for Chapter 10.

Recipe: Creating Map Annotations

Cocoa Touch defines an `MKAnnotation` protocol. You must design your own classes that conform to this protocol, which demands a `coordinate` property and `title` and subtitle instance methods. Listing 10-1 demonstrates how to do this. It builds a simple `Map-Annotation` class, providing the coordinate, title, and subtitle features demanded by the protocol. The extra `tag` integer is not required by the protocol and is added solely for convenience.

Listing 10-1 Building a Map Annotation Object

```
@interface MapAnnotation : NSObject <MKAnnotation>
- (id) initWithCoordinate: (CLLocationCoordinate2D) aCoordinate;
@property (nonatomic, readonly) CLLocationCoordinate2D coordinate;
@property (nonatomic, copy) NSString *title;
@property (nonatomic, copy) NSString *subtitle;
@property (nonatomic, assign) NSUInteger tag;
@end

@implementation MapAnnotation
- (id) initWithCoordinate: (CLLocationCoordinate2D) aCoordinate
{
        if (self = [super init])
        coordinate = aCoordinate;
        return self;
}
@end
```

Creating, Adding, and Removing Annotations

To use annotations, you create them and add them to a map view. You can do so by adding a single annotation at a time:

```
anAnnotation = [[MapAnnotation alloc] initWithCoordinate:coord];
[mapView addAnnotation:anAnnotation];
```

Alternatively, you can build an array of annotations and add them all at once:

```
[annotations addObject:annotation];
[mapView addAnnotations:annotations];
```

Delete annotations from a map by performing `removeAnnotation:` to remove just one annotation or `removeAnnotations:` to remove all items in an array.

If you need to return a map view to a no-annotations state, remove all its existing annotations. This call recovers the array of existing annotations via the `annotations` property. It then removes these from the map:

`[mapView removeAnnotations:mapView.annotations];`

This removes the automated user location and any other programmatic annotations, so to clear just the application annotations, you might want to test each annotation to ensure it's not of the `MKUserLocation` class:

```
- (void) clear
{
    NSArray *annotations =
        [NSArray arrayWithArray:mapView.annotations];
    for (id annotation in annotations)
        if (![annotation isKindOfClass:[MKUserLocation class]])
            [mapView removeAnnotation:annotation];
}
```

Annotation Views

Annotation objects are not views. The `MapAnnotation` class in Listing 10-1 does not create any onscreen elements. It is an abstract class that describes an annotation. It's the map view's job to convert that annotation description into an actual onscreen view when required. Those views belong to the `MKAnnotationView` class. You can retrieve the annotation view for an existing annotation by querying the map. Supply the annotation and request the matching view. If the annotation is not currently rendered on the map view, the following call returns nil:

`annotationView = [mapView viewForAnnotation:annotation];`

`MKPinAnnotationView` is a handy `MKAnnotationView` subclass you may want to use. These are the pins that you can drop onto maps. When tapped, they display a callout view.

Customizing Annotation Views

After adding annotations, via `addAnnotation:` or `addAnnotations:`, the map view starts building the annotation views that correspond to those annotations. When it finishes, its delegate, which must declare the `MKMapViewDelegate` protocol, receives a callback. The delegate is notified with `mapView:didAddAnnotationViews:` after the views are built and added to the map. This callback provides your application with an opportunity to customize those annotation views.

An array of annotation views is passed as the second parameter to that callback. You can iterate through this array to set features like the view's `image` or to customize its accessory buttons.

Listing 10-2 shows how you might prepare each of these annotation views for use based on their annotations.

Listing 10-2 **Preparing Annotation Views for Use**

```
- (void)mapView:(MKMapView *)mapView
   didAddAnnotationViews:(NSArray *)views
{
    // Initialize each view
    for (MKPinAnnotationView *mkaview in views)
    {
        // Only update custom items
        if (![mkaview isKindOfClass:
            [MKPinAnnotationView class]])
            continue;

        // Set the color to purple
        MapAnnotation *annotation = mkaview.annotation;
        mkaview.pinColor = MKPinAnnotationColorPurple;

        // Add buttons to each one
        UIButton *button = [UIButton
            buttonWithType:UIButtonTypeDetailDisclosure];
        mkaview.rightCalloutAccessoryView = button;
    }
}
```

This example sets a pin color and displays a button, which allows the application to respond to user interactions with annotation views. You are not limited to the built-in annotation protocol, which was minimally satisfied with the class defined in Listing 10-1. Design your annotation class with any instance variables and methods you like for more control over how you query the annotations to prepare your annotation views.

Each annotation view provides direct access to its annotation via its `annotation` property. Use that annotation data to build the exact view you need. Here are some of the annotation view properties you'll want to customize in your MapKit applications.

Each `MKPinAnnotationView` uses a color. You set this color via the `pinColor` property. MapKit provides three color choices: red (`MKPinAnnotationColorRed`), green (`MKPinAnnotationColorGreen`), and purple (`MKPinAnnotationColorPurple`). According to Apple's human-interface guidelines, red pins indicate destination points, places that the user may want to explore or navigate to. Green pins are starting points, places from which the user can begin a journey. Purple pins are user-specified. When you encourage users to add new data into the map, use purple to indicate that the user has defined them. As you saw in previous recipes, a map view-defined light-blue pin indicates the current user location.

Each annotation view offers two slots, on the left and right of the callout bubble. The `rightCalloutAccessoryView` and `leftCalloutAccessoryView` properties allow you to add buttons or any other custom subview to your callout. Listing 10-2 adds a detail disclosure right callout. You are not limited to buttons, however. You might add image views or other standard Cocoa Touch views as needed.

The `canShowCallout` property controls whether tapping a button produces a callout view. Enabled by default, you can set this property to `NO` if you do not want user taps to open callouts.

You can offset the callouts (normally, they appear directly above the pin in question) by changing the `calloutOffset` property to a new `CGPoint`. You can also change the position for the annotation view itself by adjusting its `centerOffset` property. With pin annotations, the view's art is set by default, but you can create custom annotation art by assigning a `UIImage` to the view's `image` property. Combine custom art with the center offset to produce the exact map look you want.

Responding to Annotation Button Taps

MapKit simplifies button tap management. Whenever you set a callout accessory view property to a control, MapKit takes over the control callback. You do not need to add a target and action. MapKit handles that for you. All you have to do is implement the `mapView:annotationView: calloutAccessoryControlTapped:` delegate callback, as demonstrated in Recipe 10-7.

Recipe 10-7 enables users to add pins to the current map by tapping a Tag button, which sets the pin at the center of the map. Whenever the user adjusts the map, the map view delegate receives a `mapView:regionDidChangeAnimated:` callback. The callback pulls the coordinate of the map center via its `centerCoordinate` property and stores it as the current coordinate.

When the user taps the disclosure button on an accessory view, the application calculates the distance from that pin to the current user location and displays it (in meters) at the top of the screen. Figure 10-8 shows what this might look like, if used near Apple's Infinite Loop headquarters.

Recipe 10-7 **Creating an Annotated, Interactive Map**

```
// Test this using a single set location rather than zooming
// around highways

// Update current location when the user interacts with map
- (void)mapView:(MKMapView *)aMapView
    regionDidChangeAnimated:(BOOL)animated
{
    current = [[CLLocation alloc]
        initWithLatitude:mapView.centerCoordinate.latitude
        longitude:mapView.centerCoordinate.longitude];
}
```

```objc
- (void)mapView:(MKMapView *)aMapView
    annotationView:(MKAnnotationView *)view
    calloutAccessoryControlTapped:(UIControl *)control
{
    // Calculate coordinates and distance
    CLLocationCoordinate2D viewCoord = view.annotation.coordinate;
    CLLocation *annotationLocation =
        [[CLLocation alloc] initWithLatitude:viewCoord.latitude
            longitude:viewCoord.longitude];
    CLLocation *userLocation = mapView.userLocation.location;
    float distance = [userLocation
        distanceFromLocation:annotationLocation];

    // Set the title
    self.title = [NSString stringWithFormat:@"%0f meters", distance];
}

// Set colors and add buttons
- (void)mapView:(MKMapView *)mapView
    didAddAnnotationViews:(NSArray *)views
{
    // Initialize each view
    for (MKPinAnnotationView *mkaview in views)
    {
        if (![mkaview isKindOfClass:[MKPinAnnotationView class]])
            continue;

        // Set the color to purple
        mkaview.pinColor = MKPinAnnotationColorPurple;

        // Add buttons to each one
        UIButton *button = [UIButton buttonWithType:
            UIButtonTypeDetailDisclosure];
        mkaview.rightCalloutAccessoryView = button;
    }
}

- (void) tag
{
    // Create a new annotation
    MapAnnotation *annotation =
        [[MapAnnotation alloc] initWithCoordinate:current.coordinate];

    // Label it with time and place
    NSString *locString = [NSString stringWithFormat:@"%f, %f",
        current.coordinate.latitude, current.coordinate.longitude];
    NSDateFormatter *formatter = [[NSDateFormatter alloc] init];
```

```objc
    formatter.timeStyle = NSDateFormatterLongStyle;
    annotation.title = [formatter stringFromDate:[NSDate date]];
    annotation.subtitle = locString;

    // Add it
    [mapView addAnnotation:annotation];
}

// Clear all user annotations
- (void) clear
{
    NSArray *annotations = [NSArray arrayWithArray:mapView.annotations];
    for (id annotation in annotations)
        if (![annotation isKindOfClass:[MKUserLocation class]])
            [mapView removeAnnotation:annotation];
}

- (void) loadView
{
    self.view = [[UIView alloc] init];

    // Add map
    mapView = [[MKMapView alloc] init];
    mapView.showsUserLocation = YES;
    mapView.userTrackingMode = MKUserTrackingModeFollow;
    [self.view addSubview:mapView];

    if (!CLLocationManager.locationServicesEnabled)
    {
        NSLog(@"User has opted out of location services");
        return;
    }
    else
    {
        mapView.delegate = self;
        self.navigationItem.rightBarButtonItem =
            BARBUTTON(@"Tag", @selector(tag));
        self.navigationItem.leftBarButtonItem =
            BARBUTTON(@"Clear", @selector(clear));
    }
}
```

> **Get This Recipe's Code**
>
> To find this recipe's full sample project, point your browser to https://github.com/erica/iOS-6-Advanced-Cookbook and go to the folder for Chapter 10.

Figure 10-8   These custom annotation views report the time and location when the user sets the pin. Tapping on the accessory view calculates the distance from the pin to the current user location, displaying it at in the title bar.

Summary

Core Location and MapKit go hand in hand, offering ways to locate a device's position and present related location information in a coherent map-based presentation. In this chapter, you discovered how to use Core Location to obtain real-time latitude and longitude coordinates and how to reverse geocode those coordinates into real address information. You read about working with speed and course headings both in their native and computed forms. You learned how to set up a map, adjust its region, and add a user location and custom annotations. Here are a few final thoughts for you before you navigate away from this chapter:

- Know your audience and how they will be using your application before deciding how you will approach your location needs. Some Core Location features work better for driving, others for walking and biking.
- Test, test, test, test, test, test. Core Location applications must be exhaustively tested and tuned in the field as well as at Xcode for best results in the App Store. Retrieving location data is not an exact science. Build the required slack into your application.

- "Oh, didn't I see you at -104.28393 West today?" Addresses are a lot more meaningful to most people than coordinates. Use reverse geocoding to produce human-readable information, and use forward geocoding to translate from addresses to coordinates.
- Postal codes/ZIP codes are especially API-friendly. Even if you do not plan to use a map presentation in your application, ZIP codes are ready for traditional GUI integration. A reverse-geocoded ZIP code can help retrieve nearby retail information, such as addresses, phone numbers, and information about nearby parks and attractions.
- Well-designed annotation views help bring meaningful interactivity into a map view. Don't be afraid to use buttons, images, and other custom elements that expand a map's utility.

11

GameKit

This chapter introduces various ways you can create connected game play through GameKit. GameKit offers features that enable your applications to move beyond a single-player/single-device scenario toward using Game Center and device-to-device networking.

Apple's Game Center adds a centralized service that enables your game to offer shared leaderboards and Internet-based matches. GameKit also provides an ad-hoc networking solution for peer-to-peer connectivity. Built on a technology called Bonjour, this ad-hoc connection offers simple, no-configuration communications between devices.

In this chapter, you discover how to use GameKit to build connected applications. You see how to create Game Center features. You read about adding GameKit Voice to your code for walkie-talkie-style voice chats. Ready to start? It's time to let your apps connect.

Enabling Game Center

Game Center features expand your users' experience. With Game Center, you provide matches, leaderboards, and other enhancements that bring your user out of the boundaries of a single application and into a shared communal experience. Game Center isn't just for games—although they're great for that, obviously. You can use Game Center to track progress for weight loss and fitness, to create shared design spaces, or to provide tutorial support. If you can imagine a way to share app execution across devices, Game Center is there to provide the infrastructure you need.

Before you start coding a Game Center application, there are several steps you need to take:

1. Register a unique application identifier at Apple's iOS developer portal (https://developer.apple.com/ios/manage/overview/index.action).

2. Create a 1024x1024-pixel application icon and a stand-in iPhone-sized image (640x960-pixels). You need these for iTunes Connect to build a new app listing.

3. Visit iTunes Connect (http://itunesconnect.apple.com). Register a new application using your identifier and artwork. Information need not be finalized at this time. Just add enough details so the app listing is properly saved.

4. Visit Manage Your Applications > *Application Name* > Manage Game Center, as shown in Figure 11-1. Ensure that Game Center is enabled; it should be by default.

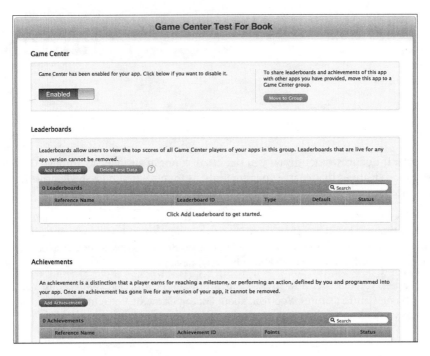

Figure 11-1   Enable Game Center on an application-by-application basis. When established, you can set up new leaderboards and achievements and allow your application to interact with other Game Center features.

You cannot test your application until you have created an iTunes Connect entry that matches your project's application identifier and enabled Game Center. If you fail to set up Game Center in advance, you cannot access standard features such as logging users and viewing achievements. A telltale warning states that your game is not recognized by Game Center. This occurs when Game Center does not recognize the identifier for the currently running application (see Figure 11-2).

The previous steps have already been performed for the recipes in this chapter. The samples should work without any further setup when using these com.sadun.cookbook apps.

Recipe: Signing In to Game Center

Figure 11-2 You can use Game Center features only with applications you have registered at iTunes Connect.

Testing for Game Center Compatibility

If you plan to deploy earlier than iOS 4.1, you may want to check to see if Game Center is available on the current device. You can easily check for class existence on, for example, `GKLocalPlayer`, to infer GameKit Game Center support:

```
if (NSClassFromString(@"GKLocalPlayer")) ...
```

Recipe: Signing In to Game Center

The easiest way to test whether your application has been properly set up for Game Center is to attempt to sign in a user. Recipe 11-1 retrieves the shared local player singleton object and provides an authentication handler for that player.

This behavior has changed in iOS 6. The `authenticateWithCompletionHandler:` method previously used is now deprecated. If you plan to deploy to iOS 5, make sure you include firmware-specific tests to run the proper code.

Authentication does not necessarily mean that the user must sign in each time the app is launched. Sessions can last quite a while. If the authentication remains valid (`player.authenticated`), your user will be greeted with a welcome-back notice rather than a password prompt.

The `GKPlayerAuthenticationDidChangeNotificationName` notification fires whenever the authentication system signs in or fails to sign in. Listen for this and update your GUI to reflect the current user state. Only present Game Center features when users have been properly authenticated:

```
[[NSNotificationCenter defaultCenter]
    addObserverForName:GKPlayerAuthenticationDidChangeNotificationName
    object:nil queue:[NSOperationQueue mainQueue]
    usingBlock:^(NSNotification *notification)
```

```objc
    {
        [weakself updateUserGUI];
    }];
```

You can best test authentication states by resetting and then quitting the iOS simulator and rerunning your application.

> **Note**
>
> As you develop, you will test in a GameKit sandbox environment. All scores and achievements made there will not leak through to the production leader boards. iTunes Connect enables you to delete test data from the app's Game Center management page.
>
> When testing on-device, visit the Game Center app, and log out of your normal account before running your beta application. This should allow you to log into the sandbox upon running your app.

Recipe 11-1 **Establishing a Game Center Player**

```objc
- (void) establishPlayer
{
    TestBedViewController __weak *weakself = self;
    [GKLocalPlayer localPlayer].authenticateHandler =
        ^(UIViewController *controller, NSError *error)
    {
        if (error)
        {
            NSLog(@"Error authenticating: %@",
                error.localizedDescription);
            alert(@"Restore game features by logging \
                in via the Game Center app.");
            return;
        }
        if (controller)
        {
            // User has not yet authenticated
            [weakself presentViewController:controller
                animated:YES completion:nil];
        }
    };
}
```

> **Get This Recipe's Code**
>
> To find this recipe's full sample project, point your browser to https://github.com/erica/iOS-6-Advanced-Cookbook and go to the folder for Chapter 11.

Designing Leaderboards and Achievements

Game Center offers a pair of items that help to engage users with games. Leaderboards enable users to view a shared list of top scores for your application. Achievements let users unlock goal milestones through game play, for example, Scored First Win, Twenty Games Played, and so on.

Both features are established at iTunes Connect. Create these (on a per-application basis or as a group that's shared among multiple apps) by clicking the Manage Game Center button in the iTunes Connect application information page. The following sections introduce the details involved in this process.

Building Leaderboards

Choose Leaderboard > Add Leaderboard > Single Leaderboard > Choose to create a new leaderboard for your application. iTunes Connect will prompt you to specify the following information:

- **Reference Name**—Enter a human-readable private name created for your own reference. This name is not used outside iTunes Connect. For example, you might want to keep track of Top Overall Point Scores or Fastest Reaction Time. Keep the reference name easy to understand and simple to search for.

- **Leaderboard ID**—This is a unique string, which is used by your application to identify the specific leaderboard. For simplicity, use reverse domain naming. Append a feature name to your application identifier (for example, com.sadun.cookbook.topPoints).

- **Score Format**—Select the way the score is counted and shown. Your choices are integer (a whole number score), fixed point (a floating point number, with one, two, or three fixed decimal places), a time interval (to the minute, second, or hundredth of a second), or currency (whole units, or units and two decimal places). Choose whether to sort from high to low (higher numbers win, for example, as in points) or low to high (lower numbers win, for example, as in shortest time elapsed).

- **Score Range (optional)**—Enter the lowest and highest possible scores for your application.

After entering this information, you must enter one or more localized descriptions, and then you can save the new leaderboard. Figure 11-3 shows a newly created leaderboard.

Localized Descriptions

The reference name you entered to create the leaderboard is not used when presenting that leaderboard to your users. Instead, you create localized descriptions that provide the text material and any other formatting options your application will use. Leaderboards must be localized to at least one language. Click Add Language and select a language.

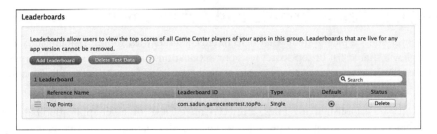

Figure 11-3 iTunes Connect enables you to create and manage leaderboards. You cannot delete leaderboards after they go live. The first leaderboard you create is your default. You can use radio buttons to select a different default if you want after adding several. You can also rearrange the order of your leaderboards and achievements.

Enter a real-world name for your leaderboard. This name will be used by your application and displayed to users. Keep it short and to the point, such as Top Scores, Best Shooters, or Quickest Reactions.

Select a score format and enter optional suffixes used for your items. (Pro tip: Add a space before your suffix so that your users can earn "50 coins" rather than "50coins".) This information can vary for each localization, enabling you to have "truckloads" in the United States and "lorry loads" in the UK. You supply an optional 512x512 pixel image to supplement each leaderboard. This image, too, is localized. You can create different images for each locale/culture you support.

Click Save to save your localization. You can revisit the description you just edited later and make any changes needed, but you cannot remove the leaderboard after it goes live.

Creating Achievements

Build achievements by adding them in iTunes Connect, much the same way as you create leaderboards. Choose Achievements > Set Up > Add New Achievement. You specify the following items:

- **Reference Name**—This is a human-readable private name that is not used outside iTunes Connect (for example, Great Start Achievement or 100 Certified Kills).

- **Achievement ID**—Provide a unique string, which is used by your application to identify the specific achievement. As with leaderboards, you'll probably want to use reverse domain naming, with your application identifier appended by some meaningful phrase (for example, com.sadun.cookbook.greatStart).

- **Hidden**—Specify whether these items are hidden on Game Center until the user has achieved them. This does not hide the number of items; it just hides the names and descriptions. Hiding allows you to surprise and delight your user with new accomplishments that they did not expect. When not hidden, you provide emotional incentives to unlock the complete set of items and a preview of what those items are.

- **Point Value**—You can assign up to 100 points per achievement and up to 1,000 points per application across all achievements. This adds a limiting factor, so you don't overwhelm your users with You Have Played This Game for 15 Minutes and 30 Seconds achievements.

Each achievement requires at least one localized description. You can use English (as shown in Figure 11-4) to add a basic write-up before adding further languages later.

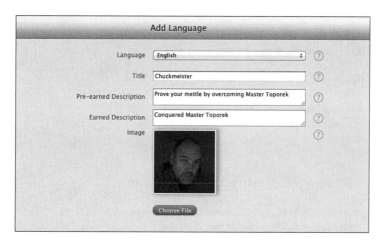

Figure 11-4 Specify the achievement elements by providing a human-readable name, a pair of descriptions for before and after the item is earned, and an image to associate with the achievement. When testing, keep in mind that the space available to display text is smaller on iPhone-family devices than iPads.

Click Add Language, choose a language, enter a human-readable title that will be displayed to your readers, and choose two descriptions. The Pre-earned Description describes the achievement before your user has unlocked it or is in the process of unlocking it; the Earned Description describes it after. For example, you might have Prove Your Mettle by Overcoming Master Toporek as the pre-earned version, and Conquered Master Toporek as the earned one. Keep the descriptions short or they'll be automatically clipped.

Nonhidden achievements always display the pre-earned description until they are achieved. Hidden achievements also display the pre-earned description to the user after you assign a (nonzero) percent-achieved value to it. If your user has gone 50% of the way toward "conquering the master," make sure that your per-earned text matches the presentation of that concept.

In addition to text, each achievement requires a (localized) 512x512 or 1024x1024 pixel art image (see Figure 11-4). Images must be jpeg, jpg, tif, tiff, or png, at least 72 DPI, and in the RGB color space.

Recipe: Accessing Leaderboards

Many applications have just one leaderboard for posting top scores. Others use leaderboards for various categories, like most accurate shooter, greatest word length, or most steps taken per week. The following method retrieves categories (the developer-defined leaderboard identifiers such as com.sadun.cookbook.topPoints) and titles (for example, Top Points) for each available leaderboard:

```
- (void) peekAtLeaderboards
{
    [GKLeaderboard loadLeaderboardsWithCompletionHandler:
        ^(NSArray *leaderboards, NSError *error)
    {
        if (error)
        {
            NSLog(@"Error retrieving leaderboards: %@",
                error.localizedFailureReason);
            return;
        }

        for (GKLeaderboard *leaderboard in leaderboards)
        {
            NSString *category = leaderboard.category;
            NSString *title = leaderboard.title;
            NSLog(@"%@ : %@", category, title);
        }
    }];
}
```

The `GKLeaderboard` Class

Each `GKLeaderboard` instance can retrieve scores from Game Center. You can search globally to report high scores or use scopes to limit your query. For example, you might specify a `playerScope` that limits the search to a player and his friends rather than returning all top scores outside of that group. The `range` property sets which scores to show. (The range count starts at 1, not at 0, by the way.) Normally, it defaults to the top 25 scores, but you can choose the top 10 instead. You can choose a time scope as well, to look at high scores for just the current day or week, and so on.

Recipe 11-2 demonstrates how you might load the data for a specific category. It requests the top ten scores within the last week, and then attempts to load the game center display names for each player. If it fails at doing so, it falls back to using the `GKScore`'s player ID property.

Recipe 11-2 **Retrieving Leaderboard Information**

```
- (void) peekAtLeaderboard: (GKLeaderboard *) leaderboard
{
    // top ten scores. Default range is 1,25
    leaderboard.range = NSMakeRange(1, 10);
    // Within last week
    leaderboard.timeScope = GKLeaderboardTimeScopeWeek;

    // Load in the scores
    [leaderboard loadScoresWithCompletionHandler:^(
        NSArray *scores, NSError *error)
    {
        if (error)
        {
            NSLog(@"Error retrieving scores: %@",
                error.localizedFailureReason);
            return;
        }

        // Retrieve player ids
        NSMutableArray *array = [NSMutableArray array];
        for (GKScore *score in scores)
            [array addObject:score.playerID];

        // Load the player names
        [GKPlayer loadPlayersForIdentifiers:array
            withCompletionHandler:
                ^(NSArray *players, NSError *error)
        {
            if (error)
            {
                // Report only with player ids
                for (GKScore *score in scores)
                    NSLog(@"[%2d] %@: %@ (%@)",
                        score.rank, score.playerID,
                        score.formattedValue, score.date);
                return;
            }

            for (int i = 0; i < scores.count; i++)
            {
                // Report with actual player names
                GKPlayer *aPlayer = [players objectAtIndex:i];
                GKScore *score = [scores objectAtIndex:i];
                NSLog(@"[%2d] %@: %@ (%@)",
                    score.rank, aPlayer.displayName,
```

```
                    score.formattedValue, score.date);
            }
        }];
    }];
}
```

> **Get This Recipe's Code**
>
> To find this recipe's full sample project, point your browser to https://github.com/erica/iOS-6-Advanced-Cookbook and go to the folder for Chapter 11.

Recipe: Displaying the Game Center View Controller

Normally, you do not access leaderboard data directly. GameKit provides a much friendlier already built view controller to show to your user (see Figure 11-5). Recipe 11-3 demonstrates how you can create, present, and dismiss the Game Center view controller. Make sure your calling class declares the `GKGameCenterControllerDelegate` protocol. This enables your code to catch the "did finish" delegate callback, so you can dismiss the view controller when the user has finished interacting with it.

Figure 11-5 GameKit offers preconfigured Game Center view controllers that you access from your application.

For those readers familiar with pre-iOS 6 Game Center, this new class duplicates features found previously in the separate leaderboard and achievement controllers. Apple writes, "The `GKGameCenterViewController` class aggregates many common Game Center features into a single user interface. It replaces `GKAchievementViewController` and `GKLeaderboardViewController` as the preferred way to show Game Center content in your game." Use the `viewState` property to set the default tab displayed by the Game Center controller.

Recipe 11-3   **Presenting the Game Center View Controller**

```
- (void)gameCenterViewControllerDidFinish:
    (GKGameCenterViewController *)gameCenterViewController
{
    // You can save the user's selection here if desired
    // and re-use it later
    // e.g. _leaderboard_Category =
    //    _gameCenterViewController.leaderboardCategory
    [self dismissViewControllerAnimated:YES completion:nil];
}

- (void) showGameCenterViewController
{
    GKGameCenterViewController *gvc =
        [[GKGameCenterViewController alloc] init];
    gvc.gameCenterDelegate = self;
    [self presentViewController:gvc
        animated:YES completion:nil];
}
```

Get This Recipe's Code

To find this recipe's full sample project, point your browser to https://github.com/erica/iOS-6-Advanced-Cookbook and go to the folder for Chapter 11.

Recipe: Submitting Scores

Submit scores by creating new `GKScore` instances and setting their value, as in Recipe 11-4. You can specify the category directly, as the following method does, or you can just call `init` and let Game Center use the default leaderboard instead.

Each user can collect many scores. These are differentiated by the category you submit. Each category stores just one value, however, and is associated with just one leaderboard. You design how you want your leaderboards to work, with various tracking statistics such as accuracy, kills, and so forth.

Recipe 11-4 **Submitting User Scores**

```
- (void) createScore
{
    NSNumber *userScore = [self requestScore];
    if (!userScore) return;

    GKScore *score = [[GKScore alloc] initWithCategory:GKCATEGORY];
    score.value = userScore.intValue;
    [score reportScoreWithCompletionHandler:^(NSError *error){
        if (error)
        {
            NSLog(@"Error submitting score to game center: %@",
                error.localizedFailureReason);
            return;
        }

        NSLog(@"Success. Score submitted.");
    }];
}
```

> **Get This Recipe's Code**
>
> To find this recipe's full sample project, point your browser to https://github.com/erica/iOS-6-Advanced-Cookbook and go to the folder for Chapter 11.

Recipe: Checking Achievements

Achievements are goals that users work toward. Each achievement can be unlocked and reset by your application as needed. Because achievements can be partially attained, you can assign a `percentComplete` value to each kind. These are displayed as partial achievements (with the pre-achieved text description) in the Game Center view controller.

Recipe 11-5 checks achievements programmatically. It calls home to Game Center to see which achievements have been activated and at what level of completion. This enables you to update the GUI if some achievement has already been unlocked. For example, advanced weapons may be available only to users who have passed some sort of initial training.

> **Note**
>
> You likely want to create a special-purpose class to handle achievement semantics rather than spreading achievement code throughout your game. It's easy to intermingle Game Center code with application semantics and end up with a giant jumble. Keep MVC design patterns in mind when working with Game Center elements, just as you would with other application development tasks.

Recipe 11-5   **Testing Achievements**

```
- (void) checkAchievement
{
    [GKAchievement loadAchievementsWithCompletionHandler:
        ^(NSArray *achievements, NSError *error)
    {
        if (error)
        {
            NSLog(@"Error loading achievements: %@",
                error.localizedFailureReason);
            return;
        }

        for (GKAchievement *achievement in achievements)
        {
            NSLog(@"Achievement: %@ : %f",
                achievement.identifier, achievement.percentComplete);
            if ([achievement.identifier isEqualToString:GKBEGINNER])
            {
                // unlock some GUI feature
            }
        }
    }];
}
```

Get This Recipe's Code

To find this recipe's full sample project, point your browser to https://github.com/erica/iOS-6-Advanced-Cookbook and go to the folder for Chapter 11.

Recipe: Reporting Achievements to Game Center

To unlock an achievement, you report it to Game Center. If the report errors out and is not successfully updated at Game Center, make sure you set up some fallback for a future repeat attempt so the user does not lose her progress.

Reward the user for completing an achievement. In Recipe 11-6, the `showsCompletionBanner` property creates the visual update shown in Figure 11-6, letting the user know that the achievement has been unlocked.

Each achievement report represents a network access at a minimum. Although reporting achievements is not especially expensive, it's not something you'll want to overuse in your applications. You don't want requests stumbling over each other, and you don't need to update an achievement many times a second as the user makes a tiny percentage of progress toward some goal. Instead, think about where achievement updates make sense rather than reflexively sticking them into code wherever the user has any state change.

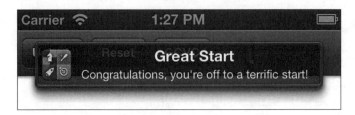

Figure 11-6 Completion banners let users know when they have unlocked achievements.

Recipe 11-6 **Unlocking Achievements**

```
- (void) unlockAchievement
{
    GKAchievement *achievement =
        [[GKAchievement alloc] initWithIdentifier: GKBEGINNER];
    if (achievement)
    {
        achievement.percentComplete = 100.0f;
        achievement.showsCompletionBanner = YES;
        [achievement reportAchievementWithCompletionHandler:^(NSError *error)
         {
             if (error)
             {
                 NSLog(@"Error reporting achievement: %@",
                     error.localizedFailureReason);

                 // Make sure to try again later in real
                 // world deployment!!!

                 return;
             }


             // Achievement is now unlocked

         }];
    }
}
```

> **Get This Recipe's Code**
> To find this recipe's full sample project, point your browser to https://github.com/erica/iOS-6-Advanced-Cookbook and go to the folder for Chapter 11.

Resetting Achievements

You may want to allow users to reset achievements and restart the game from scratch. Users can re-earn each achievement and relive the fun a second time around, even with the same Game Center ID.

The following method shows a typical way to approach this. It requests the reset and provides a completion handler to respond to the user about the success or failure of the request. Obviously, you want to confirm resets with the user—and you may want to provide a way to undo that reset by resubmitting achievements if needed. You cannot reset achievements on a category-by-category basis. A reset applies to all earned items at once:

```
- (void) resetAchievements
{
    [GKAchievement resetAchievementsWithCompletionHandler:
        ^(NSError *error)
    {
        if (error)
        {
            NSLog(@"Error resetting achievements: %@",
                error.localizedFailureReason);
            return;
        }

        // Achievements are now reset

    }];
}
```

Recipe: Multiplayer Matchmaking

GameKit enables your users to solicit matches with other players, whether those players are friends or anonymous other persons. The matchmaker view controller handles both specific invitations and general random game matches. To start with a matchmaking session, create a new match request. Specify how many players are required for gameplay and how many players can be handled total.

For Game Center gameplay, you can create matches for two to four players. Hosted matches on your own servers can allow up to 16 players at once. (This chapter does not cover hosted matches outside of Game Center.) Recipe 11-7 creates a request for a basic two-player game and presents a matchmaker controller initialized with that request.

Make sure your primary class declares the `GKMatchmakerViewControllerDelegate` protocol, and assign the `matchmakerDelegate` property. As a delegate, your controller responds to common matchmaker state updates, detailed in the next sections.

After you present the controller, it's up to the user to choose whether to invite friends or use Game Center's auto-match.

Recipe 11-7 **Requesting a Match Through the Match Maker**

```
- (void) requestMatch
{
    // Clean up any previous game
    sendingView.text = @"";
    receivingView.text = @"";

    // This is not a hosted match, which allows up to 16 players
    GKMatchRequest *request = [[GKMatchRequest alloc] init];
    request.minPlayers = 2; // Between 2 and 4
    request.maxPlayers = 2; // Between 2 and 4

    GKMatchmakerViewController *mmvc =
        [[GKMatchmakerViewController alloc]
            initWithMatchRequest:request];
    mmvc.matchmakerDelegate = self;
    mmvc.hosted = NO;
    [self presentViewController:mmvc animated:YES completion:nil];
}
```

> **Get This Recipe's Code**
>
> To find this recipe's full sample project, point your browser to https://github.com/erica/iOS-6-Advanced-Cookbook and go to the folder for Chapter 11.

Handling Matchmaker Fails

The matchmaker delegate protocol uses the following two (required) methods to handle failed match making. These deal with the user canceling the controller and with a failure to connect to Game Center. Here's how you might implement those, making sure that both methods dismiss the modal view controller. Further delegate methods are detailed in the next section:

```
- (void) matchmakerViewControllerWasCancelled:
    (GKMatchmakerViewController *)viewController
{
    [self dismissViewControllerAnimated:YES completion:nil];
}

- (void) matchmakerViewController:(GKMatchmakerViewController *)viewController
    didFailWithError:(NSError *)error
{
```

```
    [self dismissViewControllerAnimated:YES completion:nil];
    NSLog(@"Error creating match: %@",
        error.localizedFailureReason);
}
```

Recipe: Responding to the Matchmaker

Implement the (optional) `matchmakerViewController:didFindMatch:` method to start a Game Center-hosted game. (For your own server-hosted games, implement the players-found callback instead.) The match-found delegate method is called after making a successful match between your application's current player and an external player on another device.

Upon finding a match, Recipe 11-8 checks to see if you've already started playing—sometimes, race conditions mean that you've already set up another match first. Otherwise, it dismisses the matchmaker controller, saves the match to a local instance variable, and sets the match's delegate. This is a different delegate from the matchmaker delegate; a match delegate handles in-game state updates, just as the matchmaker delegate handles pre-game ones.

The delegate declares the `GKMatchDelegate` protocol. It receives data and status updates from the other parties in the game. It handles the actual gameplay after the match has been made.

Hold onto the match in a local variable. You can use that object to send data to other players. It also enables you to send queries, offer voice chatting, and disconnect an in-progress game.

Recipe 11-8   **Responding to a Found Match**

```
- (void) matchmakerViewController:
    (GKMatchmakerViewController *)viewController
    didFindMatch:(GKMatch *)aMatch
{
    // Already playing. Ignore.
    if (matchStarted)
        return;

    if (viewController)
    {
        [self dismissModalViewControllerAnimated:YES];
        match = aMatch;
        match.delegate = self;
        self.navigationItem.rightBarButtonItem = nil;
    }

    // Normal matches now wait for player connection

    // Invited connections may be ready to go now. If so, begin
    if (!matchStarted && !match.expectedPlayerCount)
```

```
    {
        // Start game!
        [self activateGameGUI];
    }
}
```

> **Get This Recipe's Code**
>
> To find this recipe's full sample project, point your browser to https://github.com/erica/iOS-6-Advanced-Cookbook and go to the folder for Chapter 11.

Starting the Game

After a match has been made, wait until the match's `expectedPlayerCount` goes down to zero before you start game play. Each time a player connects, the number decreases by one. When it hits zero, all players are connected and your game can start the actual match.

There are two points, programmatically, at which the count may hit zero, at which your game can get going. It may take place in the matchmaker's match-found callback, or it may take place in the match delegate's `match:player:didChangeState:` callback. Check in both locations.

Typically, invited games go live in the matchmaker callback and auto-match games go live in the player state callback, when the player state changes to connected. This trips up developers when they first implement matching in their apps. The sample code that accompanies this chapter demonstrates both situations.

Recipe: Creating an Invitation Handler

Your application can decide in advance how it will automatically handle invitation requests. By establishing an invitation handler, you create a code block that will be executed whenever the shared matchmaker object receives an invitation. This `inviteHandler` property of the shared matchmaker should be set early in your application's interactions with Game Center.

This handler, shown in Recipe 11-9, takes two arguments, an invitation and an array of players. When your game receives a direct invitation, the `invitation` parameter is used and set to some non-nil value. When you detect an invitation, the match has already been established; you should not create a new match request. Instead, just present a matchmaker view controller preloaded with the invitation details. Your user then waits for the host to start the game. The view controller offers a Cancel option, which allows the user to dismiss the view controller and cancel the invitation. Only the host will see a Play Now button.

> **Note**
>
> The `finishMatch` method called here may be as simple as sending `disconnect` to the current match object and resetting the game GUI to its initial state.

Recipe 11-9 **Implementing an Invitation Handler**

```
- (void) addInvitationHandler
{
    [GKMatchmaker sharedMatchmaker].inviteHandler =
        ^(GKInvite *invitation, NSArray *playersToInvite)
    {
        // This cleans up any in-progress game and changes
        // the focus to handling the invitation. YMMV.
        [self finishMatch];

        if (invitation)
        {
            GKMatchmakerViewController *mmvc =
                [[GKMatchmakerViewController alloc]
                    initWithInvite:invitation];
            mmvc.matchmakerDelegate = self;
            [self presentViewController:mmvc
                animated:YES completion:nil];
        }
        else if (playersToInvite)
        {
            GKMatchRequest *request = [[GKMatchRequest alloc] init];
            request.minPlayers = 2;
            request.maxPlayers = 2; // 2-player matches for this example
            request.playersToInvite = playersToInvite;
            GKMatchmakerViewController *mmvc =
                [[GKMatchmakerViewController alloc]
                    initWithMatchRequest:request];
            mmvc.matchmakerDelegate = self;
            [self presentViewController:mmvc
                animated:YES completion:nil];
        }
    };
}
```

Get This Recipe's Code

To find this recipe's full sample project, point your browser to https://github.com/erica/iOS-6-Advanced-Cookbook and go to the folder for Chapter 11.

Managing Match State

After establishing a match, the match delegate manages changes in the match state. The two ways things usually fall apart include failed matches (the connection to Game Center drops) and a failed player connection. (That is, the connection from Game Center to another player breaks.)

Respond to failed matches by alerting the user, cleaning up your ongoing gameplay, and reverting to a prematch ready-to-play state, as shown in the following method:

```
- (void)match:(GKMatch *) aMatch didFailWithError:(NSError *)error
{
    // Revert GUI
    [self setPrematchGUI];

    // Alert the user
    alert(@"Lost Game Center Connection: %@",
        error.localizedDescription);
}
```

For lost players, you can either treat the game as ended, as is done in the following method, or you might want to alert the user and wait for the other player to reconnect:

```
- (void)match:(GKMatch *) aMatch
    connectionWithPlayerFailed:(NSString *)playerID
    withError:(NSError *)error
{
    NSLog(@"Connection failed with player %@: %@",
        playerID, error.localizedFailureReason);
    [self setPrematchGUI];
}
```

To simplify reconnection, implement an optional delegate method. When this method returns YES, Game Center automatically reinvites the player whose connection was lost:

```
- (BOOL)match:(GKMatch *) aMatch
    shouldReinvitePlayer:(NSString *)playerID
{
    return YES;
}
```

Recipe: Handling Player State Changes

Although disconnect notices are generated when players' device go to sleep or they lose their Internet connection, players can change state in a more orderly fashion as well. For example, a player might tap a Quit button, with the application sending `disconnect` to the match object. This sends a proper state change before disconnecting and does not invoke the reinvite query that a lost connection does.

Your application also receives state changes when players first connect. This is where you should check the current match's expected player count, launching the game when the count goes down to zero.

Recipe 11-10 handles state callbacks for both connect and disconnect events, showing an example of how you might react to both of these state changes:

Recipe 11-10 **Responding to Player State**

```objc
- (void)match:(GKMatch *) aMatch
    player: (NSString *) playerID
    didChangeState: (GKPlayerConnectionState) state
{
    if (state == GKPlayerStateDisconnected)
    {
        NSLog(@"Player %@ disconnected", playerID);
        [match disconnect];
        [self setPrematchGUI];
    }
    else if (state == GKPlayerStateConnected)
    {
        if (!matchStarted && !match.expectedPlayerCount)
        {
            [GKPlayer loadPlayersForIdentifiers:@[playerID]
                withCompletionHandler:
                    ^(NSArray *players, NSError *error)
                {
                    [self activateGameGUI];
                    NSString *opponentName = playerID;
                    if (!error)
                    {
                        GKPlayer *opponent = [players lastObject];
                        opponentName = opponent.displayName;
                    }
                    alert(@"Commencing Match with %@", opponentName);
                }];
        }
    }
    else
    {
        NSLog(@"Player state changed to unknown");
    }
}
```

> **Get This Recipe's Code**
>
> To find this recipe's full sample project, point your browser to https://github.com/erica/iOS-6-Advanced-Cookbook and go to the folder for Chapter 11.

Recipe: Retrieving Player Names

Matches use player identifiers, not player names. *Identifiers* are basically numerical indicators of a particular Game Center player in the form of a string. You can retrieve an array of `GKPlayer` objects, with their human-readable `displayName` name property, on demand. This operation runs asynchronously, so if you want to display the "name" of your new opponent, you must wait for the completion block to execute. Store the opponent (or opponents) locally, as demonstrated in Recipe 11-11, so you do not need to run this operation more than once. The `PlayerHelper` class used in Recipe 11-13 and later demonstrates how to do this.

Recipe 11-11   **Loading Opponent Name**

```
// Retrieve the player information for the opponent
[GKPlayer loadPlayersForIdentifiers:@[playerID]
    withCompletionHandler:^(NSArray *players, NSError *error)
 {
     [self activateGameGUI];
     if (error) return;

     // Store the opponent object (2-player game)
     opponent = [players lastObject];

     // Announce the match
     GKPlayer *opponent = [players lastObject];
     alert(@"Commencing Match with %@", opponent.displayName);
 }];
```

> **Get This Recipe's Code**
>
> To find this recipe's full sample project, point your browser to https://github.com/erica/iOS-6-Advanced-Cookbook and go to the folder for Chapter 11.

Comparing Players

You often compare players by their IDs. The following two functions offer solutions for these checks:

```
BOOL playerEqual(GKPlayer *p1, GKPlayer *p2)
{
    return [p1.playerID isEqualToString:p2.playerID];
}

BOOL playerIDCheck(GKPlayer *p1, NSString *playerID)
{
    return [p1.playerID isEqualToString:playerID];
}
```

Retrieving the Local Player

The `GKLocalPlayer` class offers instant access to the player currently interacting with your application:

```
GKPlayer *me()
{
    return [GKLocalPlayer localPlayer];
}
```

Game Play

Games primarily involve sending data back and forth through the `GKMatch` object and updating the local GUI to reflect any remote changes. Generally, you need to send tiny bits of data quickly for real-time updates. You can bend or break this rule as needed, especially if you don't need your updates to arrive quickly. Matches use two kinds of data transmission: reliable and unreliable.

If your data must arrive coherently and cannot be lost, use reliable transmission. Reliable transmissions are delivered in the order they are sent, are guaranteed to arrive, and will be re-sent until they are fully transmitted. Think Transmission Control Protocol (TCP) transmission when working with reliable connections.

Unreliable data is sent once and may be received out-of-order by recipients. Use this approach for data that arrives in small packets but that must use near-instantaneous transmission to be useful, such as in a live shooting action game. Think User Datagram Protocol (UDP) transmission for unreliable connections.

Here's an example of a simple texting application that sends string data as it is typed into a sending text view. It uses reliable transmission ensuring that every character is transmitted in order as it is typed. As the application receives data, it converts that information back into a string and shows it on the receiving text view:

```
- (void)textViewDidChange:(UITextView *)textView
{
    NSError *error;
    NSData *dataToSend =
        [sendingView.text dataUsingEncoding:NSUTF8StringEncoding];
    BOOL success = [match sendDataToAllPlayers:dataToSend
        withDataMode:GKMatchSendDataReliable error:&error];
    if (!success)
        NSLog(@"Error sending match data: %@", error.localizedFailureReason);
}

- (void)match:(GKMatch *) aMatch didReceiveData:(NSData *)data
    fromPlayer:(NSString *)playerID
```

```objc
{
    NSString *received = [[NSString alloc]
        initWithData:data encoding:NSUTF8StringEncoding];
    receivingView.text = received;
}
```

Serializing Data

When sending data for complex gameplay, you may want to send information that's more structured than simple strings. JSON serialization provides an easy-to-use solution for compact structured data objects, both for dictionaries with key/value pairs and arrays with information sent in a key-less pre-ordained order. For example, imagine you have an application that "rolls" a local die and then transmits its value. Rolling a number might be just one action among many that a game might use. Creating a key/value dictionary pair enables your game to differentiate between rolls, moving pieces, selecting cards, and other game-specific commands:

```objc
// Roll the [1d6] die
localRoll = @((random() % 6) + 1);

// Create an info dictionary for the roll
NSDictionary *dictionary = @{@"Roll" : localRoll};

// Transmit the roll as serialized JSON data
NSData *json = [NSJSONSerialization
    dataWithJSONObject:dictionary options:0 error:nil];
[match sendDataToAllPlayers:json
    withDataMode:GKMatchSendDataReliable error:nil];
```

On the receiving side, the other player's app can deserialize the data, recover the roll value, and apply it locally. More complex actions (for example, if the player has several pieces that can be moved, not just one) require slightly more complicated structures, but the send and receive components remain the same: Convert a JSON-serializable object to data, send it, and translate it back to the object form on the receiving end:

```objc
- (void)match:(GKMatch *) aMatch
    didReceiveData:(NSData *)data
    fromPlayer:(NSString *)playerID
{
    NSDictionary *dict =
        [NSJSONSerialization JSONObjectWithData:data
            options:0 error:nil];
    NSString *key = [[dict allKeys] lastObject];
    if (!key) return;

    if ([key isEqualToString:@"Roll"])
    {
```

```
        // take action here, e.g. move n spaces
        // or extract which piece to move
    }
}
```

Top-level JSON-serializable objects must be arrays or dictionaries. Below that, they can contain further arrays, dictionaries, strings, numbers, and `NSNull` instances. Numbers cannot be NaN (the "not a number" value) or infinity. You can test whether your objects are valid JSON instances by calling `isValidJSONObject:`.

Sending Other Forms of Data

Although JSON serialization is simple to implement, it may not provide sufficient flexibility for your needs. If you need to send `NSData` or `NSDate` objects in addition to your dictionaries, arrays, strings, or numbers, you may want to investigate standard property list serialization instead.

Property lists provide a helpful abstract data type. A property list object can point to data (`NSData`), strings (`NSString`), arrays (`NSArray`), dictionaries (`NSDictionary`), dates (`NSDate`), and numbers (`NSNumber`). When working with collection objects (that is, arrays and dictionaries), all members and keys must be property list objects as well (that is, data, strings, numbers, and dates, as well as embedded arrays and dictionaries).

Although that seems limiting, you can transform most structures and objects to and from strings. For example, you can use the built-in `NSStringFromCGPoint()` or `NSStringFromClass()` functions, or you can create your own. The following pair of methods extends the `UIColor` class, providing functionality needed to send color information across a GameKit connection as strings:

```
@implementation UIColor (utilities)
- (NSString *) stringFromColor
{
    // Recover the color space and store RGB or monochrome color
    const CGFloat *c = CGColorGetComponents(self.CGColor);
    CGColorSpaceModel csm =
        CGColorSpaceGetModel(CGColorGetColorSpace(self.CGColor));
    return (csm == kCGColorSpaceModelRGB) ?
        [NSString stringWithFormat:@"%0.2f %0.2f %0.2f %0.2f",
            c[0], c[1], c[2], c[3]] :
        [NSString stringWithFormat:@"%0.2f %0.2f %0.2f %0.2f",
            c[0], c[0], c[0], c[1]];
}

+ (UIColor *) colorWithString: (NSString *) colorString
{
    // Read a color back from a string
    const CGFloat c[4];
```

```
    sscanf([colorString cStringUsingEncoding:NSUTF8StringEncoding],
        "%f %f %f %f", &c[0], &c[1], &c[2], &c[3]);
    return [UIColor colorWithRed:c[0] green:c[1] blue:c[2] alpha:c[3]];
}
@end
```

When in property list form, you can serialize your data and send it as a single chunk. On receipt, the deserialized data is ready to use. Listing 11-1 shows example `transmit` and `receivedData:` methods that handle serialization and deserialization. This code comes from a sample that stores various kinds of game state in an `NSDictionary` object.

> **Note**
> You can use the `NSKeyedArchiver` and `NSKeyedUnarchiver` classes, as well as the `NSPropertyListSerialization` class and `CFPropertyListCreateWithData` function demonstrated here.

Listing 11-1 **Serializing and Deserializing Property Lists**

```
- (void) transmit
{
    NSString *errorString;
    
    // Send a copy of the local points to the peer
    // by serializing the property list into data
    NSData *plistdata = [NSPropertyListSerialization
        dataFromPropertyList:pointData
        format:NSPropertyListXMLFormat_v1_0
        errorDescription:&errorString];
    
    if (plistdata)
        [match sendDataToAllPlayers: plistdata
            withDataMode:GKMatchSendDataReliable error:nil];
    else
        NSLog(@"Error serializing property list: %@", errorString);
}

- (void)match:(GKMatch *) aMatch
    didReceiveData:(NSData *)data
    fromPlayer:(NSString *)playerID
{
    // Deserialize the data back into a property list
    CFStringRef errorString;
    CFPropertyListRef plist =
        CFPropertyListCreateWithData (kCFAllocatorDefault,
            (CFDataRef)data, kCFPropertyListMutableContainers,
            kCFPropertyListXMLFormat_v1_0, &errorString);
```

```
    if (!plist)
    {
        NSLog(@"Error deserializating data: %@", errorString);
        return;
    }

    // Do something with the received data
}
```

Recipe: Synchronizing Data

Although Apple provides a turn-by-turn gaming option via Game Center (see the following section), you may want to develop games that perform similar features in a more immediate atmosphere, where gamers are not free to leave and return to take their turn. You can implement a turn-by-turn style game for direct matches by synchronizing data between each player. For example, one of the most common tasks for turn-based games is to choose who goes first. You can create your own roll-for-first system by generating a random number and comparing it to your opponent's.

Because each application may be slightly out of sync with the other, never assume that your roll will occur before or after the other player's. Instead, store both values locally, and wait until both items are available before checking for a winner. That means you should call your startup check both in the received-data method and in the method that performs the roll on your side.

Recipe 11-12 demonstrates how to do this. It always checks to see if it's resolved a winner. If so, the method returns because it may be called more than once after a winner has been selected. Next, it checks for the existence of both a local and remote roll. Only when those two are available does it move on to the check.

In the case of a tie, it resets both values and starts a new roll and check. Otherwise, it selects the winner (higher value wins here) and starts the game play either deferring to the other player (`opponentGoesFirst`) or not.

Recipe 11-12 Rolling for First Position

```
- (void) checkStartupWinner
{
    // Already resolved the startup winner?
    if (startupResolved)
        return;

    // Need data from both remote and local rolls in order to
    // determine a winner
    if (!remoteRoll || !localRoll)
```

```objc
    {
        [self performSelector:@selector(checkStartupWinner)
            withObject:nil afterDelay:1.0f];
        return;
    }

    // Both rolls are in. Decide a winner
    NSLog(@"Remote roll: %@, local roll: %@", remoteRoll, localRoll);
    unsigned int local = localRoll.unsignedIntValue;
    unsigned int remote = remoteRoll.unsignedIntValue;

    // Handle a tie conflict by resetting and trying again
    if (local == remote)
    {
        NSLog(@"TIE!");
        remoteRoll = nil;
        localRoll = nil;
        [self sendRoll:GKROLLFORFIRST];
        return;
    }

    startupResolved = YES;

    if (remote > local)
    {
        // they go first.
        opponentGoesFirst = YES;
    }
    else
    {
        // you go first
        opponentGoesFirst = NO;
    }

    // reset both rolls
    localRoll = nil;
    remoteRoll = nil;
}
```

Get This Recipe's Code

To find this recipe's full sample project, point your browser to https://github.com/erica/iOS-6-Advanced-Cookbook and go to the folder for Chapter 11.

Recipe: Turn-by-Turn Matchmaking

Turn-based matches enable your applications to hook into a Game Center feature that moderates games based on one-at-a-time gameplay, such as checkers, gin rummy, Scrabble, or bridge. It introduces store-and-forward game state, allowing each player to make a move and then pass that updated state back to the Game Center store.

The turn-based matchmaker enables users to connect both anonymously and by invitation, just as the standard game matchmaker does. What differs is how you establish your game and how game control moves from person to person to person by passing a virtual "baton." The person holding that baton has control of the game until she makes a move.

From there, the baton is passed to the next player—it may be the next player in order or someone based on another rule. For example, after each full round it might pass to the person with the highest score or the lowest. Or a player might choose whom to discard an item to, passing game control at the same time.

How you implement that control is up to you and the requirements of your game. How you establish a match and implement turn-by-turn follows a few basic patterns. The process to initiate a turn-by-turn match mirrors the approach for standard games. You create the same kind of standard match request, specifying the minimum and maximum number of players.

Use that request to initialize a `GKTurnBasedMatchmakerViewController` and present it. Your handler class implements the `GKTurnBasedMatchmakerView-ControllerDelegate` protocol. The delegate responds to successful and unsuccessful match attempts. Setting the `show ExistingMatches` property to `YES` enables your user to select from a new match as well as presenting existing matches. Recipe 11-13 shows the steps involved in requesting a match. This implementation may return a brand-new match or a match that's already in-progress.

The turn-by-turn matchmaker doesn't always succeed. As with standard game center controllers, respond to the user tapping the Cancel button and to any connection fail situations. Implement basic responses to these situations that dismiss the controller and then restore the interface to a state where the user can try again to establish a match.

Recipe 11-13 **Starting a Match**

```
// Build a 2-player match request
- (GKMatchRequest *) matchRequest
{
    GKMatchRequest *request = [[GKMatchRequest alloc] init];
    request.minPlayers = 2; // Between 2 and 4
    request.maxPlayers = 2; // Betseen 2 and 4
    return request;
}

// Request a match from the turn-based matchmaker
- (void) requestMatch
{
```

Chapter 11 GameKit

```objc
    GKMatchRequest *request = [self buildMatchRequest];
    GKTurnBasedMatchmakerViewController *viewController =
        [[GKTurnBasedMatchmakerViewController alloc]
            initWithMatchRequest:request];
    viewController.turnBasedMatchmakerDelegate = self;
    viewController.showExistingMatches = YES;
    [self presentViewController:viewController
        animated:YES completion:nil];
}

// User selected match
- (void)turnBasedMatchmakerViewController:
        (GKTurnBasedMatchmakerViewController *)viewController
    didFindMatch:(GKTurnBasedMatch *)aMatch
{
    // Handle dismissal
    if (viewController)
        [_delegate dismissViewControllerAnimated:YES completion:nil];

    // Add match to matches
    MatchHelper *helper = [MatchHelper helperForMatch:match];
    _matchDictionary[match.matchID] = helper;
    [helper loadData];
    [helper loadParticipants];

    // Set this match to the current match
    [_delegate chooseMatch:helper];
}


// Game Center Fail
- (void)turnBasedMatchmakerViewController:
    (GKTurnBasedMatchmakerViewController *)viewController
    didFailWithError:(NSError *)error
{
    [self dismissViewControllerAnimated:YES completion:^(){
        alert("Error creating match: %@",
            error.localizedDescription);
    }];
}

// User cancel
- (void)turnBasedMatchmakerViewControllerWasCancelled:
        (GKTurnBasedMatchmakerViewController *)viewController
{
    [self dismissViewControllerAnimated:YES completion:nil];
}
```

> **Get This Recipe's Code**
>
> To find this recipe's full sample project, point your browser to https://github.com/erica/iOS-6-Advanced-Cookbook and go to the folder for Chapter 11.

Recipe: Responding to Turn-Based Invitations

As with many Game Center features, your code should handle both in-game match requests and external invitation requests. Game updates such as invitations are processed through a handler delegate, as shown in Recipe 11-14. Declare and implement the `GKTurnBasedEventHandlerDelegate`. This protocol offers the invitation entry-point for matchmaking and allows your application to respond to turn-by-turn updates.

To subscribe, set the event handler's delegate as follows. This is a shared handler, which updates you for all matches your user is currently involved with in your game:

```
[GKTurnBasedEventHandler sharedTurnBasedEventHandler].delegate
    = turnByTurnHelper;
```

After establishing the delegate, you may receive invitation callbacks. These invitations originate from other players and are specifically targeted toward their friends. Implement the `handleInviteFromGameCenter:` method to respond to them.

As before with user-directed matchmaking, you set the matchmaking controller's match request. With invitations, however, you set the request's `playersToInvite` property to the array passed to you. Further, set the view controller's `showExistingMatches` property to `NO`. When handling an invitation, you should work only with the match you've been invited to.

Recipe 11-14 Handling Invitations

```
- (void)handleInviteFromGameCenter:
    (NSArray *)playersToInvite
{
    GKMatchRequest *request = [self matchRequest];
    request.playersToInvite = playersToInvite;
    GKTurnBasedMatchmakerViewController *viewController =
        [[GKTurnBasedMatchmakerViewController alloc]
            initWithMatchRequest:request];
    viewController.showExistingMatches = NO;
    viewController.turnBasedMatchmakerDelegate = self;
    [_delegate presentViewController:viewController
        animated:YES completion:nil];
}
```

> **Get This Recipe's Code**
>
> To find this recipe's full sample project, point your browser to https://github.com/erica/iOS-6-Advanced-Cookbook and go to the folder for Chapter 11.

Recipe: Loading Matches

Each time your game launches and your player authenticates, you should check in with game center. Set the `GKTurnBasedEventHandler` delegate to catch turn and state changes, and load in any existing matches that your player may be involved in. Turn-by-turn matches can go on for days, weeks, or months, and your player may have any number of matches in progress:

```
[[NSNotificationCenter defaultCenter] addObserverForName:
        GKPlayerAuthenticationDidChangeNotificationName
    object:nil queue:[NSOperationQueue mainQueue]
    usingBlock:^(NSNotification *notification)
{
    BOOL authenticated =
        [GKLocalPlayer localPlayer].isAuthenticated;
    if (authenticated)
    {
        turnByTurnHelper = [[TurnByTurnHelper alloc] init];
        turnByTurnHelper.delegate = self;
        [GKTurnBasedEventHandler
            sharedTurnBasedEventHandler].delegate = turnByTurnHelper;
        [turnByTurnHelper loadMatchesWithCompletion:^()
            {
                [weakself nextMatch];
            }];
    }
}];
```

Recipe 11-15 demonstrates how to request a set of current matches. This process runs asynchronously, so this implementation includes a completion handler that runs on the main thread, enabling you to set up the game interface after the match data fully arrives.

This method stores recovered matches in a local dictionary, using a custom `MatchHelper` class to manage each match. As you see, the helper's first task is to load the match data and participants, which also takes time to finish downloading.

Although match participant IDs are immediately available as soon as the match finishes loading from the game center, the actual player names (see Recipe 11-11), photos, and the data representing the state of the current match are not.

Recipe 11-15 **Loading Matches from Game Center**

```
- (void) loadMatchesWithCompletion: (CompletionBlock) completion
{
    if (!_matchDictionary)
        _matchDictionary = [NSMutableDictionary dictionary];
    
    NSLog(@"Loading matches from Game Center");
    [GKTurnBasedMatch loadMatchesWithCompletionHandler:
        ^(NSArray *theMatches, NSError *error)
     {
         if (error)
         {
             NSLog(@"Error retrieving matches: %@",
                   error.localizedDescription);
             return;
         }
         
         NSLog(@"Number of matches: %d", theMatches.count);
         for (GKTurnBasedMatch *match in theMatches)
         {
             MatchHelper *helper = [MatchHelper helperForMatch:match];
             _matchDictionary[match.matchID] = helper;
             [helper loadData];
             [helper loadParticipants];
         }
         
         [[NSOperationQueue mainQueue] addOperationWithBlock:^()
          {
              if (completion)
                  completion();
          }];
     }];
}
```

> **Get This Recipe's Code**
>
> To find this recipe's full sample project, point your browser to https://github.com/erica/iOS-6-Advanced-Cookbook and go to the folder for Chapter 11.

Recipe: Responding to Game Play

As you already know, the `GKTurnBasedEventHandlerDelegate` offers a way to respond to external invitations. The same protocol enables you to know when turns occur. These happen after users finish a turn and update the game data, and when a user quits. Your game determines what the actual scenario is upon receiving this callback and responds accordingly.

Recipe 11-16 offers a handful of methods that derive from several classes in the sample code. These are put together like this so that you can see how the flow progresses. The recipe starts with the turn callback, which calls the game's `takeTurn:` method on the main thread.

This method proceeds by checking whether the updated turn is for the displayed game. If a user is involved in several games, the update may arrive from a game that he is not currently viewing.

It also checks whether the updated match has finished. A match is done when its state updates to done. For a simple two-player game, you can also consider it finished when any match participant's state updates to done as well. What's "done" for one is also "done" for all. This is not the case for all games. You could imagine a multiplayer match where one character has died but the others continue fighting. Adapt this approach for your specific gameplay requirements.

The `takeTurn:` method works by checking each possibility. If the other player has quit but the game is still ongoing, it ends the match. If the match is ongoing, it checks which player's turn it is and updates the GUI to reflect that state. In updating the GUI, don't forget to request the latest gameplay data, so your user is ready to play only after that data arrives.

After allowing the local user to proceed with the game, complete the turn by calling the match's end turn method. You pass the updated game state data and an array of the next participants in order. For a two-player game, that array is simply the other participant.

Unlike game-ending requests, it doesn't matter whether that other participant has been matched yet. To get another player to join your new game, you must take a first turn before a second player can be found. When testing, always take the first turn on one device before trying to join the match on the second.

Recipe 11-16   **Handling Turn Events**

```
// This is a Turn Event Delegate Method
- (void)handleTurnEventForMatch:(GKTurnBasedMatch *) match
              didBecomeActive:(BOOL)didBecomeActive
{
    MatchHelper *helper = [self matchForID:match.matchID];
    helper.didBecomeActive = didBecomeActive;

    [helper loadDataWithCompletion:^(BOOL success) {
        if (!success) return;
        [[NSOperationQueue mainQueue] addOperationWithBlock:^()
        {
            [_delegate takeTurn:helper];
        }];
    }];
}
```

```objc
// This game method is called by the delegate
- (void) takeTurn: (MatchHelper *) match
{
    BOOL isCurrentMatch =
        matchEqual(currentMatch.match, match.match);
    BOOL matchEnded = match.matchIsDone;

    // Should I quit?
    if (matchEnded && match.amActive)
    {
        [match winMatch];
    }

    // Update ended match?
    if (matchEnded)
    {
        if (isCurrentMatch)
            [self chooseMatch:currentMatch];
        else
            alert(@"Match %@ has ended", match.matchID);
        return;
    }

    // Match has not ended. It is someone's turn
    if (!isCurrentMatch && match.isMyTurn)
    {
        alert(@"Your turn for match %@", match.matchID);
        return;
    }

    if (!isCurrentMatch)
    {
        NSLog(@"Non-turn activity on match %@", match.matchID);
        return;
    }

    // It is the current match and it is your turn
    [self chooseMatch:currentMatch];
}

// The following helper methods are implemented by the
// match helper class to support gameplay
- (BOOL) isMyTurn
{
    if (!_match.currentParticipant.playerID)
        return NO;
```

```objc
    return playerIDCheck(me(),
        _match.currentParticipant.playerID);
}

- (BOOL) amActive
{
    GKTurnBasedParticipant *me = myParticipantForMatch(_match);
    return (me.status == GKTurnBasedParticipantStatusActive);
}

- (BOOL) matchIsDone
{
    if (_match.status == GKTurnBasedMatchStatusEnded)
        return YES;

    // Note -- this assumes "done for one is done for all"
    for (GKTurnBasedParticipant *participant in _match.participants)
    {
        if (participant.status == GKTurnBasedParticipantStatusDone)
            return YES;
    }

    return NO;
}

// End turn at this player's end
- (void) endTurnWithTimeout: (NSTimeInterval) timeout
    withCompletion: (SuccessBlock) completion
{
    if (!myCurrentPlayerForMatch(_match))
    {
        NSLog(@"Error: You are not current player for match %@",
            _match.matchID);
        return;
    }

    NSMutableArray *participants = [NSMutableArray
        arrayWithArray:_match.participants];
    GKTurnBasedParticipant *me = myParticipantForMatch(_match);
    [participants removeObject:me];

    // Unlike ending games, it's okay to pass unmatched participants
    // to the following method
    [_match endTurnWithNextParticipants:participants
                            turnTimeout:timeout
                              matchData:_data
                      completionHandler:^(NSError *error)
```

```
    {
        if (error)
            NSLog(@"Error completing turn: %@", error.localizedDescription);

        [[NSOperationQueue mainQueue] addOperationWithBlock:^()
         {
             if (completion)
                 completion(error == nil);
         }];
    }];
}
```

Get This Recipe's Code

To find this recipe's full sample project, point your browser to https://github.com/erica/iOS-6-Advanced-Cookbook and go to the folder for Chapter 11.

Recipe: Ending Gameplay

Most games eventually come to an end. Players may quit, win, lose, draw, or come in at some place such as second, third, or fourth. Your game can produce one of these scenarios, or your user may quit directly from a Game Center view controller. The following method catches the GUI callback and redirects the quit request to my match helper class:

```
// Quit through GUI
- (void)turnBasedMatchmakerViewController:
        (GKTurnBasedMatchmakerViewController *)viewController
    playerQuitForMatch:(GKTurnBasedMatch *) match
{
    if (viewController)
        [_delegate dismissViewControllerAnimated:YES completion:nil];
    MatchHelper *helper = _matchDictionary[match.matchID];
    [helper quitMatch];
}
```

Recipe 11-17 shows how you can terminate matches, regardless of whether it's your player's turn ("in turn") or not ("out of turn").

Each match has an outcome. This is an enumeration that specifies the player's state at the end of gameplay. Game Center uses these states to display completed matches that have not yet been removed from the system. The legal outcomes supported by Game Center are as follows:

- GKTurnBasedMatchOutcomeNone
- GKTurnBasedMatchOutcomeQuit

- GKTurnBasedMatchOutcomeWon
- GKTurnBasedMatchOutcomeLost
- GKTurnBasedMatchOutcomeTied
- GKTurnBasedMatchOutcomeTimeExpired
- GKTurnBasedMatchOutcomeFirst
- GKTurnBasedMatchOutcomeSecond
- GKTurnBasedMatchOutcomeThird
- GKTurnBasedMatchOutcomeFourth
- GKTurnBasedMatchOutcomeCustomRange

When quitting in-turn, you must pass an array of participants to the nextParticipants: parameter. Never do this with unmatched players—that is, players that Game Center has not yet found for you. If there are no other legitimate players yet (check the participant PlayerIDs to be sure; if they are nil, the player is unmatched), quit out of turn instead.

Recipe 11-17 **Ending Games**

```
// Quit immediately
- (void) quitOutOfTurnWithOutcome:
    (GKTurnBasedMatchOutcome) outcome
{
    GKTurnBasedParticipant *participant =
        myParticipantForMatch(_match);
    participant.matchOutcome = outcome;

    [_match participantQuitOutOfTurnWithOutcome:outcome
                        withCompletionHandler:^(NSError *error)
     {
         if (error)
             NSLog(@"Error while quitting match out of turn %@: %@",
                 _match.matchID, error.localizedDescription);
         else
             NSLog(@"Participant quit match out of turn: %@",
                 _match.matchID);
     }];
}

// Finish the game
- (void) finishMatchWithOutcome:
    (GKTurnBasedMatchOutcome) outcome
{
    GKTurnBasedParticipant *participant =
        myParticipantForMatch(_match);
    if (!participant)
```

```objc
    {
        NSLog(@"Error: Cannot finish game. \
            You are not playing in match %@.", _match.matchID);
        return;
    }
    
    participant.matchOutcome = outcome;
    
    NSArray *participants = self.otherCurrentParticipants;
    BOOL isCurrent = myCurrentPlayerForMatch(_match);
    
    if ((participants.count == 0) || !isCurrent)
    {
        // no other valid players or out of turn
        [self quitOutOfTurnWithOutcome:outcome];
        return;
    }
    
    [_match participantQuitInTurnWithOutcome:outcome
                            nextParticipants:participants
                                  turnTimeout:GKTurnTimeoutNone
                                    matchData:_data
                            completionHandler:^(NSError *error)
     {
         if (error)
             NSLog(@"Error while quitting match %@ in turn: %@",
                 _match.matchID, error.localizedDescription);
         else
             NSLog(@"Participant did quit in turn");
     }];
}

// You quit the game
- (void) quitMatch
{
    [self finishMatchWithOutcome:GKTurnBasedMatchOutcomeQuit];
}

// You win the game
- (void) winMatch
{
    [self finishMatchWithOutcome:GKTurnBasedMatchOutcomeWon];
}

// You lose the game
- (void) loseMatch
{
    [self finishMatchWithOutcome:GKTurnBasedMatchOutcomeLost];
}
```

> **Get This Recipe's Code**
> To find this recipe's full sample project, point your browser to https://github.com/erica/iOS-6-Advanced-Cookbook and go to the folder for Chapter 11.

Recipe: Removing Matches

After a game is over, you can request its removal from the Game Center. Make sure your player has finished playing. You cannot remove a game where your player is active. The following method requests Game Center removal. This method won't error out—but it will also not succeed—if any other players are still active in the game. For this reason, take special care with your bookkeeping after requesting a match removal:

```
- (void) removeFromGameCenter
{
    [_match removeWithCompletionHandler:^(NSError * error)
    {
        if (error)
        {
            NSLog(@"Error removing match %@: %@",
                _match.matchID, error.localizedDescription);
            return;
        }

        NSLog(@"Match %@ removed from Game Center",
            _match.matchID);
        _terminated = YES;
    }];
}
```

During debugging, there are times that you simply want to clear out all ongoing matches for a test user. Recipe 11-18 takes a brute-force approach, quitting and removing all current matches. This method is obviously not meant for deployment to App Store end users but provides a handy tool in your debugging arsenal.

Recipe 11-18 Obliterating Game Center Matches for the Current Player

```
- (void) removeAllMatches
{
    // This is nuclear armageddon. Prepare!
    [GKTurnBasedMatch loadMatchesWithCompletionHandler:
     ^(NSArray *matches, NSError *error)
     {
         if (error)
         {
             NSLog(@"Error loading matches: %@",
```

```
            error.localizedFailureReason);
        return;
    }
    
    NSLog(@"Attempting to remove %d matches", matches.count);
    for (MatchHelper *helper in _matchDictionary.allValues)
    {
        GKTurnBasedMatch *aMatch = helper.match;
        GKTurnBasedParticipant *me =
            myParticipantForMatch(aMatch);
        if (me && (me.status ==
            GKTurnBasedParticipantStatusActive))
        {
            NSLog(@"Quitting match %@", aMatch.matchID);
            [helper quitOutOfTurnWithOutcome:
                GKTurnBasedMatchOutcomeQuit];
            sleep(1);
            NSLog(@"Removing Match %@", aMatch.matchID);
            [_matchDictionary removeObjectForKey:aMatch.matchID];
            [helper removeFromGameCenter];
        }
        else
        {
            NSLog(@"Removing Match %@", aMatch.matchID);
            [_matchDictionary removeObjectForKey:aMatch.matchID];
            [helper removeFromGameCenter];
        }
    }
}];
}
```

Get This Recipe's Code

To find this recipe's full sample project, point your browser to https://github.com/erica/iOS-6-Advanced-Cookbook and go to the folder for Chapter 11.

Recipe: Game Center Voice

Each Game Center match can establish voice chats. These work best as push-to-talk walkie-talkies between players. Your application can create a group chat for everyone, limit chat between team members, or create a single one-to-one channel between two players. That's because all chats are named. When players connect into the same-named chat, they can hear and participate in that chat. Your application controls the names, and API-provided hooks enable you to build some high-level GUI controls to manage the chat.

> **Note**
>
> Make sure to test Voice Chat connected to Wi-Fi, not to your cellular network data, as you may encounter errors when trying to connect to the chat.

Testing for Chat Availability

Before offering chat to your users, test whether the feature is available. The `GKVoiceChat` class offers a simple check that determines whether Voice over IP is currently permitted on the device. As Apple puts it in its documentation, "Some countries or phone carriers may restrict the availability of voice over IP services."

```
if (![GKVoiceChat isVoIPAllowed])
    return;
```

Establishing a Play and Record Audio Session

To use Game Center Voice, you need to link your project to the `AVFoundation` framework. Then, as the program launches, set the shared `AVFoundation` audio session to use a play-and-record category. Recipe 11-19 updates the shared session to both play and record audio.

Recipe 11-19   Establishing an Audio Session for Voice Chat

```
- (BOOL) establishPlayAndRecordAudioSession
{
    NSLog(@"Establishing Audio Session");
    NSError *error;
    AVAudioSession *audioSession =
        [AVAudioSession sharedInstance];
    BOOL success = [audioSession setCategory:
        AVAudioSessionCategoryPlayAndRecord error:&error];
    if (!success)
    {
        NSLog(@"Error setting session category: %@",
            error.localizedDescription);
        return NO;
    }
    else
    {
        success = [audioSession setActive: YES error: &error];
        if (success)
        {
            NSLog(@"Audio session is active");
            return YES;
        }
        else
```

```
        {
            NSLog(@"Error activating audio session: %@",
                error.localizedDescription);
            return NO;
        }
    }
    
    return NO;
}
```

> **Get This Recipe's Code**
>
> To find this recipe's full sample project, point your browser to https://github.com/erica/iOS-6-Advanced-Cookbook and go to the folder for Chapter 11.

Creating a Voice Chat

Create and hold onto a `GKVoiceChat` instance for each chat you want to use within your application, typically by assigning it to a `strong` instance variable or an array. Each voice chat channel uses a simple string to identify it. For example, you might create a "GeneralChat" for your match:

```
chat = [match voiceChatWithName:@"GeneralChat"];
```

There are no further security elements in place. All chats are limited to the current match and are controlled by your application. You programmatically provide access to the chats as required.

Starting and Stopping a Chat

To start your chat, create it, set its `active` property to `NO`, optionally initialize its `volume` property, and send it the `start` message. The chat is immediately ready to use but will not start recording or sending data yet. It won't do that until you explicitly make it active:

```
chat = [match voiceChatWithName:@"GeneralChat"];
chat.active = NO;
chat.volume = 1.0f;
[chat start];
```

To finish a chat, send it `stop`. You can start it again later or, if you reassign the variable that is holding onto it, allow it to be deallocated:

```
[chat stop];
chat = nil;
```

Chat State Monitoring

When you call `start` or `stop` on a chat, Game Center sends out notifications that the local player has joined the chat. There are four possible voice chat states: connected, disconnected, speaking, and silent. Notifications go out for each of these. So, if you have an active chat and your user pauses to take a breath, `GKVoiceChat` propagates a `GKVoiceChatPlayerSilent` event, followed by `GKVoiceChatPlayerSpeaking`, when she starts talking again.

You can use these events to update local icons, highlighting them when a user is speaking, dimming them when they are not, or removing them from view when the user is no longer part of the chat. Add a state update handler to each chat to control the GUI presentation when a player's chat state changes:

```
chat.playerStateUpdateHandler =
    ^(NSString *playerID, GKVoiceChatPlayerState state)
{
    switch (state)
    {
        case GKVoiceChatPlayerSpeaking:
            // Highlight player's picture
            break;
        case GKVoiceChatPlayerSilent:
            // Dim player's picture
            break;
        case GKVoiceChatPlayerConnected:
            // Show player name/picture
            break;
        case GKVoiceChatPlayerDisconnected:
            // Hide player name/picture
            break;
    }
};
```

Implementing a Chat Button

It helps to treat your chats as push-to-talk items, activating them only when a user opts in to speak. For example, you might create a talk button that activates the local microphone as it is pressed and deactivates it when released. Here's how you could define that button:

```
talkButton = [UIButton buttonWithType:UIButtonTypeRoundedRect];
[talkButton setTitle:@"Speak" forState:UIControlStateNormal];
talkButton.frame = CGRectMake(0.0f, 0.0f, 200.0f, 40.0f);
[self.view addSubview:talkButton];

[talkButton addTarget:self action:@selector(startSpeaking)
    forControlEvents:UIControlEventTouchDown];
[talkButton addTarget:self action:@selector(stopSpeaking)
    forControlEvents:UIControlEventTouchUpInside |
    UIControlEventTouchUpOutside];
```

The two callbacks involve nothing more than setting the chat's active property to YES (microphone enabled and live) or NO (microphone disabled). As you've already seen, always initially set this property to NO when you create the chat:

```
- (void) stopSpeaking
{
    chat.active = NO;
}

- (void) startSpeaking
{
    chat.active = YES;
}
```

Controlling Volume

The chat's `volume` property ranges between 0.0 (muted) and 1.0 (full volume). This controls how loudly the chat plays back on the local device's speakers. In addition, you can control whether you're listening to other players on a player-by-player basis. Mute individual players by calling `setMute:forPlayer:`. Like the player state update handler, this feature is meant to be integrated into a player-by-player icon GUI, allowing your users to mute others individually:

```
[chat setMute:YES forPlayer: playerID];
```

You can mute all players by iterating through the chat's list of player identifiers. This affects audio for only the selected chat. If the player is participating in more than one chat, you must mute each one or set the `volume` property for each chat to 0.0:

```
- (void) muteChat: (GKVoiceChat *) aChat
{
    NSLog(@"Muting chat %@", aChat.name);
    for (NSString *playerID in aChat.playerIDs)
        [aChat setMute:YES forPlayer:playerID];
}

- (void) unmuteChat: (GKVoiceChat *) aChat
{
    NSLog(@"Unmuting chat %@", aChat.name);
    for (NSString *playerID in aChat.playerIDs)
        [aChat setMute:NO forPlayer:playerID];
}
```

GameKit Peer Services

GameKit provides peer-to-peer connectivity between iOS devices outside of the Game Center milieu. The framework helps you create interconnected applications that exchange live data in real time. In its default implementation, GameKit works by creating and managing an ad-hoc

Bluetooth network that lets devices find each other, establish a connection, and transmit data through that connection. GameKit also allows you to find, connect, and transmit to devices on the same Wi-Fi network.

Using Bluetooth and Wi-Fi provides a fast and reliable approach to interdevice communications. Unfortunately, peer-to-peer communication using GameKit isn't fast, although it's far more reliable under iOS 6 than older firmware. Using GameKit peering can prove frustrating for end users. You should expect your support commitments to be tested accordingly if you decide to include this GameKit feature in your applications.

The standard use scenario requires that users own more than one device and use them in close quarters. No matter how nifty the peer-to-peer idea is, business-wise, it's a poor investment of time. Using Game Center is simpler, rugged, and provides a greater potential user base for those features.

With that said, here is a quick introduction to the peer-to-peer aspects of GameKit.

GameKit Bluetooth Limitations

For Bluetooth, you need proximity. Connections are limited to approximately 10 meters (30 feet). Think of your audience including people riding together on a train or in a car, in a convention-hall's meeting room, or working in the same office. Within that range, your application can establish a peer-to-peer connection.

GameKit offers excellent performance for short, tight blips of information. Apple recommends that GameKit transmissions be limited to 1,000 bytes and under. Although GameKit can handle larger blobs, up to 95 Kilobytes at a time, it's not meant for use as general device-to-device data transfer. Try to send too much data at once, and you will receive transmission errors.

If you must transfer large files, you need to break those files into manageable chunks. Make sure that you use standard handshaking and packet checksumming to ensure the reliability of your data.

GameKit's Bluetooth networking works on all modern iOS devices. The `peer-peer` key in the Info.plist `UIRequiredDeviceCapabilities` entry lets you explicitly require support for peer-to-peer connectivity over Bluetooth.

Bonjour Sessions

GameKit's peer-to-peer connections are built using Bonjour networking. Bonjour, which is Apple's trade name for zero-configuration networking, enables devices to advertise and discover network services. Built into Mac OS X since version 10.2, Bonjour offers these features without calling attention to itself. For example, Bonjour powers the features that let users find shared music for iTunes or connect to wireless printers without requiring custom configuration. These services automatically appear when they become available and disappear when they're not. It's a powerful OS feature.

GameKit provides that same Bonjour power without building the complicated Bonjour callbacks for registering and detecting services. With GameKit, you request a connection using a "peer picker" controller and then manage a "session" after the connection has been established.

GameKit's session objects provide a single focus point for data transfer management. Each session uses a unique name, which you choose, to advertise itself. When an application looks for another device to connect to, it uses this name to identify compatible services.

If you use a Bonjour browsing service to look for that name, you'll fail. Apple encodes the service name using a SHA-1 hash. For example, a service called MacBTClient Sample becomes the _11d7n7p5tob54j._udp. Bonjour service. GameKit automatically transforms the name you supply so it knows how to find matching services.

Servers, Clients, and Peers

GameKit offers three session modes: Applications can act as servers, clients, and peers. Servers advertise a service and initialize a session, allowing clients to search for and connect to them. This is the kind of behavior that a smart printer uses, letting clients find and use its capabilities. It's handy for devices that provide a fixed functionality, but it's not the best choice for most iOS applications, especially games.

Peers work as both servers and clients. They advertise and search simultaneously. When a peer selects a service, its client/server role is hidden both from the user and the developer. This makes the peer approach easy to develop for iOS. You don't have to build separate client and server applications. One peer-based application does all the work.

The Peer Connection Process

The peer picking process is handled by a class called `GKPeerPickerController`. It provides a built-in series of interactive alert dialogs that automate the task of advertising device availability and selecting a peer. Using this class is not mandatory. You can bypass it and create a custom class to search for and connect to peers.

For simple connections, however, the `GKPeerPickerController` class offers a ready-to-use interface that sidesteps the need for detecting and negotiating with peers. To use the peer picker, you allocate it, set a delegate (which must implement the `GKPeerPickerControllerDelegate` protocol), and show it.

Displaying the Peer Picker

The following code allocates and shows a new peer picker controller, setting its connection style to nearby. This skips an optional interaction step where a user selects between Online and Nearby modes. When presented, it shows the interface in Figure 11-7. You do not have to use a peer picker to establish GameKit sessions. The iOS SDK lets you create your own custom interfaces to work with the underlying GameKit connections:

```
// Create and present a new peer picker
GKPeerPickerController *picker = [[GKPeerPickerController alloc]
    init];
picker.delegate = self;
picker.connectionTypesMask = GKPeerPickerConnectionTypeNearby;
[picker show];
```

Figure 11-7   This is the first screen presented to the user for peer-to-peer Bluetooth/Wi-Fi connections.

When your mask also includes the online type (`GKPeerPickerConnectionTypeOnline`), the picker first asks the user which kind of connection to use before moving on to either the nearby connection interface or to a custom online interface that you build.

Pressing Cancel

Users may cancel out of the peer picker alert. When they do so, the delegate receives a `peerPickerControllerDidCancel:` callback. If you display a Connect button in your application, make sure to restore it at this point so that the user can try again:

```
- (void) peerPickerControllerDidCancel: (GKPeerPickerController *)picker
{
    [self setupConnectButton];
}
```

Creating the Session Object

As the picker delegate, you must supply a session object on request. Sessions, which provide an abstract class that creates and manages a data socket between devices, belong to the GKSession class and must be initialized with a session identifier. This is the unique string used to create the Bonjour service and link together two iOS devices (peers) both advertising the same service. By setting the display name to nil, the session uses the built-in device name:

```
- (GKSession *)peerPickerController:(GKPeerPickerController *)picker
      sessionForConnectionType:(GKPeerPickerConnectionType)type
{
    // Create a new session if one does not already exist
    if (!self.session) {
        self.session = [[GKSession alloc] initWithSessionID:
            (self.sessionID ? self.sessionID : @"Sample Session")
            displayName:nil sessionMode:GKSessionModePeer];

        self.session.delegate = self;
    }
    return self.session;
}
```

Although this is an optional method, you'll usually want to implement it so that you can set your session ID and mode. Upon detecting another iOS device with the same advertised service ID, the peer picker displays the peer as a compatible match, as shown in Figure 11-8.

Waiting for the peer picker list can take a few seconds (typical) or up to a few minutes (exceedingly rare under modern firmware). Apple recommends always debugging from a clean restart. If debugging delays get frustrating enough, make sure to reboot.

In normal use, connection delays usually hover approximately 45 seconds at a maximum. Warn your users to be patient. In Figure 11-9, Bear is the device name for a second iOS device running the same application. When the user taps the name Bear, this device automatically goes into client mode, and Bear goes into server mode.

Client and Server Modes

When a device changes into client mode, it stops advertising its service. The device-choice dialog shown previously in Figure 11-8 changes on the server unit. The client's peer name dims to dark gray, and the words "is not available" appear underneath. A few seconds later (and this can actually run up to 1 minute, so again warn your users about delays), both units update their peer picker display.

Figure 11-9 shows the server and client peer pickers during this process. The client waits as the server receives the connection request (left). On the server, the host user must accept or decline the connection (middle). Should they decline, an updated peer picker notifies the client (right). If they accept, both delegates receive a new callback.

Figure 11-8 The peer picker lists all devices that can act as peers.

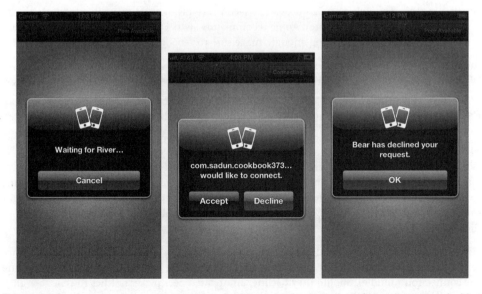

Figure 11-9 Upon choosing a partner, the client goes into wait mode (left) as the server decides whether to accept or decline the connection (middle). Should the server decline, the client receives a notice to that effect (right).

The delegate callback lets the new peers dismiss the peer picker and set their data received handler:

```
- (void)peerPickerController:(GKPeerPickerController *)picker
    didConnectPeer:(NSString *)peerID
    toSession: (GKSession *) session
{
    // Dismiss the picker, then set the data handler
    [picker dismiss];
    [self.session setDataReceiveHandler:self withContext:nil];
}
```

Sending and Receiving Data

The data handler (in this case, `self`) must implement the `receiveData:fromPeer:inSession:context:` method. The data sent to this method uses an `NSData` object; there are no hooks or handles for partial data receipt and processing. As the data arrives as a single chunk, keep your data bursts short (under 1,000 bytes) and to the point for highly interactive applications:

```
- (void) receiveData:(NSData *)data fromPeer:(NSString *)peer
    inSession: (GKSession *)session context:(void *)context
{
    // handle data here
}
```

Send data via the session object. You can send in reliable mode or unreliable mode. Reliable mode uses error checking and retrying until the data is properly sent. All items are guaranteed to arrive in the order they are sent, using TCP transmission. With unreliable mode, data is sent once using UDP transmission, with no retry, Data may arrive out of order. Use reliable mode (`GKSendDataReliable`) when you must guarantee correct delivery and unreliable mode for short bursts of data that must arrive nearly instantaneously:

```
- (void) sendDataToPeers: (NSData *) data
{
    // Send the data, checking for success or failure
    NSError *error;
    BOOL didSend = [self.session sendDataToAllPeers:data
        withDataMode:GKSendDataReliable error:&error];
    if (!didSend)
        NSLog(@"Error sending data to peers: %@",
            Error.localizedDescription);
}
```

One error you'll encounter here results from queuing too much data in reliable mode. This produces a buffer full error.

State Changes

The following session delegate callback lets you know when a peer's state has changed. The two states you want to look for are connected—that is, when the connection finally happens after the peer picker has been dismissed, and disconnected, when the other user quits the application, manually disconnects, or moves out of range:

```
- (void)session:(GKSession *)session peer:(NSString *)peerID
    didChangeState:(GKPeerConnectionState)state
{
    /* STATES: GKPeerStateAvailable, GKPeerStateUnavailable,
       GKPeerStateConnected, GKPeerStateDisconnected,
       GKPeerStateConnecting */

    if (state == GKPeerStateConnected)
    {
        // handle connected state
    }

    if (state == GKPeerStateDisconnected)
    {
        // handle disconnection
    }
}
```

To force a session to disconnect, use the `disconnectFromAllPeers` method:

```
- (void) disconnect
{
    [session disconnectFromAllPeers];
}
```

Creating a GameKit Helper

You can find a simplified peer-to-peer helper class in the sample code that accompanies this chapter. This class hides most of the GameKit details connection and data transfer details, while providing a demonstration of how to use these features. More important, it breaks down how you might look at the GameKit process, with its two key details: connection and data.

Peer-to-Peer Voice Chat

GameKit's In-Game Voice service lets applications create a walkie-talkie-style voice channel connecting two devices together. The voice additions, provided by the `GKVoiceChatService` class, sit outside normal GameKit. Chat services connect into the iPhone's audio playback and recording system, so voice chat can listen to and play back audio. Voice Chat then sends its data through GameKit and plays back the data it receives from GameKit.

`GKVoice` expects a `GKSession` with `GKPeers` to transmit its data. If you need to use voice transmission for another connection style, you must write that layer yourself.

Implementing Voice Chat

When working with voice, there's no difference in the way you start. You display a peer picker and negotiate the connection, as you would normally do with GameKit. The difference arrives after the peer connects. You need to establish the voice chat and redirect the data to and from that service.

Upon connecting to the new peer, set up the voice chat basics. Activate a play-and-record audio session (as shown earlier in this chapter), set the default chat service client, and start a new voice chat with that peer. By setting the `client` property, you ensure that your class receives the voice chat callbacks needed for negotiating data:

```
// Set the voice chat client and start voice chat
[GKVoiceChatService defaultVoiceChatService].client = self;
if (![[GKVoiceChatService defaultVoiceChatService]
    startVoiceChatWithParticipantID: peerID error: &error])
{
    NSLog(@"Error starting voice chat");
    return;
}
```

Your primary class must declare the `GKVoiceChatClient` protocol to do this. When the chat service gathers data through the microphone, it triggers the `voiceChat-Service:sendData:toParticipantID:` callback. Here, you can redirect voice data to your normal GameKit session. For a voice-only connection, just send along the data. When your application handles both voice and other data, build a dictionary and tag the data with a key, such as @"voice", or when your class receives data through the normal `receiveData:fromPeer:inSession:context:` callback, the same approaches apply. For voice only, use `receivedData:fromParticipantID:` to send the data off to the chat service. Voice Chat enables you to mix game audio with in-game voice. For voice-data hybrid applications, deserialize the data, determine whether the packet included voice or regular data, and redirect that data to the appropriate recipient:

```
- (void)voiceChatService:(GKVoiceChatService *)voiceChatService
    sendData:(NSData *)data
    toParticipantID:(NSString *)participantID
{
    // Send the next burst of data to peers
    [self.session sendData: data toPeers:[NSArray arrayWithObject:
        participantID] withDataMode: GKSendDataReliable error: nil];
}

- (void) receiveData:(NSData *)data
    fromPeer:(NSString *)peer
    inSession: (GKSession *)session context:(void *)context
```

```
{
    // Redirect any voice data to the voice chat service
    [[GKVoiceChatService defaultVoiceChatService]
        receivedData:data fromParticipantID:peer];
}
```

Creating an "Online" GameKit Connection

In the GameKit peer-to-peer world, "online" currently means any valid connection style other than Game Center or Bluetooth. You might use a local WLAN network to connect to another device on the same network or connect through WWAN (that is, the cellular service) or Wi-Fi to a remote Internet-based host. GameKit takes you only so far, as shown in Figure 11-10. By selecting Online, your user depends on you to create a custom connection to another device or service.

Figure 11-10 The Online GameKit connection means "bring your own networking."

You create this two-item dialog by supplying the online option to the peer picker mask. In all other ways, there's no change in how you create and present a standard GameKit peer picker controller:

```
- (void) startConnection
{
    if (!self.isConnected)
    {
```

```objc
    GKPeerPickerController *picker = [[GKPeerPickerController
        alloc] init];
    picker.delegate = self;
    picker.connectionTypesMask =
        GKPeerPickerConnectionTypeNearby |
        GKPeerPickerConnectionTypeOnline;
    [picker show];
    if (self.viewController)
        self.viewController.navigationItem.rightBarButtonItem =
            nil;
    }
}
```

Catch the user selection in the `peerPickerController:didSelectConnectionType:` callback. You can assume that if the user selected Nearby that all the handshaking dialogs are taken care of for you. Should the user select Online, however, it's up to you to move things to the next step. You need to dismiss the picker and display the next stage of the connection task. Here, control passes away from the peer picker to a custom class. It's up to you to produce that class. The following example method uses `BonjourHelper`, which was introduced in a previous edition of this book. Whatever class you use, its role is to begin an online connection, choose a peer to share data with, and serve that data directly to that peer:

```objc
- (void)peerPickerController:(GKPeerPickerController *)picker
    didSelectConnectionType:(GKPeerPickerConnectionType)type
{
    if(type == GKPeerPickerConnectionTypeOnline)
    {
        [picker dismiss];

        // Establish your own custom connection class here
        [BonjourHelper sharedInstance].sessionID = self.sessionID;
        [BonjourHelper sharedInstance].viewController =
            self.viewController;
        [BonjourHelper sharedInstance].dataDelegate =
            self.dataDelegate;
        [BonjourHelper connect];
    }
}
```

Summary

GameKit offers an exciting new player in the iPhone development arena. Its easy-to-use Game Center connections make it simple for you to deliver applications that communicate between remote iOS devices for satisfying information transfer and game play. In this chapter, you saw how to build those connections and produce real-time data transfers that allow games and other applications to coordinate information between separate devices. Here are a few last-minute thoughts on these technologies:

- GameKit has finally debuted OS X. Your users can now play across platforms. Consider looking into developing clients for OS X and iOS.

- Just because the name has the word "Game" doesn't mean that you cannot use GameKit and Game Center to transfer other kinds of information and data between devices. Don't limit your applications because of Apple's game branding of the service. GameKit provides a wonderful data-transfer infrastructure. The game-specific features lie on top of those. A number of terrific utilities employ these techniques.

- When working with Voice Chat locally, especially during testing, remember that nearby users may produce sound loops creating feedback distortion unless they use headsets. For peer-to-peer use, consider that people sitting 10 feet apart from each other can easily talk without the use of technology. For remote Game Center Voice Chat, be aware of how easily the technology can be abused and provide local muting options for your users.

- Don't forget to incentivize your users. Leaderboards and achievements provide wonderful ways to motivate and engage game play beyond a single device or a single session of use. Both Achievements and Leaderboards displays now enable users to quickly and easily rate your app. In iOS 6, users can tap achievements they've attained to brag about them socially.

- Game Center Matches are designed for turn-based games without timers (either per move or per game). Although you can build your own timers on top of Game Center, it's not a feature that's directly supported by Apple and could possibly impact app review.

- The turn-by-turn gaming coverage in this edition has been massively updated. However, there are still edges to smooth out and bugs to kill. If you have any suggestions for improving the code, please let me know. You can submit tickets at the github repository or just ping me by email.

12
StoreKit

StoreKit offers in-app purchasing that integrates into your software. With StoreKit, end users can use their iTunes credentials to buy unlockable features, media subscriptions, or consumable assets, such as fish food or sunlight, from within an application. They make these purchases after initially procuring and installing the application from the App Store. This chapter introduces StoreKit and shows you how to use the StoreKit API to create purchasing options for users. In this chapter, you read about getting started with StoreKit. You learn how to set up products at iTunes Connect and localize their descriptions. You see what it takes to create test users and how to work your way through various development/deployment hurdles. This chapter teaches you how to solicit purchase requests from users and how to hand over those requests to the store for payment. By the time you finish this chapter, you'll have learned about the basic StoreKit picture, from product creation to sales.

Getting Started with StoreKit

When your application demands a more complex purchase model than buy-once use-always, consider StoreKit. StoreKit offers developers a way to sell products from within an application to create additional revenue streams. There are many reasons to use StoreKit. You might support a subscription model, provide extra game levels on demand, or introduce other unlockable features.

With StoreKit, you choose the items you want to sell and you set their price. StoreKit and iTunes take care of the details. They provide the infrastructure that brings that storefront into your application through a series of API calls and delegate callbacks.

Fulfillment

On purchase, users do not download new code. All StoreKit-based applications ship with their features already built in. For example, StoreKit purchases might enable users to access parts of your application that you initially set as off limits. They can also download or unlock new data sets (which now can be hosted by Apple) or authorize access to subscription-based web feeds or upgrade items in-game for more power or longer play. StoreKit provides the way users can pay to access to these features, enabling them to go live after purchase.

StoreKit Limitations

You cannot use in-app purchasing to sell "hard" assets (such as T-shirts) or intermediate currency (such as store credit for a Web site). And, yes, real gambling is forbidden as well. Any goods sold via in-app purchase must be delivered digitally to your application. Your purchasable items must not include pornography, hate speech, or defamation.

The StoreKit Development Paradox

Unfortunately, StoreKit presents a paradox, which is this: Although you can easily add IAP features to applications that are already for sale at the App Store, you cannot fully develop and test your in-application purchasing for new apps until you have already submitted your application to iTunes. And you cannot fully submit your application to iTunes knowing that you're not done developing it. So, what's a developer to do? How do you properly develop new applications for StoreKit?

There is, fortunately, a solution. This approach is shown in Figure 12-1. To work around the StoreKit paradox, you upload a fully working but not fully fleshed-out application to iTunes Connect. You do this with the full understanding that you'll be replacing your binary at some point in the future.

Apple describes this placeholder as an approvable app, one that can be submitted to the App Store review process and be approved for sale. Should the application be rejected, your in-app items will stop working. You must have an app either already approved or in active review to develop IAP items for it.

Apple explains it this way in its technical note TN2259; the emphatic capitals are original to the note:

> DO NOT upload the development binary to iTunes Connect until the application is ready for App Review approval. If the binary is present in iTunes Connect and it is not fully functional, App Review will review the binary and likely reject the development binary. Testing In App Purchase will fail if you or App Review reject your most recent binary in iTunes Connect. The workaround in this case is to upload a binary without In App Purchase features that can get approved by App Review. Once the binary is approved, resume testing the binary with In App Purchase features.

> **Note**
> When submitting your application for testing, roll back your availability date in the iTunes Connect Pricing tab. This prevents your not-ready-for-prime-time app from inadvertently appearing for sale on the App Store until you're ready. Reset that date when you're ready to go live.

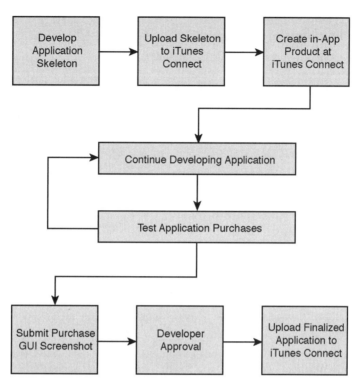

Figure 12-1   The StoreKit development process for new apps.

Developing and Testing

For new apps, after you submit your application and define at least one in-application purchase item at iTunes Connect, you can begin to fully develop and test your application and its purchases. (For existing apps, you can create that in-app purchase item without submitting an approvable skeleton.)

Use the sandbox version of StoreKit along with test user accounts to buy new items without charging a credit card. The sandbox StoreKit enables you to test your application features before, during, and after payment.

Submission

When you finish development and are ready to submit a final version to the App Store, you complete the StoreKit development process at iTunes Connect:

- You upload a screenshot showing the GUI for your application purchase.
- If you uploaded a placeholder app, and it has not yet been through the review process, you self-reject it. (If it has been approved, just upload a new version instead.)

- You upload the new fully working version of your application for review that contains the in-app purchase support.
- At iTunes Connect, you submit the in-app purchase for review.

The following sections walk you through many of the process details. You read about the StoreKit process specifics and learn how to add StoreKit to your application.

Creating Test Accounts

Test accounts play a key role in the StoreKit development scenario. Create one or more new user accounts before you begin developing new StoreKit-enabled applications. These accounts enable you to log in to iTunes to test your application payments without charging money.

Here's how you add a new user. Log in to iTunes Connect, and choose Manage Users > Test User. Click Add New User. iTunes Connect presents the form shown in Figure 12-2. When filling out this form, keep the following points in mind:

- Each e-mail address must be unique, but it doesn't need to be real. So long as the address does not conflict with any other one in the system, you'll be fine. As you might guess, other developers have already taken the easy-to-type addresses, such as abc.com, abcd.com, and so on.
- Names do not have to be real. Birthdates do not have to be real. Consider using a basic alphabetical naming system such as "a Sadun," "b Sadun," "c Sadun," and so forth. Everyone was born on January 1st.
- Apple keeps changing its password rules. Generally, passwords must be at least eight characters and include one uppercase character, one lowercase character, and a numeral. Your password cannot contain three consecutive identical characters, and it cannot be so simple that Apple flags it as needing to be "complex." As you'd imagine, this can be a pain when it comes to typing items over and over. A single easy-to-remember disposable password can be used for all your test accounts (for example, AlphaBeta1). You may want to invest in a typing utility such as Type2Phone (houdah.com). This app enables you to type over a Bluetooth connection from your computer to your phone, simplifying repetitive text entry.
- The secret question/answer fields are meaningless in this context, but they cannot be left empty. You cannot enter the same string for both fields, and the question field must be at least six characters. Consider using a question/answer pair such as "aaaaaa" and "bbbbbb" to simplify account creation.
- Selecting an iTunes Store is required. This store sets the region for your testing. If you plan to use multiple language support for various stores, make sure you create a test account in each affected region.
- You can delete user accounts and add new ones on-the-fly. If you run out of users who haven't yet purchased any items, just create new users as needed.

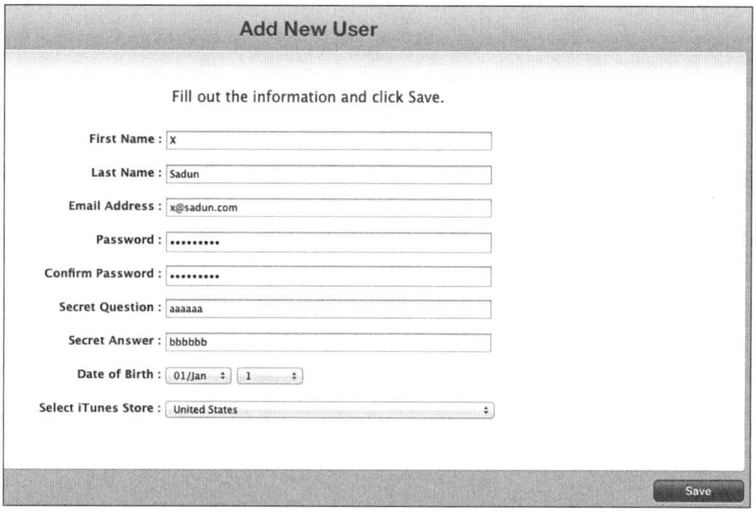

Figure 12-2 Add new test users in iTunes Connect by filling out this form.

- You never want to sign into your fake user "account" in the Settings application. If you try to do so, the iOS device forces you to consent to its standard user agreement and then tries to extract a valid credit card from you. Use Settings to log out of an account but avoid it for logging in to one.

- Get into the habit of logging out of your accounts in Settings each time you test your application.

- Test accounts are disposable. When an account has purchased an item, feel free to move onto the next one, so your testing starts fresh again. There is no practical upper limit to test accounts, and you can create as many as you need to perform the testing your application demands.

Creating New In-App Purchase Items

Each in-application purchase item must be registered at iTunes Connect. To create a new purchase, navigate to Manage Your Apps, and select any application that is either approved or in-review. Click Manage In-App Purchases, and then click Create New.

iTunes Connect prompts you to select an in-app purchase type. You can choose from the following:

- **Consumable**—Create a purchase that is used up by the user during the normal use of the app, such as a power boost, extra bullets, or fish food. Consumables may expire, and they naturally decrease over time during use.

- **Nonconsumable**—Create a one-time purchase item, such as a specialty rifle, access to additional levels or a full-featured interface, or other unlockables. Nonconsumables do not expire or decrease with use.
- **Auto-Renewable subscriptions**—These enable users to purchase content for a set period of time, which auto-charges the user at the end of each time period until the user opts out of the service. Auto-Renewable subscriptions are automatically propagated to all devices registered to the same Apple ID account.
- **Free subscriptions**—Used for putting free content into Newstand apps.
- **Nonrenewing subscriptions**—Offer time-limited access such as 1-year access to a film archive.

Select the kind of in-app purchase you want to create and then fill out the details for the new purchase item. You can start by entering basic details about the purchase, setting its pricing and availability, and entering a screenshot for review.

Filling Out the Details Section

The details section consists of a reference name, a product ID, and a list of languages. Figure 12-3 shows what this screen looks like. The reference name should describe the purchase in an easily understood manner. The Product ID provides unique identifier for looking up the product in the App Store database.

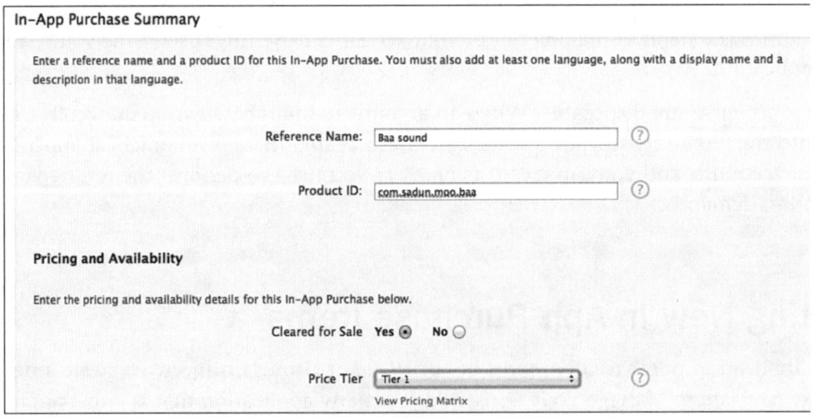

Figure 12-3   Set up the purchase details by creating a unique product ID.

Reference Name

The reference name is used to provide a name for iTunes Connect's search results and in the application's Top In-App Purchase section in the App Store. So, enter a meaningful name (for example, "Unlock Sky Master Level 3 Purchase") that helps you and others know what the item is and how it is used in your application.

Product ID

The product ID is a unique identifier, similar to the application identifier used for your app. As a rule, I use my application ID and append a purchase name to that such as `com.sadun.scanner.optionalDisclosure`. You need this identifier to query the store and retrieve details about this purchase. The same rules apply to the product ID as to application IDs. You cannot use an identifier more than once. You cannot "remove" it from the App Store. When registered, it's registered forever. If you delete it before approval, you cannot re-create it.

You need to add one or more localized language descriptions of the purchase as well.

Adding Localized Descriptions

Each purchasable item must describe itself to your application. Keep in mind that your application is the primary consumer for this information. In the localized description, you need to specify a display name (the name of the product being purchased) and description (an explanation to the user that describes what the purchase is and does).

These latter two elements are localized to specific languages. You can create data for any or all languages so long as you define at least one. You cannot submit a new purchase item without creating one or more name/description pairs. For developers who are primarily targeting the U.S. store, a single English entry should cover their needs. Although, you may consider revisiting this with the help of professional translators at a later time.

If your application is sold worldwide, you'll likely want to mirror the existing localizations you use with your app descriptions and in-app features. If your iTunes store marketing material provides a Japanese localization, for example, and your application offers a Japanese language version, you need to create a Japanese-localized in-app purchase description as well. If you do not, you can still use in-app purchases, but the language will default to whatever localizations you have provided.

> **Note**
>
> Always use native speakers to localize, edit, and proof text. Google Translate is not a substitution for proper localization. Translation bureaus like Traducto (traductoapp.com) offer both translation and proofing at various skill levels including native speakers and professional translators.

Entering Product Information

When entering the localized product information data (see Figure 12-4), keep some points in mind:

- Your application is the consumer for this information. The text you type in iTunes Connect helps create the purchase GUI that your application presents to the user.
- The user's language settings select the localization. If a localization is not available, it defaults to one of the other descriptions provided.
- If you plan to use a simple alert sheet with a Buy/Cancel choice, keep your wording tight. Limit your verbosity. If you use a more complex view, consider that as well.

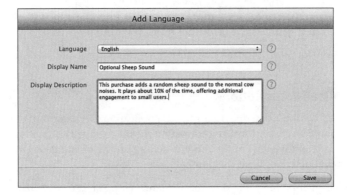

Figure 12-4 Add each language separately.

No matter how you create your GUI, remember that your description must convey the action of purchasing and a description of the item purchased—for example, "When purchased, this option unlocks this application's detail screens. These screens reveal even more data about the scanned MDNS services." A shorter description, such as "Extra detail screens" or "Unlock more details," doesn't explain to users how the purchase works and what they can expect to receive.

> **Note**
> You can edit item display details during review at iTunes Connect. You must submit new changes for review after the purchased items have been approved.

Filling Out the Pricing Section

The pricing section specifies how a purchase is priced. Leave the Cleared for Sale item checked. The Cleared for Sale check box ensures that your applications, both development and distribution, have programmatic access to the purchase item.

> **Note**
> You can change the pricing tier and Cleared for Sale check box at any time during review. You must submit new changes for review after the purchased items have been approved. You cannot edit the identifier or reuse an existing identifier, nor can you change the type of product after creating the purchase item.

Submitting a Purchase GUI Screen Shot

The For Review section appears at the bottom of the item sheet. You do not use this section until you have finished developing and debugging your application. You can leave the state of your product as Waiting for Screenshot until you finish testing it and are ready to upload it for review. When you are at that point, upload a screen shot into the provided field. The screen shot must show the in-app purchase in action, demonstrating the custom GUI you built.

Figure 12-5 displays the submission section. Valid pictures must be at least 640x960 pixels in size. The screen shot highlights how you have developed the purchase feature. Submit an image highlighting the purchase.

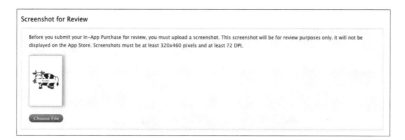

Figure 12-5   You must submit a screen shot showing your in-application purchase GUI to Apple when you are ready to have that purchase reviewed.

Submitting In-App Purchases Products for Review

After you finish your sandbox testing and are confident that the application and the in-app purchasing are ready for Apple to review, you submit the purchase for its own review process. Upload your new app, return to your App's Manage In-App Purchases, select any purchase item, and click Submit for Review.

Building a Storefront GUI

Apple's StoreKit framework does not provide a built-in GUI for soliciting user purchases. You must create your own. Retrieve localized prices and descriptions from the App Store by creating `SKProductsRequest` instances. This class asks the store for that information based on the set of identifiers you provide. Each identifier must be registered at iTunes Connect as an in-app purchase item.

Allocate a new products request instance and initialize it with that set. You can add identifiers for items you've already established and items you plan on adding in the future. Because each identifier is basically a string, you could create a loop that builds identifiers according to some naming scheme (for example, com.sadun.app.sword1, com.sadun.app.sword2, and so on) to provide for future growth. This snippet searches for a single item:

```
#define PRODUCT_ID         @"com.sadun.moo.baaa"

// Create the product request and start it
SKProductsRequest *productRequest =
    [[SKProductsRequest alloc] initWithProductIdentifiers:
        [NSSet setWithObject:PRODUCT_ID]];
productRequest.delegate = self;
[productRequest start];
```

When using a products request, your delegate must declare and implement the `SKProductsRequestDelegate` protocol. This consists of three simple callbacks. Listing 12-1 shows these callback methods for a simple application. When a response is received, this code looks for a product (only one was requested, per the preceding code snippet) and retrieves its localized price and description.

It then builds a simple alert using the description as the alert text and two buttons (the price and "No Thanks"). This alert functions as a basic purchase GUI. Figure 12-6 shows what this alert might look like.

Figure 12-6 This alert is built from the localized description retrieved from the App Store.

> **Note**
>
> StoreKit will not work if you are not connected to the network in some way. Refer to Chapter 5, "Networking," to find recipes that help check for network access.

Listing 12-1 Products Request Callback Methods

```
- (void)request:(SKRequest *)request
    didFailWithError:(NSError *)error
{
    NSLog(@"Error: Could not contact App Store properly: %@",
        error.localizedFailureReason);
}

- (void)requestDidFinish:(SKRequest *)request
{
    NSLog(@"Request finished");
}

- (void) alertView: (UIAlertView *) alertView
    clickedButtonAtIndex: (NSInteger)answer
{
    NSLog(@"User %@ buy", answer ? @"will" : @"will not");
    if (!answer) return;

    // Ready to purchase — make purchase here
    // (see next section)
}

- (void)productsRequest:(SKProductsRequest *)request
    didReceiveResponse:(SKProductsResponse *)response
{
    // Find a product
    SKProduct *product = [[response products] lastObject];
    if (!product)
    {
        NSLog(@"Error: Could not find matching products");
        return;
    }

    // Retrieve the localized price
    NSNumberFormatter *numberFormatter =
        [[NSNumberFormatter alloc] init];
    [numberFormatter setFormatterBehavior:
        NSNumberFormatterBehavior10_4];
    [numberFormatter setNumberStyle:NSNumberFormatterCurrencyStyle];
    [numberFormatter setLocale:product.priceLocale];
    NSString *priceString =
        [numberFormatter stringFromNumber:product.price];

    // Show the information
    UIAlertView *alert = [[UIAlertView alloc]
```

```
            initWithTitle:product.localizedTitle
            message:product.localizedDescription
            delegate:self
            cancelButtonTitle:@"No Thanks"
            otherButtonTitles: priceString, nil];
    [alert show];
}
```

Purchasing Items

To purchase items from your application, start by adding a transaction observer. For example, you can do this is in your application delegate's finished-launching-with-options method or in your primary view controller's load view or view did load method. Whichever observer you use, make sure that the class declares and implements the `SKPaymentTransactionObserver` protocol:

```
[[SKPaymentQueue defaultQueue] addTransactionObserver:someClassInstance];
```

With an observer in place, you can use the GUI from Listing 12-1 to begin the actual purchase:

```
// Ready to purchase the item
SKPayment *payment =
    [SKPayment paymentWithProduct:product];
[[SKPaymentQueue defaultQueue] addPayment:payment];
```

StoreKit prompts the user to confirm the in-app purchase, as shown in Figure 12-7, and then takes over the purchase process. Users may need to log in to an account before they can proceed.

Signing Out of Your iTunes Account for Testing

To use the test accounts you set up in iTunes Connect, be sure to sign out of your current, real account. Launch the Settings application, choose the Store preferences, and click Sign Out.

As mentioned earlier, *do not* attempt to sign in again with your test account credentials. Just quit out of Settings and return to your application. After clicking Buy, you are prompted to sign in to iTunes. At that prompt, choose Use Existing Account, and enter your account details.

Regaining Programmatic Control After a Purchase

The payments transaction observer receives callbacks based on the success or failure of the payment process. Listing 12-2 shows a skeleton for responding to both finished and unfinished payments. After the user finishes the purchase process, the transaction will have succeeded or failed. On success, perform whatever action the user has paid for, whether by downloading data or unlocking features.

Figure 12-7 Users must confirm the purchase after moving past your user interface into the actual App Store/StoreKit purchasing system. These screen shots were taken during sandbox testing.

Listing 12-2 **Responding to Payments**

```
- (void)paymentQueue:(SKPaymentQueue *)queue
    removedTransactions:(NSArray *)transactions
{
}

- (void) completedPurchaseTransaction:
    (SKPaymentTransaction *) transaction
{
    // PERFORM THE SUCCESS ACTION THAT UNLOCKS THE FEATURE HERE

    // For example, update user defaults, (preferably) the keychain,
    // or local variables. User defaults is *not* secure
    [[NSUserDefaults standardUserDefaults] setBool:YES forKey:@"baa"];
    [[NSUserDefaults standardUserDefaults] synchronize];
    hasBaa = YES;

    // Update GUI accordingly
    if (purchaseButton)
    {
        [purchaseButton removeFromSuperview];
        purchaseButton = nil;
    }
```

```objc
    // Provide some feedback to the user that the transaction
    // was successful, e.g.
    AudioServicesPlaySystemSound(baasound);

    // Finish transaction
    [[SKPaymentQueue defaultQueue] finishTransaction: transaction];

    // Always say thank you.
    UIAlertView *okay = [[UIAlertView alloc]
        initWithTitle:@"Thank you for your purchase!"
        message:nil delegate:nil cancelButtonTitle:nil
        otherButtonTitles:@"OK", nil];
    [okay show];
}

- (void) handleFailedTransaction: (SKPaymentTransaction *) transaction
{
    // Process any transaction error
    if (transaction.error.code != SKErrorPaymentCancelled)
    {
        UIAlertView *okay = [[UIAlertView alloc]
            initWithTitle:@"Transaction Error. Try again later."
            message:nil delegate:nil cancelButtonTitle:nil
            otherButtonTitles:@"OK", nil];
        [okay show];
    }

    // Complete the pending transaction
    [[SKPaymentQueue defaultQueue] finishTransaction: transaction];

    // Restore the GUI
}

- (void)paymentQueue:(SKPaymentQueue *)queue
    updatedTransactions:(NSArray *)transactions
{
    for (SKPaymentTransaction *transaction in transactions)
    {
        switch (transaction.transactionState)
        {
            case SKPaymentTransactionStatePurchased:
            case SKPaymentTransactionStateRestored:
                [self completedPurchaseTransaction:transaction];
                break;
            case SKPaymentTransactionStateFailed:
                [self handleFailedTransaction:transaction];
                break;
```

```
            case SKPaymentTransactionStatePurchasing:
                [self repurchase];
                break;
            default: break;
        }
    }
}
```

Registering Purchases

You can use any of a number of approaches to register purchases. You can synchronize with a web server, create local files, set user defaults (not recommended for serious business work but fine for learning how IAPs work), or add keychain entries. The solution you choose is left up to you. Just don't lose track of purchases. After a user buys an unlockable feature, subscription, or data, you must guarantee that your application supplies the promised element or elements.

Security and Piracy

As a rule, you'll want to store purchases in the keychain. Using the keychain provides the additional benefit that the data stored here can survive an application being deleted and then later reinstalled.

When you use an offsite server to register and authenticate purchases, make sure to echo those settings on the device. Users must use their applications regardless of whether they have network access. A local setting (for example, "Service enabled until 6 June 2011") enables the application to run and provides proper feedback, even when a subscribed service is inaccessible.

Never assume the user is tied to any specific device. You can track devices across installed applications using `identifierForVendor`, available in iOS 6 and later.

Restoring Purchases

Purchases may be restored on a device where an application was uninstalled and then reinstalled, or where an application was installed on a second device associated with the same iTunes/Apple ID account. If a customer's account is used on multiple devices, like a family with a total of five iPhones, iPads, and iPods, a purchase by any of the devices with those account credentials enables all the devices to download that purchase with no additional charge.

StoreKit enables you to restore purchases, which is particularly important for consumable and subscription items where you do not want to allow the user to repurchase an already valid item. For a nonconsumable item, the user can repurchase without cost ad infinitum. For these nonconsumable items, you can simply submit your purchase request. The App Store interface presents a window informing the user that he has already purchased this item, and that he can download it again for free.

To restore purchases associated with an iTunes account, call `restoreCompletedTransactions`. This works just like adding a payment and involves the same callbacks. To catch a repurchase separately from a purchase, check for `SKPaymentTransactionStateRestored` as the payment transaction state (refer to Listing 12-2):

```
- (void) repurchase
{
    // Repurchase an already purchased item
    [[SKPaymentQueue defaultQueue] restoreCompletedTransactions];
}
```

That's because purchase events provide not one, but two possible successful outcomes. The first is a completed purchase. The user has bought the item and the payment has finished processing. The second is the restored purchase described here. Make sure that your payment queue handler looks for both states.

There's a loophole here. Consider providing a consumable purchase item such as a credit to send a fax. Should the user uninstall the application and then reinstall, any repurchase functionality may restore an asset that has already been used. Applications with consumable products must be designed with more thought for the security infrastructure and demand server-side accounting that keep track of user credits and consumed assets.

Go ahead and restore purchases, but ensure that those purchases properly coordinate with your server database. As you'll read about in the following section, Apple provides a unique identifier for each purchase with a purchase receipt. A repurchased item retains that original identifier, enabling you to distinguish between new purchases and restored ones.

Purchasing Multiple Items

Users can purchase more than one copy of consumable items and subscriptions. Set the `quantity` property for a payment to request a multiple purchase. This snippet adds a payment request for three copies of a product, perhaps adding 3 months to a subscription, 3,000 hit points to a character, or so forth:

```
SKPayment *payment = [SKPayment paymentWithProduct:product];
payment.quantity = 3;
[[SKPaymentQueue defaultQueue] addPayment:payment];
```

Handling Delays in Registering Purchases

If your purchase connects with a server and you cannot complete the purchase registration process, *do not finalize the transaction*. Do not call `finishTransaction:` until you are guaranteed that all establishment work has been done for your customer.

Should you fail to set up your users with their newly purchased items before the application terminates, that's okay. The transaction remains in the purchase queue until the next time the application launches. You are given another opportunity to try to finish your work.

Validating Receipts

A successful purchase transaction contains a receipt. This receipt, which is sent in raw `NSData` format, corresponds to an encoded JSON string. It contains a signature and purchase information.

Apple strongly recommends that you validate all receipts with its servers to prevent hacking and ensure that your customers actually purchased the items they are requesting. Listing 12-3 shows how.

You must POST a request to one of Apple's two servers. The URL you use depends on the deployment of the application. Use buy.itunes.apple.com for production software and sandbox.itunes.apple.com for development.

The request body consists of a JSON dictionary. The dictionary is composed of one key (`"receipt-data"`) and one value (a Base64-encoded version of the transaction receipt data). I normally use the CocoaDev NSData Base 64 extension (from www.cocoadev.com/index.pl?BaseSixtyFour) to convert `NSData` objects into Base64-encoded strings. CocoaDev provides many great resources for Mac and iOS developers.

A valid receipt returns a JSON dictionary similar to the following. The receipt includes the transaction identifier, a product ID for the item purchased, the bundle ID for the host application, and a purchase date. Most important, it returns a status:

```
{"receipt":
    {
        "item_id":"467440745",
        "original_transaction_id":"1000000008472082",
        "bvrs":"1.0",
        "product_id":"com.sadun.moo.baa",
        "purchase_date":"2011-09-23 15:18:22 Etc/GMT",
        "quantity":"1", "bid":"com.sadun.Moo",
        "original_purchase_date":"2011-09-22 20:48:46 Etc/GMT",
        "transaction_id":"1000000008535652"},
 "status":0}
```

A valid receipt always has a 0 status. Any number other than 0 indicates that the receipt is invalid.

Simply checking for the status may not be sufficient for validation. As a rule, you should perform this check from your servers and not from the device. It's not too difficult to set up a proxy server to intercept calls to the validation server and return JSON `{"status":0}` to all requests. What's more, the receipt data sent along with the validation request can be easily deserialized.

Third-party services such as Urban Airship and Beeblex offer safe receipt verification for IAPs. Look for both time-limited tokens and strong encryption. Encryption prevents man-in-the-middle attacks; time-limited tokens prevent replay attacks. Together they make it much less likely that a simple proxy could successfully spoof an IAP receipt and fool your app into providing something for nothing.

Listing 12-3 **Checking a Receipt**

```
- (void) checkReceipt: (SKPaymentTransaction *) transaction
{
    // Retrieve the receipt data and encode it with Base 64
    NSString *receiptData =
        [transaction.transactionReceipt base64Encoding];

    // Construct a dictionary for the receipt-data
    NSDictionary *dictionary =
        [NSDictionary dictionaryWithObject:receiptData
            forKey:@"receipt-data"];

    // Translate to JSON data
    NSData *json = [NSJSONSerialization
        dataWithJSONObject:dictionary options:0 error:nil];
    if (!json)
    {
        NSLog(@"Error creating JSON receipt representation");
        return;
    }

    // Select target
    NSString *urlsting = SANDBOX ?
        @"https://sandbox.itunes.apple.com/verifyReceipt" :
        @"https://buy.itunes.apple.com/verifyReceipt";

    // Create the request
    NSMutableURLRequest *urlRequest = [NSMutableURLRequest
        requestWithURL:[NSURL URLWithString: urlsting]];
    if (!urlRequest)
    {
        NSLog(@"Error creating the URL request");
        return;
    }
    [urlRequest setHTTPMethod: @"POST"];
    [urlRequest setHTTPBody:json];

    // Post the request and retrieve the result
    NSError *error;
    NSURLResponse *response;
    NSData *result = [NSURLConnection
        sendSynchronousRequest:urlRequest
        returningResponse:&response error:&error];
```

```
    // Check any errors here, look for status information, store
    // transaction data, etc.

    // Just demonstrate by showing the result data
    NSString *resultString = [[NSString alloc]
        initWithData:result encoding:NSUTF8StringEncoding];
    NSLog(@"Receipt Validation: %@", resultString);
}
```

Summary

The StoreKit framework offers a great way to monetize your applications. As you read in this chapter, you can set up your own storefront to sell services and features from your application. Here are a few final thoughts:

- Although the entire setup and testing process for new applications may seem a little "Which came first? The chicken or the egg?," it is demonstrably possible to develop and deploy a StoreKit-based application with a minimum of headaches. It becomes easier the next time you add in-app purchases for an already approved application.

- Always remember that there are two submission steps with IAP. Ensure that both the application and the IAP items are ready for Apple to review.

- Avoid finalizing transactions until your new-purchase setup is completely, utterly, 100% done, even if that means waiting for an application relaunch. At the same time, inform the user that the purchase process is experiencing unexpected delays.

- Your methods can request only product information from in-app items that are registered to the currently running application. You cannot share requests across apps.

- Whenever possible, use off-device receipt validation. Numerous incidents in the past year have shown how easy it is to bypass security.

- Developers use IAP too much, and often without regard for the user experience. Requiring IAP in apps for kids is, in my opinion, evil—it should be strongly discouraged by Apple policy. Further, if your app requires IAP to bypass gameplay segments, your game design needs some serious reconsideration.

- Focus on providing good experiences for your paying users instead of fighting piracy. If your antipiracy protections tick off even one paying customer, you have lost the war.

- Don't forget to set up the purchase observer. More heads have been banged against desks and hair pulled out over that one step than any other StoreKit issue.

13

Push Notifications

When off-device services need to communicate directly with users, push notifications provide a solution. Just as local notifications allow apps to contact users at scheduled times, push notifications deliver messages from web-based systems. Push notifications let devices display an alert, play a custom sound, or update an application badge. In this way, off-device services connect with an iOS-based client, letting them know about new data or updates.

Unlike most other iOS development arenas, nearly all the push story takes place off the device. Developers create web-based services to manage and deploy information updates. Push notifications transmit those updates to specific devices. In this chapter, you learn how push notifications work and dive into the details needed to create your own push-based system.

Introducing Push Notifications

Push notifications, also called *remote notifications*, refer to a kind of message sent to iOS devices by an outside service. These push-based services work with any kind of application that normally checks for information updates. For example, a service might poll for new updates on Facebook, Twitter, or Google Plus; scan for nearby dangerous weather systems; respond to sensors for in-home security systems; or invite you to a shared conference. When new information becomes available for a client, the service pushes that update through Apple's remote notification system. The notification transmits directly to the device, which has registered to receive those updates.

The key to push is that these messages originate from outside the device. They are part of a client-server paradigm that enables web-based server components to communicate with iOS clients through an Apple-supplied service. With push, developers can send nearly instant updates to iOS devices that don't rely on users launching a particular application. Instead, processing occurs on the server side of things. When push messages arrive, the iOS client can respond by displaying a badge, playing a sound, and/or showing an alert box.

Moving application logic to a server limits client-side complexity, allowing many clients to share a single update system. Consider the weather danger service just mentioned. A dedicated site can monitor NOAA and other atmospheric feeds, sending remote notifications to each subscribed client as needed. Offsite processing also provides energy savings for iOS-based applications. They can now rely on push rather than using the iOS device's local CPU resources to monitor and react to important information changes.

Push's reason for being is not only tied into local resources, but it also offers a valuable solution for communicating with web-based services that goes beyond poll-and-update applications. For example, push might enable you to hook into a coaching service that transmits helpful affirmations even when an application isn't running or to a gaming service that sends you reminder notices about an upcoming tournament.

From social networking to monitoring RSS feeds, push enables iOS users to keep on top of asynchronous data feeds. It offers a powerful solution for connecting iOS clients to web-based systems of all kinds. With push, the services you write can connect to your installed iOS base and communicate updates in a clean, functional manner.

How Push Works

Push notifications are tied to specific applications and require several security checks. A push provider (the server component that generates push requests and transmits them to Apple) can communicate only with those iOS devices that host its application, that are online, and that have opted to receive remote messages. Users have the ultimate say in push updates. They can allow or disallow that kind of communication, and a well-written application lets users opt-in and opt-out of the service at will.

The chain of communication between server and client works like this. Push providers deliver message requests through a central Apple server and via that server to their iOS clients. In normal use, the server triggers on some event (like new mail or an upcoming appointment) and generates notification data aimed at a specific iOS device. It sends this message request to the Apple Push Notification Service (APNS). This notification uses JSON formatting and is limited to 256 bytes each, so the information that can be pushed through on that message is quite limited. This formatting and size ensures that APNS limits bandwidth to the tightest possible configuration.

APNS offers a centralized system that negotiates communication with iOS devices in the real world. It passes the message through to the designated device. A handler on the iOS device decides how to process the message. As Figure 13-1 shows, push providers talk to APNS, sending their message requests, and APNS talks to iOS devices, relaying those messages to handlers on the unit.

Multiple Provider Support

APNS was built to support multiple provider connections, enabling many services to communicate with it at once. It offers multiple gateways into the service so that each push service does not have to wait for availability before sending its message. Figure 13-2 illustrates the many-to-many relationship between providers and iOS devices. APNS enables providers to connect at once through multiple gateways. Each provider can push messages to many different devices.

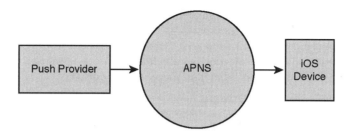

Figure 13-1 Providers send messages through Apple's centralized push notification service to communicate with an iOS device.

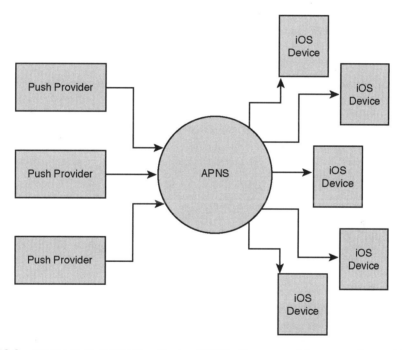

Figure 13-2 Apple's Push Notification Service (APNS) offers many gateways on its provider-facing side, enabling multiple providers to connect in parallel. Each push provider may connect to any number of iOS devices.

Security

Security is a primary component of remote notifications. The push provider must sign up for a secure sockets layer (SSL) certificate for each application it works with. Services cannot communicate with APNS unless they authenticate themselves with this certificate. They must also provide a unique key called a *token* that identifies a specific application on a single device.

After receiving an authenticated message and device token, APNS contacts the device in question. Each iOS device must be online in some way to receive a notification. They can be connected to a cellular data network or to a Wi-Fi hotspot. APNS establishes a connection with the device and relays the notification request. If the device is offline and the APNS server cannot make a connection, the notification is queued for later delivery.

Upon receiving the request, iOS performs a number of checks. Push requests are ignored when the user disables push updates for a given application; users can do so in the Settings application on their device. When updates are allowed, and only then, the iOS device determines whether the client application is currently running. If so, it sends a message directly to the running application via the application delegate. If not, it performs some kind of alert, whether displaying text, playing a sound, or updating a badge.

When an alert displays, users typically have the option to close the alert or tap View. If they choose View, the iOS device launches the application in question and sends it the notification message that it would have received while running. If the user taps Close, the notification gets ignored and the application does not launch.

This pathway, from server to APNS to device to application, forms the core flow of push notifications. Each stage moves the message along the way. Although the multiple steps may sound extensive, in real life, the notification arrives almost instantaneously. After you set up your certificates, identifiers, and connections, the actual delivery of information becomes trivial. Nearly all the work lies in first setting up that chain and then in producing the information you want to deliver.

Make sure you treat all application certificates and device tokens as sensitive information. When storing these items on your server, you must ensure that they are not generally accessible. Should this information hit the wild, it could be exploited by third parties. This would likely result in Apple revoking your SSL push certificate. This would disable all remote notifications for any apps you have sold and might force you to pull the application from the store.

Push Limitations

In real-world scenarios, push notifications are not as reliable as you might want. They can be fairly flaky. Apple does not guarantee the delivery of each notification or the order in which notifications arrive. Never send vital information by push. Reserve this feature for helpful notifications that update and inform the user, but that the user can miss without serious consequence.

Items in the push delivery queue may be displaced by new notifications. That means that notifications may compete and may get lost along the way. Although Apple's feedback service reports failed deliveries (that is, messages that cannot be properly sent through the push service, specifically to applications that have been removed from a device), you cannot retrieve information regarding bumped notifications. From the APN service point of view, a lost message was still successfully "delivered."

Push Notifications Versus Local Notifications

iOS provides both remote push notification and local notification support. Local notifications, which are discussed in the *Core iOS Developer's Cookbook*, are created and scheduled on-device. Push notification originates from a remote server and are sent to the device through APNS. This chapter exclusively covers remote notifications.

Provisioning Push

To start push development, you must visit Apple's iOS developer Provisioning Portal. This portal is located at http://developer.apple.com/ios/manage/overview/index.action. Sign in with your iOS developer credentials to gain access to the site. Here at the portal, you can work through the steps needed to create a new application identifier that can be associated with a push service.

The following sections walk you through the process. You see how to create a new identifier, generate a certificate, and request a special provisioning profile so that you can build push-enabled applications. Without a push-enabled profile, your application will not receive remote notifications.

Generate a New Application Identifier

At the iOS provisioning portal, click App IDs. You can find this option in the column on the left side of the web page. This opens a page that enables you to create new application identifiers. Each push service is based on a single identifier, which you must create and then set to allow remote notification. You cannot use a wildcard identifier with push applications; every push-enabled app demands a unique identifier.

In the App IDs section, click New App ID; this button appears at the top right of the web page. When clicked, the site opens a new Create App ID page, similar to Figure 13-3. Enter a name that describes your new identifier, such as Tutorial Push Application and a new bundle identifier.

These IDs typically use reverse domain patterns, such as com.*domainname.appname* and com.sadun.pushtutorial. The identifier must be unique and may not conflict with any other registered application identifier in Apple's system. Leave the bundle seed set to your team identifier unless you have some compelling reason to do otherwise.

Click Submit to add the new identifier. This adds the app ID irrevocably to Apple's system, where it is now registered to you. You return to the App ID page with its list of identifiers and are now ready to establish that identifier as push-compliant.

> **Note**
>
> Apple does not provide any way to remove an application identifier from the program portal after it has been created.

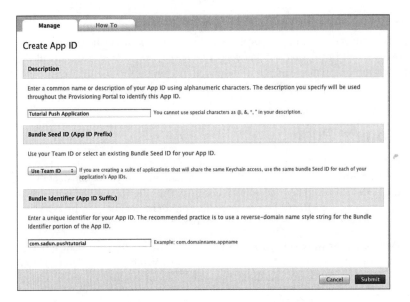

Figure 13-3   Create a new application identifier at the iOS provisioning portal.

Generate Your SSL Certificate

On the App ID page, you can see which identifiers work with push and which do not. The Apple Push Notification column shows whether push has been enabled for each app ID. The three states for this column are the following:

- Unavailable (gray) for IDs that are no longer available
- Configurable (yellow) for apps that can be used with push but that haven't yet been set up to do so
- Enabled (green) for apps that are ready for push

In the push column, you can find two dots for each application identifier: one for Development and another for Production. These options are configured separately. Locate your new app ID, make sure the yellow Configurable for Development is shown, and click Configure. This option appears in the rightmost column. When clicked, the browser opens a new Configure App ID page that permits you to associate your identifier with the push notification service.

An Enable Push Notification Services check box appears about halfway down the page. Check this box to start the certificate creation process. When checked, the two Configure buttons on the right side of the page become enabled. Click a button. A page of instructions loads, showing you how to proceed. It guides you through creating a secure certificate that will be used by your server to sign messages it sends to the APNS.

As instructed, launch the Keychain Access application. This application is located on your Macintosh in the /Applications/Utilities folder. When launched, choose Keychain Access > Certificate Assistant > Request a Certificate From a Certificate Authority. You need to perform this step again even if you've already created previous requests for your developer and distribution certificates. The new request adds information that uniquely identifies the SSL certificate.

After the Certificate Assistant opens, enter your e-mail address and add a recognizable common name, such as Push Tutorial App. This common name is important. It will come in handy for the future, so choose one that is easy to identify and accurately describes your project. The common name lets you distinguish otherwise similar looking keychain items from each other in the OS X Keychain Access utility.

After specifying a common name, choose Saved to Disk and click *Continue*. The Certificate Assistant prompts you to choose a location to save to (the Desktop is handy). Click Save, wait for the certificate to be generated, and then click *Done*. Return to your Web browser and click Continue. You are now ready to submit the certificate-signing request.

Click Choose File and navigate to the request you just generated. Select it and click Choose. Click Generate to build your new SSL push service certificate. This can take a minute or two, so be patient, and do not close the Web page. When the certificate has been generated, click Continue. Retrieve the new certificate by clicking Download. Finally, click Done. You return to the App ID page where a new, green Enabled indicator should appear next to your app ID. Apple also e-mails you a confirmation that your certificate request was approved.

> **Note**
> Should you ever need to download the public key portion of your SSL certificate again, click Configure to return to the Configure App ID page. There, you can click Download to request another copy. As a rule, you should export the private+public portions of the certificate into a p12 file for safekeeping. If you lose the private key, you'll have to regenerate the entire certificate.

If you plan to run your Push Server from your Macintosh (handy for development before deploying a server, or before the server piece is fully developed), add the new item to your keychain. It will appear in your Certificates. You can identify the certificate by clicking the small triangle next to it to reveal the common name you used when creating the certificate request.

Push-Specific Provisions

You can't use wildcard team provisioning profiles for push-enabled applications. Instead, you must create a single profile for just that application. This means that if you intend to create development, ad hoc, and distribution versions of your app, you must request three new mobile provisioning profiles in addition to whatever provisioning profiles you have already created for other work.

In the latest versions of Xcode, you can build your profile from the Organizer (Command-Shift-2 > Devices > Provisioning Profiles > New). Enter a profile name, choose the already-configured App ID, and select the devices you want to work with. There's a Select All button to make this easier. Click Next and let Xcode do the rest of the work for you.

Alternatively, go to the Provisioning section of the developer portal. Select Development or Distribution profile by clicking the appropriate tab. Click New Profile to begin creating your new provision. A Create Provisioning Profile page opens:

- **Development Provision**—For development, enter a profile name, such as Push Tutorial Development. Check the certificate you will be using and choose your application identifier from the pop-up list. Select the devices you will be using and click Submit.

- **Distribution Provision**—For distribution, select App Store or Ad Hoc. Enter a name for your new provision such as Push Tutorial Distribution or Push Tutorial Ad Hoc. Choose your application identifier from the pop-up list. For ad-hoc distribution only, select the devices to include in your provision. Click Submit to finish.

It may take a moment or two for your profile to generate. Wait a short while and reload the page. The provision status should change from Pending to Active. Download your new provision and add it to Xcode through the Organizer > Devices > Provisioning Profiles. Click Import, select the provision from your downloads folder, and click Open.

Creating a Push-Compatible Application

After installing your push-enabled provisioning profile into Xcode, you are ready to create a push-compatible application. Start by ensuring that your new bundle identifier matches the provision you just created in TARGETS > Info > Bundle identifier. Then, in Build Settings > Code Signing Identity, choose the developer identity for the new provisioning profile. You use separate identities and profiles for debug and release builds.

Registering Your Application

Signing an application with a push-compatible mobile provisioning profile is just the first step to working with push notifications. The application must request to register itself with the iOS device's remote notification system. You do this with a single UIApplication call, as follows. The application: didFinishLaunchingWithOptions: delegate method provides a particularly convenient place to call this:

```
[[UIApplication sharedApplication]
    registerForRemoteNotificationTypes:types];
```

This call tells iOS that your application wants to accept push messages. The types you pass specify what kinds of alerts your application will receive. iOS offers three types of notifications:

- **UIRemoteNotificationTypeBadge**—This notification adds a red badge to your application icon on SpringBoard.
- **UIRemoteNotificationTypeSound**—Sound notifications let you play sound files from your application bundle.
- **UIRemoteNotificationTypeAlert**—This style displays a text alert box in SpringBoard or any other application with a custom message using the alert notification.

Choose the types you want to use and or them together. They are bit flags, which combine to tell the notification registration process how you want to proceed. For example, the following flags allow alerts and badges, but not sounds:

types = UIRemoteNotificationTypeBadge | UIRemoteNotificationTypeAlert;

Performing the registration updates user settings and allows users to customize their push preferences from Settings > Notifications. There, users can choose how to present the notification alert style (banner, alert, and none), which types to allow (badge, alert, and sound), whether to show the alert on the lock screen, and so forth.

To remove your application from active participation in push notifications, send unregister-ForRemoteNotifications. This unregisters your application for all notification types and does not take any arguments:

[[UIApplication sharedApplication] unregisterForRemoteNotifications];

Retrieving the Device Token

Your application cannot receive push messages until it generates and delivers a device token to your server. It must send that device token to the offsite service that pushes the actual notifications. Recipe 13-1, which follows this section, does not implement server functionality. It provides only the client software.

A token is tied to one device. In combination with the SSL certificate, it uniquely identifies the iOS device and can be used to send messages back to the device in question. Be aware that device tokens can change after you restore device firmware.

Device tokens are created as a byproduct of registration. Upon receiving a registration request, iOS contacts the Apple Push Notification Service. It uses a SSL request. Somewhat obviously, the unit must be connected to the Internet. If it is not, the request will fail. With a live connection, iOS forwards the request to APNS and waits for it to respond with a device token.

APNS builds the device token and returns it to iOS, which in turn passes it back to the application via an application delegate callback, namely `application:didRegisterForRemoteNotificationsWithDeviceToken:`. Your application must retrieve the device token and pass it to the provider component of your service, where it needs to be stored securely. Anyone who gains access to a device token and the application's push credentials could spam messages to devices. You must treat this information as sensitive and protect it accordingly.

> **Note**
>
> At times, the token may take time to generate. Consider designing around possible delays into your application by registering at each application run. Until the token is created and uploaded to your site, you cannot provide remote notifications to your users.

Handling Token Request Errors

At times, APNS cannot create a token or your device may not send a request. For example, you cannot generate tokens from the simulator. A `UIApplicationDelegate` method `application: didFailToRegisterForRemoteNotifications-WithError:` lets you handle these token request errors. For the most part, you need to retrieve the error and display it to the user:

```
// Provide a user explanation for when the registration fails
- (void)application:(UIApplication *)application
    didFailToRegisterForRemoteNotificationsWithError:(NSError *)error
{
    NSString *status = [NSString stringWithFormat:
        @"%@\nRegistration failed.\n\nError: %@", pushStatus(),
        error.localizedFailureReason];
    tbvc.textView.text = status;
}
```

Responding to Notifications

iOS uses a set chain of operations (see Figure 13-4) to respond to push notifications. When an application runs, the notification is sent directly to a `UIApplicationDelegate` method, `application:didReceiveRemoteNotification:`. The payload, which is sent in JSON format, is converted automatically into an `NSDictionary`, and the application is free to use the information in that payload however it wants. As the application is already running, no further sounds, badges, or alerts are invoked.

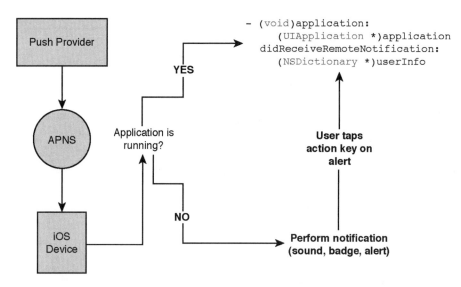

Figure 13-4 Visible and audible notification are presented only when the application is not running. Should the user tap on the alert in Notification Center, the application launches, and the payload is sent as a notification to the application delegate.

```
// Handle an actual notification
- (void)application:(UIApplication *)application
    didReceiveRemoteNotification:(NSDictionary *)userInfo
{
    NSString *status = [NSString stringWithFormat:
        @"Notification received:\n%@",userInfo.description];
    tbvc.textView.text = status;
    NSLog(@"%@", userInfo);
}
```

When an application is not running, iOS performs all requested notifications that are allowed by registration and by user settings. These notifications may include playing a sound, badging the application, and displaying an alert. Playing a sound can also trigger iPhone vibration when a notification is received.

By default, notifications are added to the Notification Center (see Figure 13-5). (Users can disable this via Settings > Notification > Application Name > Notification Center.) When the user taps the notification in the list, the application launches with the notification as a launch option.

Figure 13-5 Remote alerts can appear on the Lock Screen, in SpringBoard, in applications, and in the Notification Center. Users may slide the alert on the lockscreen or tap it in Notification Center to switch to the notifying application. In this case, that application is HelloWorld, whose name is clearly shown on the alert.

Upon launch, the application delegate receives the same remote notification callback that an already running application would have seen. Alerts appear on the lock screen when the device is locked and in the Notification Center, unless the user overrides the default settings.

Recipe: Push Client Skeleton

Recipe 13-1 introduces a basic client that allows users to register and unregister for push notifications. The interface (shown in Figure 13-6) uses three switches that control the services to be registered. When the application launches, it queries the app's enabled remote notification types and updates the switches to match. Thereafter, the client keeps track of registrations and unregistrations, adjusting the switches to keep sync with the reality of the settings.

Two buttons at the top left and right of the interface let users unregister and register their application. Unregistering disables all services associated with the app. It provides a clean sweep. In contrast, registering apps requires flags to indicate which services are requested.

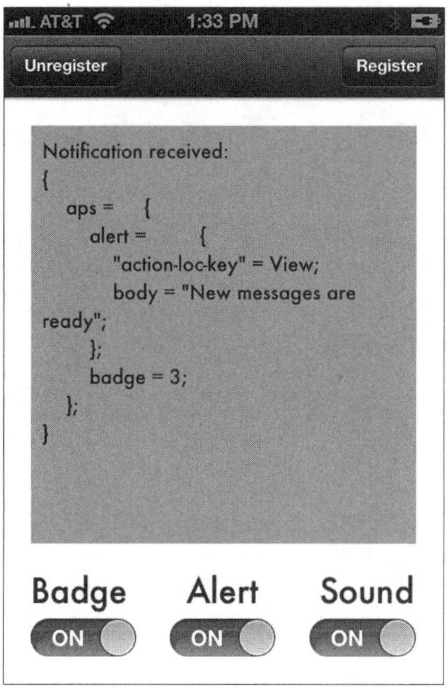

Figure 13-6   The Push Client skeleton introduced in Recipe 13-1 lets users specify which services they want to register.

When requesting new services, the user is always prompted to approve. Figure 13-7 shows the dialog that appears. The user must confirm by explicitly granting the application permission. If the user does not, by tapping Don't Allow, the flags remain at their previous settings.

Unfortunately, the `confirmation` dialog does not generate a callback when it is dismissed, regardless of whether the user agreed. To catch this event, you can listen for a general notification (`UIApplicationDidBecomeActiveNotification`) that gets generated when the dialog returns control to the application. It's a hack, but it lets you know when the user responded and how the user responded. In Recipe 13-1, the `confirmationWasHidden:` method catches this notification and updates the switches to match any new registration settings.

Being something of a skeletal system, this push client doesn't actually respond to push notifications beyond showing the contents of the user info payload that gets delivered. Figure 13-8 illustrates the actual payload that was sent in Figure 13-6. This display is performed in the `application:didReceiveRemoteNotification:` method in the application delegate.

> **Note**
>
> The three sound files included in the online sample project (ping1.caf, ping2.caf, and ping3.caf) let you test sound notifications with real audio.

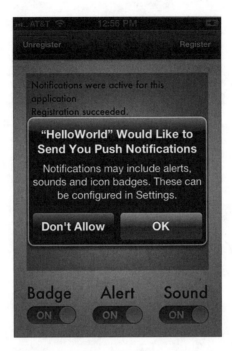

Figure 13-7 Users must explicitly grant permission for an application to receive remote notifications.

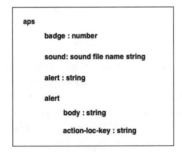

Figure 13-8 The `aps` dictionary may include badge requests, sound file names, and/or an alert string. The action-loc-key is less important now due to the Notification Center. If the user disables the Notification Center (this is rare), they will see the old-style alert, which displays the action button.

Recipe 13-1 **Push Client Skeleton**

```
// Important: Set up your own app id and provisioning profile
// before attempting to play with this code

@implementation TestBedViewController
@synthesize textView;

// Basic status
NSString *pushStatus ()
{
    return [[UIApplication sharedApplication]
        enabledRemoteNotificationTypes] ?
        @"Notifications were active for this application" :
        @"Remote notifications were not active for this application";
}

// Fetch the current switch settings
- (NSUInteger) switchSettings
{
    NSUInteger settings = 0;
    if (badgeSwitch.isOn) settings =
        settings | UIRemoteNotificationTypeBadge;
    if (alertSwitch.isOn) settings =
        settings | UIRemoteNotificationTypeAlert;
    if (soundSwitch.isOn) settings =
        settings | UIRemoteNotificationTypeSound;
    return settings;
}

// Change the switches to match reality
- (void) updateSwitches
{
    NSUInteger rntypes = [[UIApplication sharedApplication]
        enabledRemoteNotificationTypes];
    badgeSwitch.on = (rntypes & UIRemoteNotificationTypeBadge);
    alertSwitch.on = (rntypes & UIRemoteNotificationTypeAlert);
    soundSwitch.on = (rntypes & UIRemoteNotificationTypeSound);
}

// Register application for the services set out by the switches
- (void) registerServices
{
    if (![self switchSettings])
    {
        textView.text = [NSString stringWithFormat:
            @"%@\nNothing to register. Skipping.", pushStatus()];
        [self updateSwitches];
```

```objc
        return;
    }
    
    NSString *status = [NSString stringWithFormat:
        @"%@\nAttempting registration", pushStatus()];
    textView.text = status;
    [[UIApplication sharedApplication]
        registerForRemoteNotificationTypes:[self switchSettings]];
}

// Unregister application for all push notifications
- (void) unregisterServices
{
    NSString *status = [NSString stringWithFormat:@"%@\nUnregistering.",
        pushStatus()];
    textView.text = status;
    
    [[UIApplication sharedApplication] unregisterForRemoteNotifications];
    [self updateSwitches];
}

- (void) loadView
{
    self.view = [[UIView alloc] init];
    
    // Workaround for catching the end of user choice
    TestBedViewController __weak *weakself = self;
    [[NSNotificationCenter defaultCenter]
        addObserverForName:UIApplicationDidBecomeActiveNotification
        object:nil queue:[NSOperationQueue mainQueue]
        usingBlock:^(NSNotification *notification)
    {
        [[UIApplication sharedApplication]
            registerForRemoteNotificationTypes:
                [weakself switchSettings]];
         [weakself updateSwitches];
     }];
    
    self.navigationItem.rightBarButtonItem =
        BARBUTTON(@"Register", @selector(registerServices));
    self.navigationItem.leftBarButtonItem =
        BARBUTTON(@"Unregister", @selector(unregisterServices));
    [self updateSwitches];
}
@end

#pragma mark - Application Delegate
```

```objc
@implementation TestBedAppDelegate
// Retrieve the device token
- (void)application:(UIApplication *)application
    didRegisterForRemoteNotificationsWithDeviceToken:(NSData *)deviceToken
{
    NSUInteger rntypes = [[UIApplication sharedApplication]
        enabledRemoteNotificationTypes];
    NSString *results = [NSString stringWithFormat:
        @"Badge: %@, Alert:%@, Sound: %@",
        (rntypes & UIRemoteNotificationTypeBadge) ? @"Yes" : @"No",
        (rntypes & UIRemoteNotificationTypeAlert) ? @"Yes" : @"No",
        (rntypes & UIRemoteNotificationTypeSound) ? @"Yes" : @"No"];

    // Show the retrieved Device Token
    NSString *status = [NSString stringWithFormat:
        @"%@\nRegistration succeeded.\n\nDevice Token: %@\n%@",
        pushStatus(), deviceToken, results];
    tbvc.textView.text = status;
    NSLog(@"deviceToken: %@", deviceToken);

    // Write token to file for easy retrieval in iTunes.
    // Handy for learning push dev - but not a great idea for
    // App Store deployment. Set UIFileSharingEnabled to YES
    [deviceToken.description writeToFile:[NSHomeDirectory()
        stringByAppendingPathComponent:@"Documents/DeviceToken.txt"]
        atomically:YES encoding:NSUTF8StringEncoding error:nil];
}

// Respond to failed registration
- (void)application:(UIApplication *)application
    didFailToRegisterForRemoteNotificationsWithError:(NSError *)error
{
    NSLog(@"Error registering for remote notifications: %@",
        error.localizedFailureReason);
    NSString *status = [NSString stringWithFormat:
        @"%@\nRegistration failed.\n\nError: %@",
        pushStatus(), error.localizedFailureReason];
    tbvc.textView.text = status;
}

// Handle an actual notification
- (void)application:(UIApplication *)application
    didReceiveRemoteNotification:(NSDictionary *)userInfo
{
    NSString *status = [NSString stringWithFormat:
        @"Notification received:\n%@", userInfo.description];
    tbvc.textView.text = status;
```

```objc
    NSLog(@"%@", userInfo);
}

// Little work-around for showing text
- (void) showString: (NSString *) aString
{
    tbvc.textView.text = aString;
}

// Report the notification payload when launched by alert
- (void) launchNotification: (NSNotification *) notification
{
    // Workaround allows the text view to be created if needed first
    [self performSelector:@selector(showString:)
        withObject:[[notification userInfo] description]
        afterDelay:1.0f];
}

- (BOOL)application:(UIApplication *)application
    didFinishLaunchingWithOptions:(NSDictionary *)launchOptions
{
    window = [[UIWindow alloc] initWithFrame:[[UIScreen mainScreen] bounds]];
    tbvc = [[TestBedViewController alloc] init];
    UINavigationController *nav =
        [[UINavigationController alloc] initWithRootViewController:tbvc];
    window.rootViewController = nav;
    [window makeKeyAndVisible];

    // Listen for remote notification launches
    [[NSNotificationCenter defaultCenter]
        addObserver:self selector:@selector(launchNotification:)
        name:@"UIApplicationDidFinishLaunchingNotification" object:nil];

    NSLog(@"Launch options: %@", launchOptions);
    return YES;
}
@end
```

Get This Recipe's Code

To find this recipe's full sample project, point your browser to https://github.com/erica/iOS-6-Advanced-Cookbook and go to the folder for Chapter 13.

Building Notification Payloads

Delivering push notification through APNS requires three things: your SSL certificate, a device ID, and a custom payload with the notification you want to send. The payload uses JSON formatting. You've already read about generating the certificate and producing the device identifiers, which you need to pass up to your server. Building the JSON payloads basically involves transforming a small well-defined dictionary into JSON format.

JavaScript Object Notation (JSON) is a simple data interchange format based on key-value pairs. The JSON Web site (www.json.org) offers a full syntax breakdown of the format, which allows you to represent values that are strings, numbers, and arrays. The APNS payload consists of up to 256 bytes, which must contain your complete notification information.

Notification payloads must include an `aps` dictionary. This dictionary defines the properties that produce the sound, badge, and/or alert sent to the user. In addition, you can add custom dictionaries with any data you need to send to your application, so long as you stay within the 256-byte limit. Figure 13-8 shows the hierarchy for basic (non-localized) alerts.

The `aps` dictionary contains one or more notification types. These include the standard types you've already read about: badges, sounds, and alerts. Badge and sound notifications each take one argument. The badge is set by a number, and the sound is set by a string that refers to a file already inside the application bundle. If that file is not found (or the developer passes `default` as the argument), a default sound plays for any notification with a sound request. When a badge request is not included, iOS removes any existing badge from the application icon.

Localized Alerts

When working with localized applications, construct your `aps > alert` dictionary with two additional keys. Use `loc-key` to pass a key that is defined in your application's Localizable.strings file. iOS looks up the key and replaces it with the string found for the current localization.

At times, localization strings use arguments like `%@` and `%n$@`. Should that hold true for the localization you are using, you can pass those arguments as an array of strings via `loc-args`. As a rule, Apple recommends against using complicated localizations because they can consume a major portion of your 256-byte bandwidth.

Transforming from Dictionary to JSON

After you design your dictionary, you must transform it to JSON. The JSON format is simple but precise. The `NSJSONSerialization` class creates a JSON string. Here's the basic call:

```
NSData *jsonData = [NSJSONSerialization
    dataWithJSONObject:mainDict options:0 error:nil];
```

Custom Data

So long as your payload has room left, keeping in mind your tight byte budget, you can send additional information in the form of key-value pairs. These custom items can include arrays and dictionaries as well as strings, numbers, and constants. You define how to use and interpret this additional information. The entire payload dictionary is sent to your application, so whatever information you pass along will be available to the `application:didReceiveRemoteNotification:` method via the user dictionary.

A dictionary containing custom key-value pairs does not need to present an alert for end-user interaction; doing so allows your user to choose to open your application if it isn't running. If your application is already launched, the key-value pairs arrive as a part of the payload dictionary.

Receiving Data on Launch

When your client receives a notification, tapping on it launches your application. Then, after launching, iOS sends your application delegate an optional callback. The delegate recovers its notification dictionary by implementing a method named `application:didFinishLaunchingWithOptions:`.

iOS passes the notification dictionary to the delegate method via the launch options parameter. For remote notifications, this is the official callback to retrieve data from an alert-box launch. The `didReceiveRemoteNotification:` method is not called when iOS receives a notification and the application is not running.

This "finished launching" method is actually designed to handle many different launch circumstances. Push notification is just one. Others include opening by local notifications, by URL scheme handling, and so forth. In any case, the method must return a Boolean value. As a rule, return `YES` if you processed the request or `NO` if you did not. This value is actually ignored in the case of remote notification launches, but you must still return a value.

> **Note**
>
> When your user taps Close and later opens your application, the notification is not sent on launch. You must check in with your server manually to retrieve any new user information. Applications are not guaranteed to receive alerts. In addition to tapping Close, the alert may simply get lost. Always design your application so that it doesn't rely solely on receiving push notifications to update itself and its data.

Recipe: Sending Notifications

The notification process involves several steps (see Figure 13-9). First, you build your JSON payload, which you just read about. Next, you retrieve the SSL certificate and the device token for the unit you want to send to. How you store these is left up to you, but you must remember that these are sensitive pieces of information. Open a secure connection to the APNS server. Finally, you handshake with the server, send the notification package, and close the connection.

This is the most basic way of communicating and assumes you have just one payload to send. In fact, you can establish a session and send many packets at a time; however, that is left as an exercise for the reader as is creating services in languages other than Objective-C. The Apple

Developer Forums (devforums.apple.com) host ongoing discussions about push providers and offer an excellent jumping off point for finding sample code for PHP, Python, Ruby, Perl, and other languages.

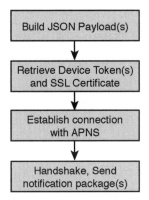

Figure 13-9   *The steps for sending remote notifications.*

Be aware that APNS may react badly to a rapid series of connections that are repeatedly established and torn down. If you have multiple notifications to send at once, go ahead and send them during a single session. Otherwise, APNS might confuse your push deliveries with a denial-of-service attack.

Recipe 13-2 demonstrates how to send a single payload to APNS, showing the steps needed to implement the fourth and final box in Figure 13-9. The recipe is built around code developed by Stefan Hafeneger and uses Apple's ioSock sample source code. It's been cleaned up a bit, since the previous edition of this Cookbook, to deal with deprecated APIs.

The individual server setups vary greatly depending on your security, databases, organization, and programming language. Recipe 13-2 demonstrates a minimum of what is required to implement this functionality and serves as a template for your own server implementation in whatever form this might take.

Sandbox and Production

Apple provides both sandbox (development) and production (distribution) environments for push notification. You must create separate SSL certificates for each. The sandbox helps you develop and test your application before submitting to App Store. It works with a smaller set of servers and is not meant for large-scale testing. The production system is reserved for deployed applications that have been accepted to App Store:

- The sandbox servers are located at gateway.sandbox.push.apple.com, port 2195.
- The production servers are located at gateway.push.apple.com, port 2195.

> **Note**
>
> Recipe 13-2 is meant to be compiled for and used on the Macintosh as a command-line utility.

Recipe 13-2 **Pushing Payloads to the APNS Server**

```
// Adapted from code by Stefan Hafeneger
- (BOOL) push: (NSString *) payload
{
    otSocket socket;
    SSLContextRef context;
    SecKeychainRef keychain;
    SecIdentityRef identity;
    SecCertificateRef certificate;
    OSStatus result;

    // Ensure device token
    if (!deviceTokenID)
    {
        fprintf(stderr, "Error: Device Token is nil\n");
        return NO;
    }

    // Ensure certificate
    if (!certificateData)
    {
        fprintf(stderr, "Error: Certificate Data is nil\n");
        return NO;
    }

    // Establish connection to server.
    PeerSpec peer;
    result = MakeServerConnection("gateway.sandbox.push.apple.com",
        2195, &socket, &peer);
    if (result)
    {
        fprintf(stderr, "Error creating server connection\n");
        return NO;
    }

    // Create new SSL context.
    result = SSLNewContext(false, &context);
    if (result)
    {
        fprintf(stderr, "Error creating SSL context\n");
        return NO;
    }
```

```objc
// Set callback functions for SSL context.
result = SSLSetIOFuncs(context, SocketRead, SocketWrite);
if (result)
{
    fprintf(stderr, "Error setting SSL context callback functions\n");
    return NO;
}

// Set SSL context connection.
result = SSLSetConnection(context, socket);
if (result)
{
    fprintf(stderr, "Error setting the SSL context connection\n");
    return NO;
}

// Set server domain name.
result = SSLSetPeerDomainName(context,
    "gateway.sandbox.push.apple.com", 30);
if (result)
{
    fprintf(stderr, "Error setting the server domain name\n");
    return NO;
}

// Open keychain.
result = SecKeychainCopyDefault(&keychain);
if (result)
{
    fprintf(stderr, "Error accessing keychain\n");
    return NO;
}

// Create certificate from data
CFDataRef data = (__bridge CFDataRef) self.certificateData;
certificate = SecCertificateCreateWithData(
    kCFAllocatorDefault, data);
if (!certificate)
{
    printf("Error creating certificate from data\n");
    return nil;
}

// Create identity.
result = SecIdentityCreateWithCertificate(keychain, certificate,
    &identity);
if (result)
```

```objc
{
    fprintf(stderr, "Error creating identity from certificate\n");
    return NO;
}

// Set client certificate.
CFArrayRef certificates = CFArrayCreate(NULL,
    (const void **)&identity, 1, NULL);
result = SSLSetCertificate(context, certificates);
if (result)
{
    fprintf(stderr, "Error setting the client certificate\n");
    return NO;
}

CFRelease(certificates);

// Perform SSL handshake.
do {result = SSLHandshake(context);}
    while(result == errSSLWouldBlock);


// Convert string into device token data.
NSMutableData *deviceToken = [NSMutableData data];
unsigned value;
NSScanner *scanner = [NSScanner
    scannerWithString:self.deviceTokenID];
while(![scanner isAtEnd]) {
    [scanner scanHexInt:&value];
    value = htonl(value);
    [deviceToken appendBytes:&value length:sizeof(value)];
}

// Create C input variables.
char *deviceTokenBinary = (char *)deviceToken.bytes;
char *payloadBinary = (char *)[payload UTF8String];
size_t payloadLength = strlen(payloadBinary);

// Prepare message
uint8_t command = 0;
char message[293];
char *pointer = message;
uint16_t networkTokenLength = htons(32);
uint16_t networkPayloadLength = htons(payloadLength);

// Compose message.
memcpy(pointer, &command, sizeof(uint8_t));
```

```
    pointer += sizeof(uint8_t);
    memcpy(pointer, &networkTokenLength, sizeof(uint16_t));
    pointer += sizeof(uint16_t);
    memcpy(pointer, deviceTokenBinary, 32);
    pointer += 32;
    memcpy(pointer, &networkPayloadLength, sizeof(uint16_t));
    pointer += sizeof(uint16_t);
    memcpy(pointer, payloadBinary, payloadLength);
    pointer += payloadLength;

    // Send message over SSL.
    size_t processed = 0;
    result = SSLWrite(context, &message, (pointer - message),
        &processed);
    if (result)
    {
        fprintf(stderr, "Error sending message via SSL.\n");
        return NO;
    }
    else
    {
        printf("Message sent.\n");
        return YES;
    }
}
```

Get This Recipe's Code

To find this recipe's full sample project, point your browser to https://github.com/erica/iOS-6-Advanced-Cookbook and go to the folder for Chapter 13.

Feedback Service

Apps don't live forever. Users add, remove, and replace applications on their devices all the time. From an APNS point of view, it's pointless to deliver notifications to devices that no longer host your application. As a push provider, it's your duty to remove inactive device tokens from your active support list. As Apple puts it, "APNS monitors providers for their diligence in checking the feedback service and refraining from sending push notifications to nonexistent applications on devices." Big Brother *is* watching.

Apple provides a simple way to manage inactive device tokens. When users uninstall apps from a device, push notifications begin to fail. Apple tracks these failures and provides reports from its APNS feedback server. The APNS feedback service lists devices that failed to receive notifications. As a provider, you need to fetch this report on a periodic basis and weed through your device tokens.

The feedback server hosts sandbox and production addresses, just like the notification server. You find these at feedback.push.apple.com (port 2196) and feedback.sandbox.push.apple.com (port 2196). You contact the server with a production SSL certificate and shake hands in the same way you do to send notifications. After the handshake, read your results. The server sends data immediately without any further explicit commands on your side.

The feedback data consists of 38 bytes. This includes the time (4 bytes), the token length (2 bytes), and the token itself (32 bytes). The timestamp tells you when APNS first determined that the application no longer existed on the device. This uses a standard UNIX epoch, namely seconds since Midnight, January 1, 1970. The device token is stored in binary format. You need to convert it to a hex representation to match it to your device tokens if you use strings to store token data. At the time of writing this book, you can ignore the length bytes. They are always 0 and 32, referring to the 32-byte length of the device token:

```
// Retrieve message from SSL.
size_t processed = 0;
char buffer[38];
do
{
    // Fetch the next item
    result = SSLRead(context, buffer, 38, &processed);
    if (result) break;

    // Recover Date from data
    char *b = buffer;
    NSTimeInterval ti = ((unsigned char)b[0] << 24) +
        ((unsigned char)b[1] << 16) +
        ((unsigned char)b[2] << 8) +
        (unsigned char)b[3];
    NSDate *date = [NSDate dateWithTimeIntervalSince1970:ti];

    // Recover Device ID
    NSMutableString *deviceID = [NSMutableString string];
    b += 6;
    for (int i = 0; i < 32; i++) [
        deviceID appendFormat:@"%02x", (unsigned char)b[i]];

    // Add dictionary to results
    [results addObject:
        [NSDictionary dictionaryWithObject:date
        forKey:deviceID]];

} while (processed > 0);
```

> **Note**
>
> Search your Xcode Organizer Console for "aps" to locate APNS error messages.

Designing for Push

When designing for push, keep scaling in mind. Normal computing doesn't need to scale. When coding is done, an app runs on a device using the local CPU. Should a developer deploy an extra 10,000 copies, there's no further investment involved other than increased technical support.

Push computing does scale. Whether you have 10,000 or 100,000 or 1,000,000 users matters. That's because developers must provide the service layer that handles the operations for every unit sold. The more users supported, the greater the costs will be. Consider that these services need to be completely reliable and that consumers will not be tolerant of extended downtimes.

On top of reliability, add in security concerns. Many polled services require secure credentials. Those credentials must be uploaded to the service for remote use rather than being stored solely on the device. Even if the service in question does not use that kind of authentication, the device token that allows your service to contact a specific device is sensitive in itself. Should that identifier be stolen, it could let spammers send unsolicited alerts. Any developer who enters this arena must take these possible threats seriously and provide secure solutions for storing and protecting information.

Third-party provider Urban Airship (urbanairship.com) offers ready-to-use push infrastructure, which is widely used in the iOS developer community. Newcomer Parse (parse.com), an up-and-coming competitor, simplifies push composition and deployment as well.

Summary

In this chapter, you saw push notifications both from a client-building point of view and as a provider. You learned about the kinds of notifications you can send and how to create the payload that moves those notifications to the device. You discovered registering and unregistering devices and how users can opt in and out from the service.

Much of the push story lies outside this chapter. It's up to you to set up a server and deal with security, bandwidth, and scaling issues. The reality of deployment is that there are many platforms and languages that you can use that go beyond the Objective-C sample code shown here, typically via Ruby, Python, and PHP. Regardless, the concepts discussed and recipes shown in this chapter give you a good stepping-off point. You know what the issues are and how things must work. Now it's up to you to put them to good use:

- The big wins of notifications are their instant updates and immediate presentation. Like SMS messages, they're hard to overlook when they arrive on your device. There's nothing wrong in opting out of push if your application does not demand that kind of immediacy.
- Guard your SSL certificate and device tokens. It varies how Apple will respond to security breaches, but experience suggests that it will be messy and unpleasant.

- Don't leave users without service when you have promised to provide it to them. Build a timeline into your business plan that anticipates what it will take to keep delivering notifications over time and how you will fund this. Consumers will not be tolerant of extended downtimes; your service must be reliable.
- Apple provides a verbose logging tool for push notification, which you can read about at its Developer Forums site.
- Don't spam users. Push notifications are not meant to sell products or promote sales. Respect your user base.
- Build to scale. Although your application may not initially have tens of thousands of users, anticipate a successful app launch as well as a modest one. Create a system that can grow along with your user base.

Index

A

ABStandin class, 299-302
Accelerate, rotating images, 215-216
acceleration
    catching events, 11
    moving onscreen objects, 16-19
accelerometers
    retrieving current angle synchronously, 13-16
    scroll view, 19-21
    sliding onscreen objects based on feedback, 17-19
achievements, Game Center
    checking, 382-383
    creating, 376-377
    reporting, 383-385
    resetting, 385
activity item sources, Activity View controller, 55
Activity View controller, 54-62
    activity item sources, 55
    adding services, 58-62
    excluding activities, 62
    HTML e-mail support, 62
    item providers, 56

item source callbacks, 56-57
items, 62-63
services, 62-63
activityImage method, 58
activityTitle method, 58
activityType method, 58
activityViewController method, 59
adding contacts, Address Book, 331-333
Address Book, 297, 338
    ABStandin class, 299-302
    addresses, 313-315
    contacts
        accessing contact image data, 325-326
        adding and removing, 331-333
        modifying and viewing, 334-335
        picking people, 326-331
        searching for, 318-319, 322-325
        sorting, 319
    custom dictionaries, 311-312
    databases, 298
    date properties, 306-307
    frameworks, 297-322
        ABContact, 322
        AddressBook UI, 298
        instant-message properties, 313-315
        multivalue items, 312-313
        retrieving and setting strings, 304-306
        wrapping, 303
    groups, 319-322
    images, 315-316
        displaying in table cells, 326
    multivalue data, storing, 311-312

querying, 302-303
records, 298-299
    creating, 317
    functions, 304
    removing, 317-318
social profile, 313-315
Unknown Person Controller, 335-337
addresses, Address Book, 313-315
alerts, localizing, 465
altitude property (Core Location), 347
ambient audio, creating, 270-272
annotation property (document interaction controller), 71
annotations
    maps, 363-368
    user locations, 360-363
API providers, request tokens, 188-189
APNS (Apple Push Notification Service), 448-451, 473-474. See also **push notifications**
    feedback service, 471-472
    handling token request errors, 456
    multiple provider support, 448
    notification payloads, building, 465-466
    responding to notifications, 456-458
    retrieving device token, 455-456
    security, 449-450
    sending notifications, 466-471
App IDs, generating new, 451-454
Apple Push Notification Service (APNS). See **APNS (Apple Push Notification Service)**
application bundles, images, 197

apps
    in-app purchase items, creating, 431-435
    developing and testing, 429
    fonts, adding custom to, 118
    registering, 454-458
    submitting, StoreKit, 429-430

assets library, reading images from, 202-203

attitude property (Core Motion), 22

attribute stacks, Core Text, 100-105

attributed strings, Core Text, 89-93
    drawing with, 109-111
    fonts, 116-117
    mutable, 95-98
    paragraph styles, 92-93
    Text View, 93-94

attributed text
    Bezier paths, drawing along, 151-154
    creating, pseudo-HTML, 105-109
    drawing into PDFs, 120-122

attributes
    Core Text, 87-88
    layering via iterated ranges, 97-98

audio, 261, 294-295
    Game Center, sessions, 411-415
    interruptions, handling, 272-274
    looping, 269-272
    Media Queries, creating, 288-290
    MPMusicPlayerController, 290-294
    picking, MPMediaPickerController, 286-288
    playing with AVAudioPlayer, 261-269
    recording, 274-280
        audio queues, 280-286

authentication, handling challenges, 176-177

authorization, Core Location, 339-344

available disk space, checking, 35-36

AVAudioPlayer
    audio, recording, 274-286
    audio interruptions, handling, 272-274
    monitoring audio levels, 265-269
    playback progress, 264-269
    playing audio, 261-269
    scrubbing, 264

AVFoundation cameras, accessing, 235-242

B

battery state, devices, monitoring, 6-8

Bezier paths, 166
    attributed text, drawing along, 151-154
    bounding, 137-142
    elements, 144-148
    fitting, 142-144
    moving items along, 148-151
    points
        extracting, 127-129
        retrieving, 149-151
        thinning, 129-132

478 bitmap images

bitmap images
   analyzing, 257-259
   representations, 210-214
blocks, handler, Core Motion, 23-26
Bluetooth limitations, GameKit, 416
Bonjour sessions, GameKit, 416-417
bounding Bezier paths, 137-142
boxes, bounding, 138-141
building simple web servers, 181-184

C

C-based Core Text, 88-89
cameras, 229
   AVFoundation, accessing, 235-242
   CI (Core Image) filtering, 248-251
      face detection, 251-257
   enabling flashlights, 233-235
   Exchangeable Image File Format (EXIF), 242-247
   image helper, 241-242
   image orientations, 247-248
   photographs, snapping, 229-233
   previews, 240
      laying out, 241
   querying, 236-237
   retrieving, 236-237
   sampling live feeds, 257-260
   sessions, establishing, 237-239
   switching, 239-240
canPerformWithActivityItems method, 58
catching, acceleration events, 11
Catmull-Rom, splines, 133-134

chat, Game Center, 411-415, 423-424
   testing availability, 412
CI (Core Image) filtering, 248-251
   face detection, 251-257
classes
   GKLeaderboard, 378-380
   UIDevice, 1-2, 5-9, 12-13
   UIImage, 199-200
   UIScreen, 8-9, 29, 31
code listings, 147-148. *See also* **recipes**
   Adding Camera Previews (7-6), 241
   Adding Images to Core Text Flow (3-4), 112-114
   Application Activities (2-1), 59-62
   Attributed String Core Text View (3-3), 109
   Attributed String View (3-2), 109
   Bezier Elements (4-1), 145-146
   Building a Map Annotation Object (10-1), 363
   Cameras (7-1), 236-237
   Capturing Output (7-3), 238-239
   Checking a Receipt (12-3), 444-445
   Converting Between RGB and HSB (7-9), 259-260
   Converting Geometry from EXIF to Image Coordinates (7-8), 252-254
   Creating a Session (7-2), 237-238
   Drawing an Image into a PDF File (6-2), 222-223
   Embedding and Retrieving Previews (7-5), 240
   Fitting Element-Based Bezier Paths (4-2), 147-148

Making a Screen Shot of a View (6-1), 222

Preparing Annotation Views for Use (10-2), 365

Products Request Callback Methods (12-1), 437-438

Recovering File System Size and File System Free Size (1-1), 36

Responding to Payments (12-2), 439-441

Retrieving Image Metadata (7-7), 243-244

Returning a Font from Its Traits (3-1), 99-100

Selecting from Available Cameras (7-4), 239-240

Serializing and Deserializing Property Lists (11-1), 396-397

conformance

retrieving lists, UTIs (Uniform Type Identifiers), 43-45

testing, UTIs (Uniform Type Identifiers), 42-43, 44-45

conformance arrays, 44-45

contacts, Address Book

accessing contact image data, 325-326

adding and removing, 331-333

modifying and viewing, 334-335

picking people, 326-331

searching, 322-325

searching for, 318-319

sorting, 319

controllers

Activity View, 54-62

activity item sources, 55

adding services, 58-62

excluding activities, 62

HTML e-mail support, 62

item providers, 56

item source callbacks, 56-57

items, 62-63

services, 62-63

document interaction, 69-75

checking Open menu, 72-75

creating instances, 69-71

properties, 71

Quick Look support, 71-72

MPMediaPickerController, picking audio, 286-288

MPMusicPlayerController, 292-294

Quick Look, 62-69

adding actions, 66-69

document interaction controllers, providing support, 71-72

Unknown Person Controller, Address Book, 335-337

converting between coordinate systems, 210-211

convex hulls, bounding, 138-141

convolution, images, 216-219

coordinate property (Core Location), 347

coordinate systems, converting between, 210-211

Core Image (CI) filtering, 248-251

face detection, 251-257

Core Location, 339, 344-347, 369-370

authorizing, 339-344

geocoding, 353-355

Geofencing, 348-350

location and privacy, resetting, 341

location properties, 346-347

maps, creating annotations, 363-368
speed, tracking, 347-348
testing, 339-341, 343-344
tracking north, 350-353
user locations
    annotations, 360-363
    viewing, 355-360
user permissions, checking, 343

Core Motion, 21-26
handler blocks, 23-26
properties, 21-22
testing for sensors, 22

Core Text, 87, 125-126
adding images to, 112-114
attributed strings, 89-93
    mutable, 95-98
    paragraph styles, 92-93
    Text View, 93-94
attributed text
    creating using pseudo-HTML, 105-109
    drawing into PDFs, 120-122
attributes, 87-88
    stacks, 100-105
C-based, 88-89
creating image cut-outs, 112-114
drawing into scroll view, 114-116
drawing with, 109-111
fonts, 116-117
    adding custom to apps, 118
large text, 122-125
multipage, 119-120
Objective-C, 88-89

responder styles, 98-100
UIKit, 89
counting groups, Address Book, 319
course property (Core Location), 347
credentials
entering, 171-176
secure storage, 167-171
current angle, accelerometers, retrieving synchronously, 13-16
curves, 144-148
custom document types, creating, 77-78

D

data
serializing, GameKit, 396-397
uploading, 177-181
date properties, Address Book, 306-307
databases, Address Book, 298
design, push notifications, 473
detecting
screens, 29-30
shakes, motion events, 27-28
development
apps, 429
push notifications, 467-471
devices
accessing basic information, 1-2, 9-10
adding capability restrictions, 2-5
battery state, monitoring, 5-8
orientation, 12-13
permissions, 3
proximity sensor, enabling/disabling, 5

required capabilities, 4
Retina support, detecting, 8-9
retrieving tokens, 455-456
user permission descriptions, 3
disabling proximity sensor, 5
display links, screens, adding, 31
distribution, push notifications, 467-471
document file sharing, enabling, 49
document interaction controllers, 69-75
    checking Open menu, 72-75
    creating instances, 69-71
    properties, 71
    Quick Look support, 71-72
documents
    creating custom types, 77-78
    declaring support, 75-82
    Documents folder, scanning for new, 50-53
Documents folder
    monitoring, 48-53
    scanning for new documents, 50-53
    user controls, 49-50
    Xcode access, 50
drawing, Core Text, 109-111
drawings, smoothing, 132-135

E

Ecamm's Printopia, 55
emitters, 226-228
enabling proximity sensor, 5
ending game play, GameKit, 407-410
entering credentials, 171-176

events
    acceleration, catching, 11
    motion, detecting shakes, 27-28
EXIF (Exchangeable Image File Format), 242-247
external screens, 29-35

F

face detection, CI (Core Image) filtering, 251-257
feedback service, APNS (Apple Push Notification Service), 471-472
file extensions, UTIs (Uniform Type Identifiers), 40-41
file system, recovering size, 36
files, PDFs, drawing into, 222-223
filtering images, CI (Core Image), 248-251
fitting Bezier paths, 142-144
flashlights, cameras, enabling, 233-235
folders, Documents, monitoring, 48-53
fonts
    apps, adding custom to, 118
    Core Text, 116-117
frameworks Address Book, 297-322
    ABStandin class, 299-302
    AddressBook UI, 298
    databases, 298
    date properties, 306-307
    images, 315-316
    multivalue items, 307-309, 312-313
    querying, 302-303
    record functions, 304
    records, 298-299

frameworks Address Book

retrieving and setting strings, 304-306
storing multivalue data, 311-312
wrapping, 303

G

Game Center, 371, 425-426. *See also* **GameKit**
    achievements
        checking, 382-383
        reporting, 383-385
        resetting, 385
    enabling, 371-373
    loading matches from, 403
    removing matches, 410-411
    scores, submitting, 381-382
    signing in to, 373-374
    view controller, displaying, 380
    voice, 411-415

game play
    ending, GameKit, 407-410
    responding to, GameKit, 403-407

GameKit371, 425-426
    achievements
        checking, 382-383
        creating, 376-377
        reporting, 383-385
        resetting, 385
    audio sessions, establishing, 412-413
    Bluetooth limitations, 416
    chat
        creating, 413
        implementing buttons, 414-415
        starting and stopping, 413

        state monitoring, 414
        testing availability, 412
        volume control, 415
    Game Center view controller, displaying, 380
    game play, 393-394
        ending, 407-410
        responding to, 403-407
    handling player state changes, 390-391
    invitation handlers, creating, 388-389
    leaderboards
        accessing, 378-380
        building, 375-376
    matches
        loading, 402-403
        removing, 410-411
    multiplayer matchmaking, 385-387
        managing match state, 390
        responses, 387-388
        turn-by-turn, 399-401
    peer services, 415-425
        Bonjour, 416-417
        creating helper, 422
        online connections, 424-425
        peer connection process, 417-421
        peer-to-peer voice chat, 422-423
        sending and receiving data, 421
        state changes, 422
        voice chat, 423-424
    player names, retrieving, 392-393
    scores, submitting, 381-382
    serializing data, 394-397
    session modes, 417

starting games, 388

synchronizing data, 397-398

turn-based invitations, responding to, 401-402

games, starting, GameKit, 388

geocoding, Core Location, 353-355

Geofencing, Core Location, 348-350

geometry, 127, 166

    Bezier paths

        drawing attributed text along, 151-154

        fitting, 142-144

        moving items along, 148-151

        points, 127, 129-132

        retrieving, 127-129

    curves, 144-148

    drawings, smoothing, 132-135

    transforms, 154-161

    velocity-based stroking, 136-137

    view intersections, testing, 161-165

GKLeaderboard class, 378-380

graphics. *See* images

gravity property (Core Motion), 22

groups, Address Book, 319-322

GUIDs (Globally Unique Identifiers), 35

H

Hafeneger, Stefan, 467

handler blocks, Core Motion, 23-26

handling

    authentication challenges, 176-177

    player state changes, GameKit, 390-391

hints, naming, 200-201

horizontalAccuracy property (Core Location), 347

HSB (hue, saturation, brightness), converting RGB (red, green, blue) to, 259-260

I

iCloud, images, 198

icons property (document interaction controller), 71

IIDs (Interface Identifiers), 35

image cut-outs, creating, Core Text, 112-114

ImageIO framework, 242-243

images, 197, 228

    Address Book, 315-316

    applying aspect, 205-207

    assets library, reading from, 202-203

    bitmap representations, 210-214

    capturing view-based screen shots, 221-222

    CI (Core Image) filtering, 248-251

        face detection, 251-257

    converting data to and from bitmap data, 212-214

    convolution, 216-219

    emitters, 226-228

    Exchangeable Image File Format (EXIF), 242-247

    fitting and filling, 203-207

    loading from URLs, 202

    metadata, exposing, 245-246

484 images

orientations, 247-248
processing, 215-216
  core, 219-221
reading data, 199-203
reflections, 223-226
rotating, 208-210
  Accelerate, 215-216
sandbox, finding, 201-202
snapping, 229-233
sources, 197
UIImage, wrapping, 244-247
in-app purchase items, creating, 431-435
Info.plist file, 3-5
inheritance, UTIs (Uniform Type Identifiers), 40
instances, document interaction controller, creating, 69-71
Internet, images, 198
interruptions, audio, handling, 272-274
invitation handlers, GameKit, creating, 388-389
invitations, responding to, GameKit, 401
iPhone files, serving through Web service, 181-184
item providers, Activity View controller, 56
item source callbacks, Activity View controller, 56-57
items, Activity View controller, 62-63
iterated ranges, layering attributes via, 97-98
iTunes accounts, signing out, 438

J-L

JSON (JavaScript Object Notation)
  serialization, 394
  transforming from dictionary to, 465

laying out camera previews, 241
leaderboards
  accessing, GameKit, 378-380
  building, GameKit, 375-376
levels, audio, monitoring, 265-269
listings. See code listings; recipes
lists, conformance, retrieving, 43-45
live feeds, sampling, 257-260
loading
  images from URLs, 202
  matches, GameKit, 402-403
local notifications versus push notifications, 451
location services, testing for, 339-341
locations, Core Location
  annotations, 360-363
  properties, 346-347
  resetting, 341
  viewing, 355-360
looping audio, 269-272

M

magneticField property (Core Motion), 22
MapKit, 339, 369-370
maps, annotations, Core Location, 363-368
match state, GameKit, managing, 390

matches, GameKit

   loading, 402-403

   removing, 410-411

matchmaker fails, handling, 386-387

Media Queries, creating, 288-290

metadata

   images, exposing, 245-246

   querying, 243-244

monitoring

   battery state and proximity, 6-8

   Documents folder, 48-53

motion events, detecting shakes, 27-28

MPMediaPickerController, picking audio, 286-288

MPMusicPlayerController, 292-294

multipage Core Text, 120

multiplayer matchmaking, GameKit, 385-387

   managing match state, 390

   responses, 387-388

   turn-by-turn, 399-401

multivalue items, Address Book, 311-313

N

naming hints, 200-201

networking, 167, 196

   authentication, handling challenges, 176-177

   credentials

      entering, 171-176

      secure storage, 167-171

   OAuth utilities, 184-196

   uploading data, 177-181

   web servers, building, 181-184

north, tracking, Core Location, 350-353

notifications (push), 447-449, 473-474

   APNS (Apple Push Notification Service), 448-451

   App IDs, generating new, 451-454

   building payloads, 465-466

   designing for, 473

   limitations, 450

   versus local notifications, 451

   multiple provider support, 448

   production, 467-471

   provisioning push, 451-454

   push client skeletons, 459-464

   registering apps, 454-458

   responding to, 456-458

   sandbox, 467-471

   security, 449-450

   sending, 466-471

NSFileManager class, 36

O

OAuth utilities, 184-196

Objective-C, Core Text, 88-89

onscreen objects, sliding based on accelerometer feedback, 17-19

Open menu, document interaction controllers, 72-75

orientations

   devices, 12-13

      calculating from accelerometers, 14-15

   images, 247-248

overscanning compensation, screens, 31

P

paragraph styles, Core Text, 92-93
passively updating, pasteboard, 47-48
pasteboard, 45-48
   images, 198
   passively updating, 47-48
   properties, 46-47
   retrieving data, 47
   storing data, 46
paths, Bezier, 166
   drawing attributed text along, 151-154
   elements, 144-148
   fitting, 142-144
   moving items along, 148-151
   points
      extracting, 127-129
      thinning, 129-132
   retrieving points and slopes, 149-151
payloads, push notifications, building, 465-466
payments, responding to, StoreKit, 438-441
PDF (Portable Document Format) files, drawing into, 222-223
   attributed text, 120-122
peer services, GameKit, 415-425
   Bonjour, 416-417
   creating helper, 422
   online connections, 424-425
   peer connection process, 417-421
   peer-to-peer voice chat, 422-423
   sending and receiving data, 421
   state changes, 422
   voice chat, 423-424
permissions, devices, 3
photo album, 197
photographs, snapping, 229-233. *See also* images
picking audio, MPMediaPickerController, 286-288
pictures. *See* images
player state changes, GameKit, handling, 390-391
predicates, 288-290
prepareWithActivityItems method, 58
previews, cameras, 240
   laying out, 241
Printopia, 55
privacy, Core Location, resetting, 341
processing images, 215-216, 219-221
production, push notifications, 467-471
properties
   Core Motion, 21-22
   document interaction controllers, 71
   system pasteboard, 46-47
proximity sensor, enabling/disabling, 5
pseudo-HTML, creating attributed text, 105-109
purchases
   registering, 441, 442
   restoring, 441-442
purchasing items, 438-442
   multiple, 442
push client skeletons, 459-464

push notifications, 447-449, 473-474
  APNS (Apple Push Notification Service), 448-451
  App IDs, generating new, 451-454
  building payloads, 465-466
  designing for, 473
  limitations, 450
  versus local notifications, 451
  multiple provider support, 448
  production, 467-471
  provisioning push, 451-454
  push client skeletons, 459-464
  registering apps, 454-458
  responding to, 456-458
  sandbox, 467-471
  security, 449-450
  sending, 466-471

Q

queries, Media Queries, creating, 288-290
querying
  Address Book, 302-303
  cameras, 236-237
  metadata, 243-244
Quick Look controller, 62-66
  adding actions, 66-69
  document interaction controllers, providing support, 71-72

R

reading image data, 199-203
receipts, validating, 443-445

recipes. *See also* code listings
  Activity View Controller (2-4), 56-57
  Adding a Simple Core Image Filter (7-5), 250-251
  Adding Emitters (6-8), 226-228
  Analyzing Bitmap Samples (7-7), 257-259
  Applying Image Aspect (6-1), 205-207
  Audio Recording with AVAudioRecorder (8-4), 276-280
  Authentication with NSURLCredential Instances (5-3), 176-177
  Basic Core Motion (1-6), 23-25
  Basic OAuth Signing Utilities (5-6), 185-188
  Big Text. Really Big Text (3-9), 123-125
  Bounding Boxes and Convex Hulls (4-5), 138-141
  Building Attributed Strings with an Objective-C Wrapper (3-3), 102-104
  Catching Acceleration Events (1-3), 11
  Catmull-Rom Splining (4-3), 133-134
  Choosing Display Properties (9-4), 330-331
  Controlling Torch Mode (7-2), 234-235
  Converting to and from Image Bitmaps (6-3), 213-214
  Convolving Images with the Accelerate Framework (6-5), 217-218
  Core Image Basics (6-6), 220-221

Core Text and Scroll Views (3-5), 114-116

Creating a Font List (3-6), 117

Creating Ambient Audio Through Looping (8-2), 270-272

Creating an Annotated, Interactive Map (10-7), 366-368

Creating an Automatic Text-Entry to Pasteboard Solution (2-2), 48

Creating Reflections (6-7), 224-226

Credential Helper (5-1), 169-171

Detecting Faces (7-6), 255-257

Detecting the Direction of North (10-3), 351-352

Displaying Address Book Images in Table Cells (9-2), 326

Document Interaction Controllers (2-7), 73-75

Drawing to PDF (3-8), 121

Ending Games (11-17), 408-409

Establishing a Game Center Player (11-1), 374

Establishing an Audio Session for Voice Chat (11-19), 412-413

Exposing Image Metadata (7-4), 245-246

Extending Device Information Gathering (1-2), 9-10

Extracting Bezier Path Points (4-1), 128-129

Fitting Paths into Custom Rectangles (4-6), 143-144

Handling Incoming Documents (2-8), 79-81

Handling Invitations (11-14), 401

Handling Turn Events (11-16), 404-407

Helper Class for Cameras (7-3), 241-242

Implementing an Invitation Handler (11-9), 389

Layering Attributes Via Iterated Ranges (3-2), 97-98

Laying Out Text Along a Bezier Path (4-8), 152-154

Loading Matches from Game Center (11-15), 403

Loading Opponent Name (11-11), 392

Monitoring Proximity and Battery (1-1), 6-8

Multipage Core Text (3-7), 120

OAuth Process (5-7), 192-195

Obliterating Game Center Matches for the Current Player (11-18), 410-411

Password Entry View Controller (5-2), 172-175

Picking People (9-3), 328-329

Playing Back Audio with AVAudioPlayer (8-1), 265-269

Presenting the Game Center View Controller (11-3), 381

Presenting User Location Within a Map (10-5), 357-360

Providing URL Scheme Support (2-9), 84

Pseudo HTML Markup (3-4), 106-109

Push Client Skeleton (13-1), 461-464

Pushing Payloads to the APNS Server (13-2), 468-471

Quick Look (2-5), 65-66

Quick Look (2-6), 67-68

Recording with Audio Queues: The Recorder.m Implementation (8-5), 280-285

Recovering Address Information from Coordinates and Descriptions (10-4), 354-355

Requesting a Match Through the Match Maker (11-7), 386

Responding to a Found Match (11-8), 387-388

Responding to Player State (11-10), 391

Retrieving Leaderboard Information (11-2), 379-380

Retrieving Points and Slopes from Bezier Paths (4-7), 149-151

Retrieving Transform Values (4-10), 162-165

Rolling for First Position (11-12), 397-398

Rotating an Image (6-2), 209-210

Rotating Images with the Accelerate Framework (6-4), 215-216

Selecting and Displaying Contacts with Search (9-1), 323-325

Selecting Music Items from the iPod Library (8-6), 287-288

Serving iPhone Files Through a Web Service (5-5), 181-184

Simple Media Playback with the iPod Music Player (8-7), 292-294

Sliding an Onscreen Object Based on Accelerometer Feedback (1-4), 17-19

Snapping Pictures (7-1), 232-233

Starting a Match (11-13), 399-400

Storing the Interruption Time for Later Pickup (8-3), 272-274

Submitting User Scores (11-4), 382

Testing Achievements (11-5), 383

Testing Conformance (2-1), 44-45

Thinning Bezier Path Points (4-2), 131-132

Tilt Scroller (1-5), 19-21

Tracking the Device Through the MapView (10-6), 361-362

Transformed View Access (4-9), 159-161

Unlocking Achievements (11-6), 384

Uploading Images to imgur (5-4), 178-181

Using a kqueue File Monitor (2-3), 51-53

Using Basic Attributed Strings with a Text View (3-1), 93-94

Using Core Location to Geofence (10-2), 349-350

Using Core Location to Retrieve Latitude and Longitude (10-1), 345-346

Using Device Motion Updates to Fix an Image in Space (1-7), 26-27

Using the New Person View Controller (9-5), 331-333

Velocity-Based Stroking (4-4), 136-137

VIDEOkit (1-8), 32-35

Working with the Unknown Controller (9-6), 336-337

recording audio, 274-280

    audio queues, 280-286

records, Address Book, 298-299

    creating, 317

    functions, 304

    removing, 317-318

reflections, images, 223-226
registering
    apps, 454-458
    purchases, 441-442
relative angles, calculating, 15-16
removing contacts, Address Book, 331-333
reporting, Game Center achievements, 383-385
request tokens, API providers, 188-189
resetting achievements, GameKit, 385
responder styles, Core Text, 98-100
Retina support, detecting, 8-9
retrieving
    cameras, 236-237
    current angle synchronously, accelerometers, 13-16
    data, system pasteboard, 47
    strings, Address Book, 304-306
RGB (red, green, blue) color codes, converting to HSB, 259-260
rotating images, 209-210, 215-216
rotationRate property (Core Motion), 22

S

sandbox
    images, 198
        finding, 201-202
    push notifications, 467-471
schemes, URL, declaring, 82-83
scores, submitting, GameKit, 381-382
screen shots, capturing view-based, 221-222

screens
    detecting, 29-30
    display links, adding, 31
    external, 29-35
    overscanning compensation, 31
    retrieving resolutions, 30
    Video Out, setting up, 30-31
    VIDEOkit, 31-35
scroll views
    accelerometer-based, 19-21
    Core Text, drawing into, 114-116
searches, contacts, Address Book, 318-319, 322-325
sending push notifications, 466-471
sensors
    proximity, 5
    testing for, Core Motion, 22
serializing data, GameKit, 394-397
services, Activity View controller, 62-63
    adding, 58-62
session modes, GameKit, 417
sessions, cameras, establishing, 237-239
setting strings, Address Book, 304-306
shakes, detecting, motion events, 27-28
shared data, images, 198-199
signing in to Game Center, 373-374
slopes, Bezier paths, retrieving, 149-151
smoothing drawings, 132-135
social profiles, Address Book, 313-315
sorting contacts, Address Book, 319
sound. *See* audio
sources, images, 197-199

speed, tracking, Core Location, 347-348
storage, credentials, secure, 167-171
storefront GUI, building, 435-438
StoreKit, 427-430, 445
   apps
      developing and testing, 429
      submitting, 429-430
   development paradox, 428
   in-app purchase items, creating, 431-435
   purchases
      registering, 441-442
      restoring, 441-442
   purchasing items, 438-442
   storefront GUI, building, 435-438
   test accounts, creating, 430-431
   validating receipts, 443-445
storing data, system pasteboard, 46
strings
   Address Book, retrieving and setting, 304-306
   attributed, Core Text, 89-98
support, documents, declaring, 75-82
switching, cameras, 239-240
synchronizing data, GameKit, 397-398
system pasteboard, 45-48
   passively updating, 47-48
   properties, 46-47
   retrieving data, 47
   storing data, 46

T

TCP (Transmission Control Protocol), 393
test accounts, StoreKit, creating, 430-431
testing
   apps, 429
   conformance, UTIs (Uniform Type Identifiers), 42-45
   Core Location, 343-344
   Game Center achievements, 382-383
   for location services, 339-341
   URLs, 83
   view intersections, 161-165
text displays, large text, 122-125
Text View, attributed strings, 93-94
thinning Bezier path points, 129-132
Tilt Scroller, 19-21
timestamp property (Core Location), 347
tokens, retrieving and storing, 189
torch mode, controlling, 234-235
tracking speed, Core Location, 347-348
tracking users, 35
transforms, 154-161
   basic, 154-155
   values
      retrieving, 156-157
      setting, 157-158
   view point locations, retrieving, 158-161
Transmission Control Protocol (TCP), 393
turn-based invitations, responding to, GameKit, 401-402
turn-by-turn matchmaking, GameKit, 399-401
turn events, handling, 404-407

U

UDP (User Datagram Protocol), 393
UIDevice class, 1-2, 5-9, 12-13
UIImage class, 199-200
    orientations, 249-248
    wrapping, 244-247
UIKit, Core Text, 89
UIScreen class, 8-9, 29, 31
Uniform Type Identifiers (UTIs). See UTIs (Uniform Type Identifiers)
Unknown Person Controller, Address Book, 335-337
uploading data, 177-181
URL-based services, creating, 82-84
URL property (document interaction controller), 71
URLs (uniform resource locators)
    declaring schemes, 82-83
    images, loading from, 202
    testing, 83
user acceleration property (Core Motion), 22
user controls, Documents folder, 49-50
User Datagram Protocol (UDP), 393
user locations, Core Location
    annotations, 360-363
    viewing, 355-360
user permissions, Core Location, checking, 343
users, tracking, 35
UTI property (document interaction controller), 71
utilities, OAuth, 184-196

UTIs (Uniform Type Identifiers), 39-45
    file extensions, 40-41
    inheritance, 40
    producing preferred extensions or MIME types, 41-42
    testing conformance, 42-43
UUIDs (Universally Unique Identifiers), 35

V

validating receipts, 443-445
values, transforms
    retrieving, 156-157
    setting, 157-158
velocity-based stroking, 135-137
verticalAccuracy property (Core Location), 347
Video Out, setting up, 30-31
VIDEOkit, 31-35
view-based screen shots, capturing, 221-222
View Controller, Address Book contacts, 331-333
view intersections, testing, 161-165
view point locations, transforms, retrieving, 158-161
viewing user locations, Core Location, 355-360
views, accelerometer-based scroll, 19-21
voice, Game Center, 411-415

W-Z

web servers, building, 181-184

wrapping

    Address Book framework, 303

    UIImage, 244

Xcode access, Documents folder, 50